Lecture Notes in Networks and Systems

Volume 753

AF870041

The series "Lecture Notes in Networks and Systems" publishes the latest developments in Networks and Systems—quickly, informally and with high quality. Original research reported in proceedings and post-proceedings represents the core of LNNS.

Volumes published in LNNS embrace all aspects and subfields of, as well as new challenges in, Networks and Systems.

The series contains proceedings and edited volumes in systems and networks, spanning the areas of Cyber-Physical Systems, Autonomous Systems, Sensor Networks, Control Systems, Energy Systems, Automotive Systems, Biological Systems, Vehicular Networking and Connected Vehicles, Aerospace Systems, Automation, Manufacturing, Smart Grids, Nonlinear Systems, Power Systems, Robotics, Social Systems, Economic Systems and other. Of particular value to both the contributors and the readership are the short publication timeframe and the world-wide distribution and exposure which enable both a wide and rapid dissemination of research output.

The series covers the theory, applications, and perspectives on the state of the art and future developments relevant to systems and networks, decision making, control, complex processes and related areas, as embedded in the fields of interdisciplinary and applied sciences, engineering, computer science, physics, economics, social, and life sciences, as well as the paradigms and methodologies behind them.

Indexed by SCOPUS, INSPEC, WTI Frankfurt eG, zbMATH, SCImago.

All books published in the series are submitted for consideration in Web of Science.

For proposals from Asia please contact Aninda Bose (aninda.bose@springer.com).

Dalia Magdi · Ahmed Abou El-Fetouh ·
Mohamed Mamdouh · Amit Joshi
Editors

Green Sustainability: Towards Innovative Digital Transformation

Proceedings of ITAF 2023

Springer

Editors
Dalia Magdi
School of Computer Science
Canadian International College
Cairo, Egypt

Mohamed Mamdouh
Faculty of Computer Science
and Information Technology
Ahram Canadian University
Cairo, Egypt

Ahmed Abou El-Fetouh
Delta Higher Institute for Management
and Accounting Information System
El Mansora, Egypt

Amit Joshi
Global Knowledge Research Foundation
Ahmedabad, Gujarat, India

ISSN 2367-3370 ISSN 2367-3389 (electronic)
Lecture Notes in Networks and Systems
ISBN 978-981-99-4763-8 ISBN 978-981-99-4764-5 (eBook)
https://doi.org/10.1007/978-981-99-4764-5

This Springer imprint is published by the registered company Springer Nature Singapore Pte Ltd.
The registered company address is: 152 Beach Road, #21-01/04 Gateway East, Singapore 189721, Singapore

Paper in this product is recyclable.

Preface

This volume contains the papers presented at ITAF 2022–2023: Green Sustainability: Towards Innovative Digital Transformation, held at Cairo, Egypt, 4 and 5 February 2023, collaborated by Global Knowledge Research Foundation. The associated partners were Springer, InterYIT IFIP. The 3rd ITAF congress featured two days of focused networking and information sharing at the IoT cutting edge. The third edition brings together researchers, leading innovators, business executives, and industry professionals to examine the latest advances and applications for commercial and industrial end users across sectors within the emerging Internet of Things ecosphere. It targeted the state of the art as well as emerging topics related to Internet of Things such as Digital Transformation, E-Government, Big Data Research, Emerging Services and Analytics, Internet of Things (IoT) Fundamentals, Electronic Computation and Analysis, Big Data for Multi-discipline Services, Security, Privacy and Trust, IoT Technologies, and Open and Cloud Technologies.

The main objective of the conference was to provide opportunities for the researchers, academicians, industry persons, students, and expertise from all over the world to interact and exchange ideas and experience in the field of Internet of Things. It also focused on innovative issues at international level by bringing together the experts from different countries.

It introduced emerging technological options, platforms, and case studies of digital transformation implementation in areas by researchers, leaders, engineers, executives, and developers who will present the digital transformation industry which are dramatically shifting business strategies and changing the way we live, work, and play.

The ITAF Conference incited keynotes, case studies, and breakout sessions, focusing on smart solutions leading Egypt in digital transformation technologies into 2030 and beyond.

The conference started by the welcome speech of **Assoc. Prof. Dalia Magdi**, Conference Chair, ITAF 2023, Vice dean of School of Computer Science, Canadian International College; then the speech of **Prof. Ahmed Abou El-Fetouh**, Dean of Delta Academy; **Dr. Amit Joshi**, Organizing Secretary, ITAF 2023, Director of

Global Knowledge Research Foundation; and **Mr. Aninda Bose**, Executive Editor at Springer Nature Group.

On behalf of ITAF 2023 board, we thank all respectable keynote speakers—**Assoc. Prof. Jagdish Chand Bansal**, South Asian University, New Delhi, India; Pro Engineering College and Research Centre, Jaipur, India; **RD. Vijay Singh Rathore**, Professor and Director—outreach, Jaipur Engineering College, India; **Dr. Istvan Vassanyi**, University of Pannonia medical Informatics Research and Development Center, Hungary; **Dr. Attila Magyar**, University of Pannonia Electrical Energy Systems Research Laboratory, Hungary; and **Eng. Jamal Mekhaemar**, CTIO CORELLIA, a RICOH Company.

A lot of researches were submitted in various advanced technology areas, and 41 researches were reviewed and accepted by the committee members to be presented. There were 7 technical sessions in total, and talks on academic and industrial sector were focused on both the days.

Finally, on the behalf of ITAF 2023, we would like to thank all our partners, keynote speakers, researchers, attendees, and guests who have participated and shared our two successful days, looking forward to see you next year.

Cairo, Egypt

Assoc. Prof. Dalia Magdi
Conference Chair, and Editor

Contents

Editors and Contributors

About the Editors

Dalia Magdi is Vice Dean of Computer Science School at the Canadian International College. She acted as Dean of Faculty of Management and Information Systems, she was Head of Information System Department, Faculty of Management and Information Systems, Vice Director of CRI (Centre de Recherche Informatique), x Coordinator of Univrsite de Paris-Sud, French University in Egypt, X Coordinator of University of New Brunswick program at Sadat Academy for Management Sciences, Member of the Editorial Board of many international journals, and Reviewer of many international journals such as *SCIREA Journal of Information Science, SCIREA Journal of Computer Sciences, Internet of Things* and *Cloud Computing (IOTCC) Journal, Horizon Journal of Library and Information Science*, and *Journal of Computer Science and Security* (JCSS) in the areas of Computer Science and Security. She was invited as Keynote Speaker and participated in many national and international conferences. She was Chair of Internet of things application and future international conference 2019 and 2022.

Ahmed Abou El-Fetouh is Professor of Information Systems majored in Intelligent Information systems, Decision Support Systems, and Geographic Information Systems and Remote Sensing at the Faculty of Computers and Information, Mansoura University, Egypt. He holds both the Bachelor's degree and the Master's degree in Accounting Information Systems. He also obtained the Ph.D. degree in Accounting Information Systems. Prof. Ahmed is former Dean of the Faculty of Computers and Information, Mansoura University, Egypt. He is currently Dean of the Higher Institute for Administrative and Accountable Information Systems, Delta Academy affiliated to the Ministry of Higher Education, Egypt.

Mohamed Mamdouh is Assistant Professor of Software Engineering at Faculty of Computer Science and Information Technology, Ahram Canadian University, Egypt, and earned a Ph.D. in Information Systems at Mansoura University, Egypt. He has more than 14 years of teaching experience in different educational institutes. His research interests are artificial intelligence, machine learning, data science, bioinformatics, and Internet of things (IoT). He has published extensively as Author and Co-author of many papers in highly regarded, peer-reviewed journals and international conferences. He is Researcher in Centre for Research and Interdisciplinary CRI at French University in Egypt, an environment for experimental research in the above-mentioned areas, and is involved in several open-source software projects. He is Active Member in International Rough Set Society IRSS. He was Co-chair of Internet of things application and future international conference 2019 and 2022.

Amit Joshi is currently Director of Global Knowledge Research Foundation, also Entrepreneur and Researcher who has completed his graduation (B.Tech.) in Information Technology and M.Tech. in Computer Science and Engineering and completed his research in the areas of cloud computing and cryptography in medical imaging with a focus on analysis of the current Government Strategies and World forums needs in different sectors on security purposes. He has an experience of around 10 years in academic and industry in prestigious organizations. He is Active Member of ACM, IEEE, CSI, AMIE, IACSIT-Singapore, IDES, ACEEE, NPA, and many other professional societies. Further currently he is also International Chair of InterYIT at International Federation of Information Processing (IFIP, Austria). He has presented and published more than 50 papers in national and international journals/conferences of IEEE and ACM. He has also edited more than 20 books which are published by Springer, ACM, and other reputed publishers. He has also organized more than 40 national and international conferences and workshops through ACM, Springer, IEEE across 5 countries including India, UK, Thailand, Europe.

Contributors

Elham Mohamed Abdellatif Department of Accounting, Faculty of Commerce, Mansoura University, Mansoura, Egypt

Mohamed Abdelsalam Siemens EDA, Cairo, Egypt

Mohammed Abdelsalam Business Information Systems Department, Faculty of Commerce and Business Administration, Helwan University, Cairo, Egypt

Amira Mohamed Naguib Abdulaziz Accounting Dept, Delta Higher Institute for Management and Accounting Information Systems, Talkha, Egypt

Ayman Aboualnour Arab Academy for Science, Technology, and Maritime Transport, Cairo, Egypt

Nader Alber Faculty of Business, Ain Shams University, Cairo, Egypt

Enjy Khaled Ali Software Engineering Department, Ahram Canadian University Giza, Giza Governorate, Egypt

Yasmin M. Alsakar Faculty of Computers and Information, Mansoura University, Mansoura, Egypt

Mona M. Arafa Information Systems Department, Faculty of Computers and Artificial Intelligence, Benha University, Benha, Egypt

Galyna Azarenkova V. N. Karazin, Kharkiv National University, Kharkiv, Ukraine

Vitalina Babenko V. N. Karazin Kharkiv National University, Kharkiv, Ukraine;
Kharkiv National Automobile and Highway University, Kharkiv, Ukraine;
National University of Life and Environment Science of Ukraine, Kyiv, Ukraine

Sherif Barakat Department of Information System, Faculty of Computers and Information, Mansoura University, Mansoura, Egypt

Igor Barannik Simon Kuznets Kharkiv National University of Economics, Kharkiv, Ukraine

Olena Chukurna State University of Intelligent Technologies and Telecommunications, Odessa, Ukraine

Oleksii Diachenko Odessa State Agrarian University, Odessa, Ukraine

M. M. Eissa Software Engineering Department, Ahram Canadian University Giza, Giza Governorate, Egypt

Ahmed Mahmoud Elbrashy Accounting Dept, Delta Higher Institute for Management and Accounting Information Systems, Talkha, Egypt

Ahmed Saleh A. El Fetouh Delta Higher Institute for Management and Accounting Information Systems, Mansoura, Egypt

Noha E. El-Attar Information Systems Department, Faculty of Computers and Artificial Intelligence, Benha University, Benha, Egypt

Fatma El-Zahraa A. El-Gamal Faculty of Computers and Information, Mansoura University, Mansoura, Egypt

Amir El-Ghamry Faculty of Computers and Information, Mansoura University, Mansoura, Egypt;
School of Engineering and Computer Science, Hosted By Global Academic Foundation, University of Hertfordshire, Garden City, Egypt;
Faculty of Computer Science and Engineering, New Mansoura University, Mansoura, Egypt

Ahmed El-Halafawy Faculty of Business, Ain Shams University, Cairo, Egypt

Sherif Elhennawy Faculty of Computers and Artificial Intelligence, Cairo University, Giza, Egypt

M. Watheq El-Kharashi Department of Computer and Systems Engineering, Ain Shams University, Cairo, Egypt

Mohammed Elmogy Faculty of Computers and Information, Mansoura University, Mansoura, Egypt

Tarek Abd El-Rahman Faculty of Engineering, Computer and System Engineering Department, Ain Shams University, Cairo, Egypt

Nora El-rashidy Machine Learning and Information Retrieval Department, Faculty of Artificial Intelligence, Kafrelsheiksh University, Kafrelsheiksh, Egypt

Emad Elsamahy Arab Academy for Science, Technology, and Maritime Transport, Cairo, Egypt

Hoda T. Elsayed Faculty of Computers and Artificial Intelligence, Cairo University, Giza, Egypt

Nesma E. ElSayed Delta Higher Institute for Management and Accounting Information Systems, Mansoura, Egypt

Mohamed H. El-Shafey Faculty of Engineering, Computer and System Engineering Department, Ain Shams University, Cairo, Egypt

Tarek Elshishtawy Information Systems Department, Faculty of Computers and Artificial Intelligence, Benha University, Benha, Egypt

Sara S. Emam Information Systems Department, Faculty of Computers and Artificial Intelligence, Benha University, Benha, Egypt

Dina Fakhry Department of Computer and Systems Engineering, Ain Shams University, Cairo, Egypt

Tetiana Filimonova State University of Trade and Economics, Kyiv, Ukraine

Maha Fouad Department of Information System, Faculty of Computers and Information, Mansoura University, Mansoura, Egypt

Mohamed Gamal Faculty of Computers and Artificial Intelligence, Cairo University, Giza, Egypt

Volodymyr Gamaliy State University of Trade and Economics, Kyiv, Ukraine

Hamed Nabil Hamed Department of Accounting, Faculty of Commerce, Mansoura University, Mansoura, Egypt

Karam M. Hassan Faculty of Computers and Information, Mansoura University, Mansoura, Egypt

Iman M. A. Helal Faculty of Computers and Artificial Intelligence, Cairo University, Giza, Egypt

Olena Iastremska Simon Kuznets Kharkiv National University of Economics, Kharkiv, Ukraine

Mai Ramadan Ibraheem Information Technology Dept Faculty of Computers and Information, Kafrelsheiksh University, Kafr Elsheiksh, Egypt

Rogushina Julia Institute of Software Systems of the National Academy of Sciences, Kyiv, Ukraine

Irina Kryukova Odessa State Agrarian University, Odessa, Ukraine

Dalia Ahmed Magdi Faculty of Computers and Information Specialties, Sadat Academy for Management Sciences, School of Computer Science, Canadian International College, Cairo, Egypt

Lyudmyla Malyarets Simon Kuznets Kharkiv National University of Economics, Kharkiv, Ukraine

Oleksandra Mandych State Biotechnological University, Kharkiv, Ukraine

Adolfo Maza Universidad de Cantabria, Santander, Spain

Sherif A. Mazen Faculty of Computers and Artificial Intelligence, Cairo University, Giza, Egypt

Ibrahim Fathy Moawad Faculty of Computer and Information Sciences, Ain Shams University, Abbassia, Cairo, Egypt;
Faculty of Computer Science and Engineering, New Mansoura University, New Mansoura City, Egypt

Mohamed Hegazy Mohamed Business Information Systems Department, Helwan University, Cairo, Egypt

Aya Mohammed A. Moussa Business Information Technology Program, Faculty of Computer and Information, Mansoura University, Mansoura, Egypt

Olexandr Nazarenko State University of Intelligent Technologies and Communications, Odessa, Ukraine

Maryna Nehrey National University of Life and Environment Science of Ukraine, Kyiv, Ukraine

A. Fatma Omara Faculty of Computers and Information, Cairo University Giza, Giza, Egypt

Kateryna Oriekhova V. N. Karazin, Kharkiv National University, Kharkiv, Ukraine

Boris Pokhodenko Kharkiv National Automobile and Highway University, V. N. Karazin Kharkiv National University, Kharkiv, Ukraine

Oleg Pursky State University of Trade and Economics, Kyiv, Ukraine

Olga Pushko Sumy State University, Sumy, Ukraine

Amira Rezk Department of Information System, Faculty of Computers and Information, Mansoura University, Mansoura, Egypt

Walid M. Saad The Egyptian Technical Research and Development Centre, Cairo, Egypt

Mona Safar Department of Computer and Systems Engineering, Ain Shams University, Cairo, Egypt

Nehal A. Sakr Faculty of Computers and Information, Mansoura University, Mansoura, Egypt

Amgad A. Salama The Egyptian Technical Research and Development Centre, Cairo, Egypt

Samir Aboul Fotouh Saleh Department of Accounting, Faculty of Commerce, Mansoura University, Mansoura, Egypt

Islam Samih Business Information Systems Department, Faculty of Commerce and Business Administration, Helwan University, Cairo, Egypt

Mohamed Shalaby Egyptian Armed Forces, Cairo, Egypt

Victor Zamlynskyi State University of Intelligent Technologies and Telecommunications, Odessa, Ukraine

Digital Transformation—Ecosystems, Banking and Financial Services

Assessment of the Level of Environmental and Economic Sustainability of Subjects of the Public Sector of the Country's Economy

Lyudmyla Malyarets, Vitalina Babenko, Olena Iastremska, and Igor Barannik

Abstract The article outlines the methodological foundations for assessing the environmental and economic sustainability of business entities in the public sector of the country's economy. The substantive essence of ecological and economic stability has been clarified. Its main differences are: multidimensionality, multicriteria, continuity, structuredness, close interconnection of economic and environmental components, direct dependence on the factors of influence of external and internal environments, conditionality by the potential of the corresponding economic entity, and the ability to flexible structure of the potential. In military conditions, the environmental and economic sustainability of business entities is ensured by a sufficient level of sustainability and its reserve, effective management for the preservation of the external environment, and an effective structural policy of economic activity. Using the data of Ukraine as an example, the solution of the analytical problem of determining the level of environmental and economic sustainability of economic entities in the public sector of the economy in the regional context is given.

Keywords Environmental and economic sustainability · Level of sustainability · Regional profile · Methodological basis for assessment

L. Malyarets (✉) · O. Iastremska · I. Barannik
Simon Kuznets Kharkiv National University of Economics, Kharkiv 61166, Ukraine
e-mail: malyarets@ukr.net

O. Iastremska
e-mail: Iastremska_om@hneu.net

I. Barannik
e-mail: Barannik@ukr.net

V. Babenko
V.N. Karazin, Kharkiv National University, Kharkiv 61022, Ukraine
e-mail: vitalinababenko@karazin.ua

D. Magdi et al. (eds.), *Green Sustainability: Towards Innovative Digital Transformation*, Lecture Notes in Networks and Systems 753,
https://doi.org/10.1007/978-981-99-4764-5_1

1 Introduction

The military conditions of economic entities activity have raised the problem of environmental and economic sustainability in a new way. Previously, scientists and practitioners, when defining economic sustainability, spoke of a certain "state" of an enterprise when its stable functioning is ensured. At the same time, they are pointing to the main characteristics and homeostasis with the external environment, the ability to counteract the impact of negative factors on its activities were pointed out. But in the modern activities of enterprises, attention is focused on considering their environmental and economic sustainability in critical, force majeure conditions, such as military, post-war, natural disasters, social disasters (COVID-19), and global world crises.

2 Literature Review

Many works of scientists and practitioners are devoted to solving the problems of environmental and economic sustainability. The concept of sustainability is closely related to the basic concepts of development and growth [1]. This is very well and thoroughly substantiated in the fundamental monograph Sustainable development: crisis or regulation? [one]. This monograph contains recommendations for solving many conceptual problems of sustainability, namely: the relationship between the economic and environmental crisis; contradiction to the development of economy and ecology; institutionalization of relations between economy and ecology; interpretation of sustainable development; the importance of knowledge in sustainable development; and the importance of social responsibility in ensuring sustainable development. Naturally, this monograph is of great importance in the formation of the modern theory of sustainable development. Regarding the definition of economic sustainability, scientists Engelbert Stockhammer, Harald Hochreiter, Bernhard Obermayr, and Klaus Steiner in their work recommend using the index of sustainable economic well-being as an appendix to GDP. They believe that a holistic reporting system for measuring economic sustainability should be created [2]. The substantiation of the approach in modeling the sustainability of the activity of a corporation by scientists Payman et al. [3] differs. They developed the sustainability model on the basis of a probabilistic approach, taking into account environmental issues. The idea of multicriteria in assessing stability is supported by Hassani et al. linking the process of its determination with the solution of optimization problems [4]. Trianni et al. in their work have demonstrated that they are adherents of a structural approach to the definition of sustainability. In Ukraine, at the state level, the problems of economic sustainability were fundamentally dealt with by scientists from the Institute of Economics and Forecasting of the National Academy of Sciences of Ukraine, the National Institute for Strategic Studies of the National Academy of Sciences of Ukraine, and others.

Experts of the Organization for Economic Cooperation and Development distinguish four groups of conditions for ensuring the stability of the national economy, namely: (1) the country's foreign trade openness, a high share of competitive commodity markets in the economy, and an efficiently functioning labor market increase the ability of the economy to absorb negative external influences and overcome their consequences; (2) developed and effectively regulated capital markets support the stability of the economy by overcoming the trend of increasing external debt, supporting project co-financing, diversifying financial instruments, and developing small and medium-sized businesses; (3) an effective tax policy and a system of social protection of the population ensure an increase in the level of economic stability through the promotion of comprehensive economic growth and the reduction of compromise solutions to stimulate it; and (4) developed state institutions (institutions and organizations) ensure the sustainability of the economy through the formation and implementation of an effective policy to counteract external negative impacts [5].

Many scientists define the sustainability of an enterprise as its ability to withstand destabilizing influences and changes in the external and internal environment due to the efficient use of resources, as well as the ability of an enterprise to adapt to these changes while maintaining its sustainable potential, structural integrity, profitability, and liquidity in the long term [6, 7].

Speaking of sustainability, determine the factors influencing it. Savina and Skibina [8] propose to include the following factors influencing environmental and economic sustainability: production (the level of obsolescence and physical depreciation of equipment; production capabilities of equipment; availability of raw materials and materials; reserve capacity, quality control etc.); economic (percentage of equity capital; attraction of borrowed funds; volume of receivables and payables; level of profitability; etc.) [9]; environmental (environmental damage, environmental tax, etc.); innovative (innovations in management, production, personnel, etc.); and organizational and structural (level of personnel qualification; enterprise development strategy; intellectual potential of the enterprise; labor productivity; organizational structure; management methods; etc.) [10]. In addition to analyzing the factors influencing environmental and economic sustainability, scientists advise analyzing the degree to which it is ensured. Ensuring environmental and economic sustainability is a permanent process, the basis of which is formed by a cognitive-subjective pattern of formation and modification of the abilities of managers that determine the range of reflexive actions on the turbulence of business conditions of a proactive or reactive nature [11].

3 Materials and Methods

Summarizing various approaches to the definition of environmental and economic sustainability, it is necessary to highlight its main modern differences: multidimensionality, multicriteria, continuity, structuredness, close relationship between economic and environmental components, direct dependence on the influence of external and internal environments, conditionality by the potential of the corresponding one, and the ability to flexible change the structure potential [12]. To ensure environmental and economic sustainability in modern conditions of limited activity of business entities, it is necessary to form precisely these characteristics of it. Thus, in modern military conditions, ensuring the environmental and economic sustainability of business entities should be carried out in the following areas: maintaining a sufficient level of stability and its reserve; effective management at the enterprise to preserve the external environment; and effective structural policy of economic activity. All this ensures the environmental and economic sustainability of business entities in force majeure conditions, such as war and natural disasters. Consequently, environmental and economic sustainability is a characteristic of economic entities to maintain homeostasis with the external environment in various conditions of their activity, including in limited conditions, to counteract the negative impact of destabilizing factors that destroy the normal life of people [13–15].

Let us consider an important analytical task of determining the level of environmental and economic sustainability of economic entities in the public sector of the economy in the regional context for 2021. The official recommendations for assessing the economic activity of business entities indicate its criteria, namely: the absence or reduction of wage arrears, the rate of change in the size of the average monthly wage, the implementation of the financial plan, the degree of depreciation of fixed assets, the change in the amount of net profit/loss, the coverage ratio, financial stability ratio, solvency ratio, and the results of the audit report. Thus, the economic component of the environmental and economic sustainability of economic entities in the public sector of the economy in the regional context should be determined on the basis of the main indicators of their financial and economic activities, namely: net income (x_1, million UAH), net financial result (x_2, million UAH); accounts receivable (x_3, mln UAH), accounts payable (x_4, mln UAH), total assets value (x_5, mln UAH), own capital (x_6, mln UAH), average number of employees (x_7, thousand people), and wage arrears (x_8, mln. UAH) [5]. The environmental component of environmental and economic sustainability in the regional context will be determined on the basis of capital investments on environmental protection, by region (x_9, at current prices, thsd. UAH), current expenditures on environmental protection, by region (x_{10}, in current prices, thsd. UAH), and indices of increase/decrease of air emissions and greenhouse gases emissions from stationary pollution sources per capita by regions (x_{11}, in % to previous year) [5].

To determine the level of environmental and economic sustainability of economic entities in the regional context, it is necessary to collapse the entire system of indicators into one value, which is an integral indicator. Based on the advantages of calculating the integral indicator using the taxonomic development indicator of V. Plyuta, namely the simplicity of the algorithm for calculating the method, a clear interpretation of the value of the integral indicator, it is he who is recommended to determine the level of environmental and economic sustainability of economic entities in the public sector of the economy in the regional context [16]. When constructing a taxonomic indicator according to the method of V. Pluty, the following computational problems are solved:

(1) definition of stimulators, destimulators, nominators among the indicators of the signs of the socio-economic system:
$X = (x_{ij}),\ i = \overline{1, m},\ j = \overline{1, n}$, where *i*-th sign for *j*-th period or object;
(2) formation of the standard:
(a) according to the MiniMax criterion; (b) reference values are established;
(3) normalization or standardization of indicators:
$$Z = (z_{ij});\ z_{ij} = \frac{x_{ij} - \overline{x_i}}{\sigma_i},\ \overline{x_i} = \frac{1}{n}\sum_{j=1}^{n} x_{ij}, \sigma_i = \sqrt{\frac{\sum_{j=1}^{n}(x_{ij} - \overline{x_i})^2}{n}};$$
(4) calculation of the values of the generalizing indicator:
$$d_j = \left(\sum_{i=1}^{m}(z_{ij} - z_{i0})^2\right)^{\frac{1}{2}};\ \overline{d} = \frac{1}{n}\sum_{j=1}^{n} d_j;\ \delta = \overline{d}\ \text{або}\ \delta = M_e;$$
$$s_d = \left(\frac{1}{n}\sum_{j=1}^{n}(d_j - \delta)^2\right)^{\frac{1}{2}};\ d = \delta + a s_d;\ a = 3;\ d = \delta + 3 s_d;$$
$$I_j = \frac{d_j}{d};\ I_j^* = 1 - I_j.$$

Here, z_{ij}—standardized indicator values; $\overline{x_i}$—average values of indicators; σ_i—standard deviations of indicators; d_j—distance of standardized indicator values to a standardized reference; $\overline{d_j}$—average distances; and s_d—mean square distances. The problems of computing values when calculating a taxonomic indicator consist of calculating the values a and δ. Value a—number of standard deviations in fractions σ, which can be equal to 2 if the distribution of the trait is symmetrical, or 3 in the general case. Most often, it is taken equal to 3. Undoubtedly, if a certain accuracy is to be achieved in the task, then all laws of the distribution of indicator values should be diagnosed for symmetry.

This diagram depicts the main points of the implementation of the mathematical method for constructing a taxonomic indicator in solving various problems. First of all, it should be noted the differences in the methods of forming the standard. When using the MiniMax criterion in the formation of the standard, the levels of integral indicators are comparable locally in this sample, and it is not objective to equalize them with others. When we form a standard, setting the values of indicators and relying on normative values or planned, expert ones, we have an assessment in a global comparison, and it is possible to compare objects from different groups. In solving economic problems, this can be interpreted as follows: a comparative analysis with respect to the standard according to the MiniMax criterion allows setting and achieving the solution of local tasks of the operational management of the enterprise and with respect to the established standard—global tasks of the strategic management of the enterprise.

A characteristic property of the integral indicator I_j is that its value is in the range from 0 to 1. According to the calculations, the interpretation of the taxonomic indicator does not agree with intuitive ideas (the taxonomic indicator increases with the removal of the values of the indicators from the reference one and decreases as they approach it). Therefore, the taxonomic indicator was brought to the form $I_j^* = 1 - I_j$. The interpretation of this indicator is as follows: it takes high values when the values of indicators in the system are close to the standard and low values when they are far [16].

4 Results

The results of the implementation of this algorithm in determining the level of environmental and economic sustainability of economic entities in the regions of Ukraine in 2021 are presented in Fig. 1. In Fig. 1, the following designations of regions are: Vinnytsia (1), Volyn (2), Dnipropetrovsk (3), Donetsk (4), Zhytomyr (5), Transcarpathia (6), Zaporozhye (7), Ivano-Frankivsk (8), Kyiv (9), Kirovograd (10), Lukhansk (11), Lviv (12), Mykolayiv (13), Odesa (14), Poltava (15), Rivne (16), Sumy (17), Ternopil (18), Kharkiv (19), Gerson (20), Khmelnichsky (21), Cherkasy (22), Chernivtsi (23), and Chernikhov (24).

The regions of Kyiv, Kharkiv, Dnipropetrovsk, Zaporozhye, and Odesa have the highest level of environmental and economic sustainability of business entities in the public sector of the economy. Kirovograd, Luhansk, and Ivano-Frankivsk have the minimum level of environmental and economic sustainability in Ukraine.

To develop managerial solutions to improve the level of environmental and economic sustainability of economic entities in the public sector of the economy, it is necessary to identify clusters according to the system of indicators of this sustainability. To solve this analytical problem, cluster analysis is recommended, namely the Ward method, since it is he who gives the natural classification of objects in the

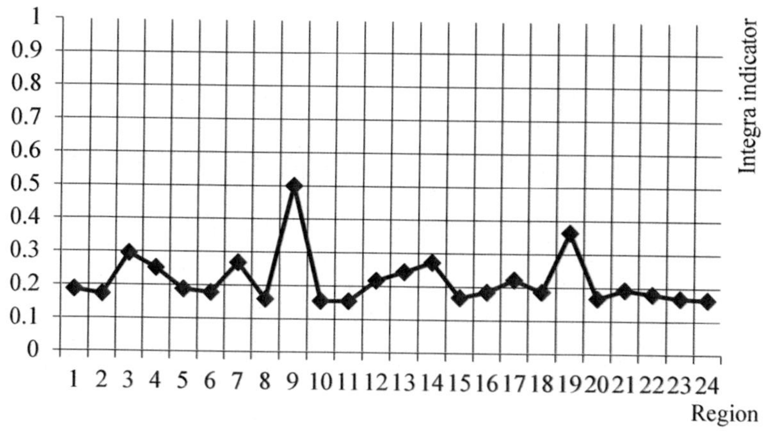

Fig. 1 Levels of environmental and economic sustainability of business entities of the public sector of the economy in the regions of Ukraine for 2021

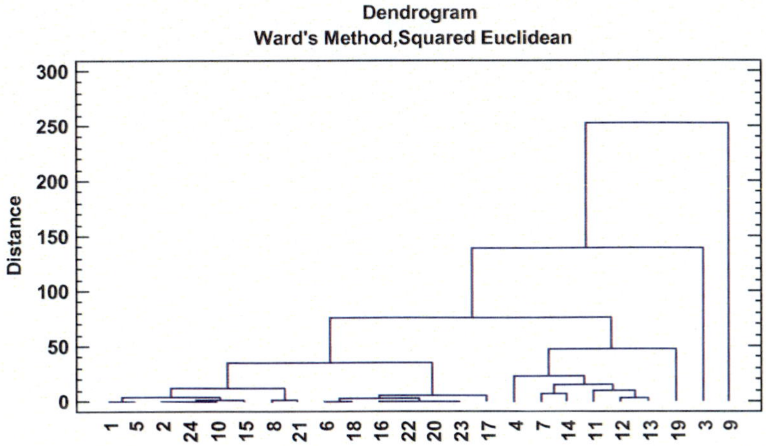

Fig. 2 Clusters of regions of Ukraine in 2021 according to the environmental and economic sustainability of business entities in the public sector of the economy

aggregate [16]. Figure 2 shows a dendrogram that allows you to identify clusters of environmental and economic sustainability of the subjects of the public sector of the regional economy in Ukraine.

In Fig. 2, there are three large clusters of regions, and three regions have a special status. Regions have a special status of environmental and economic sustainability of subjects of the public sector of the economy Dnipropetrovsk (3), Kyiv (9), and Kharkiv (19).

5 Summary and Conclusion

The conceptual principle for assessing the level of environmental and economic sustainability of economic entities in the state sector of the country's economy in military conditions is to maintain a sufficient level of sustainability and its reserve; effective management at the enterprise to preserve the external environment; and effective structural policy of economic activity.

The main modern differences in assessing the level of environmental and economic sustainability are: multidimensionality, multicriteria, continuity, structuredness, close relationship between economic and environmental components, direct dependence on the influence of external and internal environments, conditionality by the potential of the relevant business entity, and the ability to flexible change the structure of potential.

The methodological framework for assessing the environmental and economic sustainability of business entities in the public sector of the country's economy consists of the relevant elements, namely:

Definition of environmental and economic sustainability as a property of business entities to maintain homeostasis with the external environment in various conditions of their activities, including in conditions of limitedness, to counteract the negative impact of destabilizing factors that destroy the normal life of people;

Information support, which includes indicators that determine the economic and environmental components of sustainability, and the integral indicator determines the overall level of sustainability;

Analytical support, including an adjusted mathematical method for calculating the taxonomic indicator of development by V. Plyuta, as well as cluster analysis using the Ward method;

The logic of the stages of the assessment, implemented in the following sequence: (1) substantiation of indicators of environmental and economic sustainability; (2) calculation of the integral indicator of the stability level; (3) definition of clusters of environmental and economic sustainability of business entities in the regional context; and (4) analysis of the environmental and economic sustainability of economic entities in each cluster.

Solving the analytical task of determining the level of environmental and economic sustainability of economic entities in the regional context using the method of calculating the taxonomic indicator of development and its subsequent use to justify the clusters of regions makes it possible to objectively approach the development of state programs for the simultaneous management of the economy and the environment and reasonably conduct environmental and economic policy in the country.

References

1. Aregbeshola RA (2017) Import substitution industrialization and economic growth: evidence from the group of BRICS countries. Future Bus J 3:138–158
2. Stockhammer E, Hochreiter H, Obermayr B, Steiner K (1997) The index of sustainable economic welfare (ISEW) as an alternative to GDP in measuring economic welfare. Ecol Econ (21)1:19–34
3. Ahi P, Searcy C, Jaber MY (2018) A quantitative approach for assessing sustainability performance of corporations. Ecol Econ 152:336–346
4. Hassani L, Daneshvarkakhki M, Sabouhisabouni M, Ghanbari R (2019) The optimization of resilience and sustainability using mathematical programming models and metaheuristic algorithms. J Clean Prod 228:1062–1072
5. Reference data [Electronic resource]. Access mode: https://www.me.gov.ua/?lang=ru-UA [in Ukrainian]
6. Lyba VA, Revenko DS (2013) Economic sustainability of the enterprise: basic concepts and components of the system. Econ Manage Enterprises Mach Room: Prob Theory Prac VIP. 1(21):56–64
7. Bachev H (2016) Sustainability of farming enterprise—understanding, governance, evaluation. Bull Taras Shevchenko Nat Univ Kyiv (2):6–15
8. Savina GG, Skibina TI (2016) Factors of external and internal influence on the level of efficiency of enterprise management of the complex of public services. Eff Econ (12). http://www.economy.nayka.com.ua/?op=1&z=5300 Accessed 10 Feb 2018

9. Babenko V, Sidorov V, Koniaieva Y, Kysliuk L (2019) Features in scientific and technical cooperation in the field of non-conventional renewable energy. Global J Environ Sci Manage 5(SI):105–112. http://dx.doi.org/https://doi.org/10.22034/gjesm.2019.05.SI.12
10. Kaganovska T, Muzyka A, Hulyk A, Tymchenko H, Javadov H, Grabovskaya O (2022) Introduction of information technologies as the newest concept of optimization of civil proceedings. J Inf Technol Manage 14(3):1–25. https://doi.org/10.22059/jitm.2022.87260
11. Mizina OV, Shirokova IM (2010) Assessment of the economic sustainability of an industrial enterprise at the tactical and strategic levels. Sci Works DonNTU. Ser Econ (39–2):168–173
12. Malyarets L, Otenko V, Otenko I, Chepeliuk M (2021) Assessment the development of the commodity structure a country's exports and imports (case study of Ukraine). Montenegrin J Econ 17(4):7–16
13. Sustainable development: crisis or regulation?: Scientific monograph (2017) Podgorica, p 538
14. Trianni A, Cagno E, Neri A, Howard M (2019) Measuring industrial sustainability performance: Empirical evidence from Italian and German manufacturing small and medium enterprises. J Clean Prod 229:1355–1376
15. Malyarets L, Barannik I, Sabadash L, Grynko P (2019) modeling the economic sustainability of the macro system (for Example Ukraine). Montenegrin J Econ 14(3)23–35. http://www.mnje.com/sites/mnje.com/files/023-035_-ludmila_malarec_et_al.pdf
16. Malyarets LM (2006) Measurement of signs of objects in the economy: methodology and practice: scientific publication. Kharkov: Ed. KhNEU 384

A Suggested Index of Green Finance Availability: The Case of Egypt

Nader Alber and Ahmed El-Halafawy

Abstract The green economy is one of the most controversial topic for the economy of the twenty-first century, as world has changed its development direction from unfair development that ignores environmental and social dimensions to green development that takes into account the environmental dimension, reduces the carbon dioxide emissions, and protects our planet from the negative effect of the rise of global temperature, through the mitigation and adaptation with the climate change. The study aims at creating a new index of green finance availability in Egypt. This index has been constructed using factor analysis, according to each of (1) legal framework, (2) number of procedures, (3) time taken for financing, (4) cost, and (5) encouraging incentives. Results indicate that the proposed index of green finance availability seems to be affected positively by each of "investment attraction" and "required procedures to encourage green invest" and negatively by "obstacles facing green investments", without any evidence regarding the effect of "access to green finance".

Keywords Green economy · Green investments · Green finance · Sustainable development

1 Introduction

Competition among different countries and political groups around the world has been increased to attract more Foreign Direct Investments (FDIs), due to their importance for the economic growth sustainable development. Regarding the case of Egypt, Fig. 1 shows the developments of FDI from 2007 to 2019, as follows:

N. Alber (✉) · A. El-Halafawy
Faculty of Business, Ain Shams University, Cairo, Egypt
e-mail: naderalberfanous@yahoo.com

D. Magdi et al. (eds.), *Green Sustainability: Towards Innovative Digital Transformation*, Lecture Notes in Networks and Systems 753,
https://doi.org/10.1007/978-981-99-4764-5_2

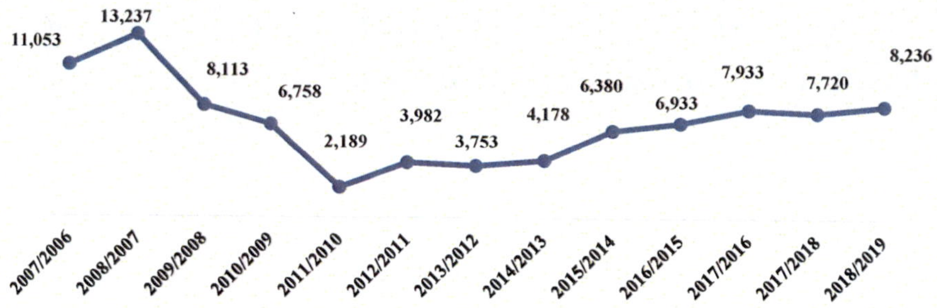

Fig. 1 Development of FDI in Egypt ($ 1,000,000). http://www.cbe.org.eg

As a result of the reforms adopted by the Egyptian economy, Fig. 1 indicates that the FDI has been increased from 3.9 billion dollars in the fiscal year 2004/2005 to 13.2 billion dollars in 2007/2008, then decreased as a result of the effects of the global financial crisis 2008, where the decline has been continued to reach 3.7 billion dollars in the fiscal year 2012/2013 as a result of political turmoil (the revolutions of January 25, 2011 and June 30, 2013). As a result of enhancement in the political stability, FDI has begun to rise, reaching $8236 million in 2018/2019 according to published data from the Central Bank of Egypt. Besides, the growth in the real sector may enhance the development of the financial sector in terms of its profit efficiency [1], financial stability [2], asset allocation [3], and financial inclusion [4].

Alber and El-Halafawy [5] aim at investigating the impact of ranking on "Doing Business Report" on the FDI inflow in Egypt during the period from 2007 to 2019. Results support the significance of the effects of each of "Registering property", "Getting credit" and "Paying taxes" on "FDI inflow" with explanation power of 92.9%. Therefore, it's important to work on matching the Egyptian business environment with the Doing Business methodology, in terms of reducing the number of procedures regarding property registration, getting credit and paying taxes. This could be conducted through encouraging the automation if the procedures that take place between investors and various administrative authorities, such as the experience of one-stop shop in GAFI. Besides, it is important to focus on reviewing and auditing laws and legal legislative related to the business climate, to improve the competitiveness and efficiency of the Egyptian economy, especially through PPP in fields of infrastructure, new and renewable energy, transport and logistics, garbage and solid recycling, Fin tech, education, health care and development of the Suez Canal Economic Zone.

The study aims at creating a new index of green finance availability in Egypt. This index has been constructed using factor analysis, according to each of (1) legal framework, (2) number of procedures, (3) time taken for financing, (4) cost, and (5) encouraging incentives.

The paper is arranged as follows: after this introduction, Sect. 2 reviews research literature. Section 3 explains how to develop hypotheses and measure variables. Section 4 is for empirical work, presenting results, discussing how these results answer research questions. Section 5 summarizes the paper and provides remarks about conclusions.

2 Literature Review

Green finance is still a controversial topic, due to its importance for sustainable development in one side and to the needed costs to avoid environmental damages in the other one. Literature has covered many aspects, related to legal framework, required procedures, estimated costs, and encouraging incentives. While Volz and Bohnke [15] concerns with reducing negative emissions, [14] addresses the role of green investments in protecting the environment from these emissions. Berensmann and Lindenberg [6] tried to evaluate the effectiveness of the Paris Climate Agreement, while several papers have addressed some possible solutions (e.g., [8, 7, 10, 13, 16]).

Volz and Böhnke [15] have indicated that Indonesian government has pledged to reduce carbon dioxide emissions by 26% by 2020, while emphasizing that economic and financial policies may be consistent with achieving sustainable development, shifting toward a green economy to preserve the environment and face climate change, improve energy efficiency and better use of natural resources. Thus, Indonesian government should demonstrate the bottlenecks facing the banking sector, to allocate more financing for green investments. The study has showed that the most important bottlenecks are: (1) weak financing available to banks to finance green investments, (2) the very low prices of traditional energy from fossil fuel when compared to renewable energy, (3) weak monitoring, supervision, and investigations from environmental standards for investment projects that pollute the environment, and (4) Uncertainty in the cash flows of green projects compared to traditional investments. Besides, these bottlenecks can be addressed through: (1) allocate a percentage of banks' financing activities to finance green investments; (2) making some amendments to the organizational regulations to make them more obligatory; (3) granting additional incentives for green investment projects, and (4) spreading thought and awareness about the importance of moving toward a green economy to face climate changes, and that the banking sector must deal with climate changes as systemic risks that must be faced.

Coping with this environmental risks, [14] indicates that the green financing policy includes a set of institutional measures that aim at attracting green investments to promote and protect the environment from risks, threats, climate changes, and transform to green energy. The investments required for green projects and industries in China are estimated at about 2 trillion Chinese yuan at the time of the study, equivalent approximately 320 billion dollars (3% of the Chinese GDP) during the five years (2016–2020). The government implements 10–15% of these investments, while the private sector implements the largest part which is 85–90%. The real challenge facing green investments in China is the lack of sufficient incentives to attract this type of investment, as the current incentives are granted to encourage "Green Investment" and to protect environment from "Brown Investment".

Berensmann and Lindenberg [6] have illustrated how to provide the necessary financing for the process of transformation toward green investments, and the role of various parties in mobilizing green financing from banks, local and international financing institutions, stock exchanges, donors, central banks around the world, regulators and investors, which is known as greening the financial system. Results show that many countries of the world that are signatories to the Paris Climate Agreement do not have a clear and announced plan of action on the steps of transition and transformation toward a green economy, with the need to give a greater role to the private sector, and that the global government sector cannot finance all of green projects; with more control and investigation in order to avoid sources of funding from illegal operations, which is known green money laundering.

It's important to investigate how financing green projects have been developed, Green banks and green bonds have some potential to support clean energy development. The advantages of green banks include offering better credit conditions for clean energy projects, the ability to aggregate small projects to achieve a commercially attractive scale, creation of innovative financial products, and market expansion through dissemination of information about the benefits of clean energy. Supporters of green bonds believe that they can provide long-term, reasonably priced capital to refinance a project once it has passed through the construction phase and is operating successfully [12]. Global energy investment has totaled USD 1.8 trillion in 2017, a 2% decline in real terms from the previous year according to the 2018 World Energy Investment report [9].

Fossil fuels still dominate energy investment. A major concern in the transition to low-carbon energy provision, therefore, is how to steer investments toward RE [10]. One possible solution is to stimulate non-bank financial institutions, including pension funds and insurance companies, and to invest in green projects. In the context of current systems of carbon pricing, "carbon price risk" has emerged as a new form of political risk for both companies and investors. This risk is related to the probability of the emergence of future international climate agreements and changes in carbon-related national policies [8]. In addition, central banks are in a powerful position to support the development of green finance models and enforce adequate pricing of environmental and carbon risk by financial institutions. The important consideration is the financial governance policies through which central banks, as well as other relevant financial regulatory agencies, can address environmental risk and promote sustainable finance [7].

HIT funds have expanded from Japan to Cambodia, Viet Nam, and Peru. They are also attracting attention from the government of Thailand, Malaysia's central bank and Mongolia. The venture capital market is generally not well developed in many countries including in many Asian economies and the financial system of many developing countries remains dominated by banks, but Internet sales are gradually expanding and the use of alternative financing vehicles, such as HIT funds, will help risky sectors to grow [16].

New financial technologies "Fintech", such as blockchain, the Internet of Things and big data, could unlock green finance over the same timeframe as the Paris Agreement and the SDGs. According to [11], there are three possible broad applications of Fintech to green finance: blockchain applications for sustainable development; blockchain use-cases for renewable energy, decentralized electricity markets, carbon credits, and climate finance; and innovation in financial instruments, including green bonds.

Basel capital requirements place several restrictions on lending, and this is why banks are reluctant to lend these projects. Besides, banks' resources come from short-to-medium-term deposits, whereas green infrastructure projects require long-term finance, resulting in a maturity mismatch for banks. So, banks are not able to provide all required credit for green projects, and we need to look for new channels of finance for this sector to fill the financing gap [13].

Compared with relevant previous work, this paper is different in three aspects. While the most of literature focused on possible solutions, this paper concerns with suggesting a realistic index which may reflect the effectiveness of these possible solutions. In addition, this paper supports its findings by a statistical significance, while many other papers concerned with the theoretical elaboration. Besides, this paper focuses on the case Egypt, providing a practical potential solutions.

3 Data Description and Developing Hypotheses

The study aims at creating a new index of green finance availability in Egypt. This index has been constructed using factor analysis, according to each of (1) legal framework, (2) number of procedures, (3) time taken for financing, (4) cost, and (5) encouraging incentives.

Data have been collected from a sample of 370 companies established during the first half of 2019. During this period, more than 10,000 companies have been registered. So, the research sample (370 companies) represents the research population (10,919 companies) at a confidence level of 95%. Responses of the 370 CEOs have been collected according to the questionnaire shown in the appendix. Reliability has been verified using Cronbach's alpha, providing an overall coefficient of 0.839.

Regarding each pillar of questionnaire, reliability has been tested, providing results shown in Table 1, as follows:

Table 1 shows that all the pillars of the questionnaire are characterized by strong stability, as all values are greater than 0.5, which indicates relying on the questionnaire's sentences and trusting its results. Besides, descriptive statistics of these pillars are shown in Table 2, as follows:

Descriptive statistics indicate homogeneity of respondents' opinions. The study aims at creating a new index of green finance availability in Egypt. This index has been constructed using factor analysis, according to each of (1) legal framework, (2) number of procedures, (3) time taken for financing, (4) cost, and (5) encouraging incentives. Table 3 shows that the value of the weighted mean for the total

Table 1 Reliability test of the green finance availability drivers

No	Pillars of green finance availability	Cronbach's alpha	No. of phrases
1	Investment attraction factors	0.824	8
2	Factors affecting access to green finance	0.725	2
3	Obstacles facing green investments	0.652	2
4	Required procedures to encourage green investment	0.724	7
	Green finance availability pillars	0.824	19

Table 2 Descriptive statistics of the green finance availability drivers

No	Pillars of green finance availability	Mean	Std. dev	Trend
1	Investment attraction factors	3.777	0.4450	Agree
2	Factors affecting access to green finance	4.274	0.4145	Strongly agree
3	Obstacles facing green investments	3.675	0.4038	Agree
4	Required procedures to encourage green invest	4.251	0.4159	Strongly agree

green finance pillar = 4.27, which indicates the tendency of the sample's opinions in the direction of "Strongly Agree" according to the Likert scale, and that the weighted mean of all components of the green finance pillars range from 4.13 to 4.44, with a standard deviation of 0.41, which indicates the homogeneity of respondent's opinions.

Table 4 indicates the correlation matrix among green finance index components, as follows:

The previous table shows that most of the correlation coefficients are significant (between medium and strong), but the correlation coefficients between the incentive variable and the rest of the variables are "weak" or not significant, which indicates that the incentive variable may not have a significant role in formulating the suggested indicator.

Besides, the value of the determinant is 0.091, greater than 0.00001, which means that there is no multicollinearity problem between the variables. So, multicollinearity problem does not exist, and this is why there is no need of excluding the variables that have a correlation coefficient greater than (0.8 $\pm$).

Table 3 Descriptive statistics of the green finance index components

No	Pillars of green finance availability	Mean	Std. dev
1	Legal framework	4.18	0.44
2	Number of procedures	4.29	0.60
3	Time taken for financing	4.32	0.61
4	Cost	4.44	0.49
5	Encouraging incentives	4.13	0.78

Table 4 Correlation matrix among green finance index components

Item	Y1	Y2	Y3	Y4	Y5
Y1 (Legal framework)	1				
Y2 (Number of procedures)	0.656***	1			
Y3 (Time taken for financing)	0.631**	0.847**	1		
Y4 (Cost)	0.547**	0.284**	0.522**	1	
Y5 (Encouraging incentives)	0.219**	0.085	− 0.017	0.105	1

Source outputs of data processing using, where * denotes significance level of 0.10, ** denotes significance level of 0.05, and *** denotes significance level of 0.01

To test whether the sample size is appropriate and reliable in conducting the factor analysis or not, "Kaiser–Meyer–Olkin" test should be conducted to mediate the difference between 0.5 and the correct one. Table 5 indicates this test as follows:

The value of the "Kaiser–Meyer–Olkin" test = 0.737 that mediates the difference between 0.5 and the correct one. This means that the sample size is appropriate and reliable in conducting the factor analysis.

The value of "Bartlett's" test shows that the probability value (*P*-value) is less than 0.05, and therefore, we reject the (null hypothesis: the correlation matrix between the variables is the identity matrix); and therefore, there is a real correlation matrix. So, we can form the indicator of the availability of green finance.

Table 6 indicates that the variance explained in is 57.5%, with communalities of 0.716, 0.799, 0.793, 0.539, and 0.030 for each of "łegal framework", "number of procedures", "time taken for financing", "cost", and "encouraging incentives", respectively.

Table 5 KMO and Bartlett's test

KMO	Bartlett's test		
	(χ^2)	Df	*P*-value
0.737	471.484	10	0.000

Table 6 Formulation of green finance availability index using factor analysis

Item	Communalities	Loading	Importance
Legal framework	0.716	0.846	0.294
Number of procedures	0.799	0.894	0.311
Time taken for financing	0.793	0.891	0.310
Cost	0.539	0.734	0.255
Encouraging incentives	0.030	0.172	0.060
Total variance explained	0.575		

According to the loading values, we have estimated the importance of each variable in constructing the index of availability of green finance, as follows:

Index of green finance availability = 0.294 legal framework + 0.311 procedures + 0.310 time + 0.255 costs + 0.06 incentives.

This paper tries to test the following four hypotheses:

1. There is not a significant effect of "investment attraction" on the proposed index of green finance availability.
2. There is not a significant effect of "access to green finance" on the proposed index of green finance availability.
3. There is not a significant effect of "obstacles facing green investments" on the proposed index of green finance availability.
4. There is not a significant effect of "required procedures to encourage green investment" on the proposed index of green finance availability.

4 Testing Hypotheses

To test the research hypotheses, we have used multiple regression technique, to assess the determinants of the proposed "Green Finance Availability Index", as shown in Table 7.

Regression analysis supports the significance of the effects of each of $X1$ (Investment attraction), $X3$ (Obstacles facing green investments), and $X4$ (Procedures required to encourage green investment) on FDI inflow in Egypt with explanation power of 37.6%. Besides, results don't support any effects of each of $X1$, $X2$, $X5$, $X7$, $X8$, and $X9$ on FDI inflow. Besides, multicollinearity problem doesn't exist, as the value of the VIF coefficient is less than 10.

So, for the first, third, and fourth hypotheses, we can reject the null hypotheses and accept the alternative ones. Besides, for the second hypothesis, we can accept the null hypotheses and reject the alternative ones. Based on the findings, it is important

Table 7 Determinants of the "green finance availability index"

Variable	Coefficient	t	VIF
C	– 6.199	– 7.262***	
$X1$ (Investment attraction)	0.313	2.075**	1.42
$X2$ (Access to green finance)			
$X3$ (Obstacles facing green investments)	– 0.343	– 1.805*	1.85
$X4$ (Procedures required to encourage green investment)	1.476	8.244***	1.74
F	39.40***		
Standard error	0.769		
R^2	0.376		

Source: outputs of data processing using, where * denotes significance level of 0.10, ** denotes significance level of 0.05, and *** denotes significance level of 0.01

to propose issuing of new legislation to regulate green investments, considering the positive effects of each of "investment attraction" and "procedures required to encourage green investment". Besides, the government should pay a great attention to the "obstacles facing green investments", due to their negative effects.

5 Results and Concluded Remarks

The green economy is one of the most controversial topic for the economy of the twenty-first century, as world has changed its development direction from unfair development that ignores environmental and social dimensions to green development that takes into account the environmental dimension, reduces the carbon dioxide emissions, and protects our planet from the negative effect of the rise of global temperature, through the mitigation and adaptation with the climate change.

The study aims at creating a new index of green finance availability in Egypt. This index has been constructed using factor analysis, according to each of (1) legal framework, (2) number of procedures, (3) time taken for financing, (4) cost, and (5) encouraging incentives.

Results indicate that the proposed index of green finance availability seems to be affected positively by each of "investment attraction" and "required procedures to encourage green invest" and negatively by "obstacles facing green investments", without any evidence regarding the effect of "access to green finance".

Appendix: Pillars of the Research Dimensions

1. Investment attraction
1. Trained labor at an appropriate cost
2. Production requirements are available at competitive prices
3. Availability of infrastructure
4. Stability of exchange rates
5. Political stability
6. The speed of entry and exit from the market
7. Effective laws and diversified and attractive investment policies
8. Encouraging customs and tax incentives
2. Access to green finance
1. Ease access to finance
2. Ease of export and import procedures
3. Obstacles facing green investments
1. The absence of a legal and framework specific in green investments

(continued)

(continued)

2. Needs to legislation effectiveness framework to the protection of intellectual property
4. Required procedures to encourage green invest
1. Flexible legislative framework to facilitate procedures
2. Granting more incentives for projects in the slum areas
3. Expanding the establishment of environmentally friendly smart industrial cities
4. Fighting bureaucracy and shifting toward electronic government
5. Removing the restrictions on some of investments in specific zones
6. Paying attention to technical education to provide trained workers
7. Interest in research and development (R&D)
Green finance availability index
1. Legal framework
2. Number of procedures
3. Time taken for financing
4. Cost to get green finance
5. Encouraging incentives to get green finance

References

1. Alber N (2011) The effect of banking expansion on profit efficiency of Saudi Banks. Corp Ownersh Control 8(2):51–58
2. Alber N (2017) banking efficiency and financial stability: which causes which? a panel analysis. In: Springer proceedings in business and economics, pp 91–98
3. Alber N (2018) Asset allocation, capital structure, theory of the firm and banking performance: a panel analysis. In: Springer proceedings in business and economics, pp 34–51
4. Alber N (2019) Determinants of financial inclusion: the case of 125 Countries from 2004 to 2017. In: Springer proceedings in business and economics, pp 1–10
5. Alber, El-Halafawy (2023) Does ranking on doing business report affect the FDI inflow? the case of Egypt. In: The third world conference on internet of things: applications & future (ITAF 2023), Cairo, Egypt
6. Berensmann K, Lindenberg N (2016) Green Finance: actors, challenges and policy recommendations, German Development Institute/Deutsches Institut für Entwicklungspolitik (DIE) Briefing Paper 23/2016, Available at SSRN: https://ssrn.com/abstract=2881922
7. Dikau S, Volz U (2018) Central banking, climate change and green finance. 867. ADBI Working Paper, Asian Development Bank Institute, Tokyo
8. Gianfrate G, Lorenzato G (2018) Stimulating Non-Bank Financial Institutions' Participation in Green investments. 860. ADBI Working Paper, Asian Development Bank Institute, Tokyo
9. International Energy Agency (2018) World Energy Investment 2018. International Energy Agency, Paris
10. Mazzucatoa M, Semieniukb G (2017) Financing renewable energy: who is financing what and why it matters. Technol Forecast Soc Change. dx.doi.org/https://doi.org/10.1016/j.techfore.2017.05.021
11. Nassiry D (2018) The role of fintech in unlocking green finance: policy insights for developing Countries. 883. ADBI Working Paper, Asian Development Bank Institute, Tokyo

12. Natural Resources Defense Council (2016) Clean energy finance outlook: opportunities for green banks and green bonds in Chile. Natural Resources Defense Council, New York, NY
13. Sachs J, Woo W, Yoshino N, Taghizadeh-Hesary F (2019) Why is green finance important. ADBI Working Paper Series (917)
14. UN Environment Program (2015) Establishing China's Green financial system, background paper a: theoretical framework of green finance. Retrieved from: https://www.unep.org/resources/report/establishing-chinas-green-financial-system
15. Volz U, Böhnke J (2015) Vanessa financing the green transformation, how to make green finance work in Indonesia. Eidt, Laura Knierim, Katharina Richer and Greta-Maria Roeber
16. Yoshino N, Taghizadeh-Hesary F, Nakahigashi M (2018) Modelling the social funding and spill-over tax for addressing the green energy financing gap. Econ Model. https://doi.org/10.1016/j.econmod.2018.11.018

Corporate Financial Performance Prediction Using Artificial Intelligence Techniques

Elham Mohamed Abdellatif, Samir Aboul Fotouh Saleh, and Hamed Nabil Hamed

Abstract Performance prediction is critical for aligning the company's operations with its strategic direction. A company's financial performance is one of the most important factors that determine whether or not it is successful in achieving its goals or implementing an appropriate strategy. A company must know the level of its financial health compared to other companies in the market, as this is a powerful tool that helps make the right decision. An analysis of a company's financial performance should include multiple standards, and this multidimensional view of performance indicates the need to different models or patterns of the relationship between the company's performance and its determinants. Financial performance prediction models evolved from using statistical methods to artificial intelligence methods such as neural networks. These models were not able to explain the relationships between past and current financial performance. This may be due to the omission of important qualitative information that can be extracted from annual financial reports, such as corporate governance. Therefore, this paper proposes the use of machine learning to support the prediction of corporate financial performance and obtain accurate results that outperform traditional analysis methods.

Keywords Financial performance · Artificial intelligence · Digital transformation · Machine learning

E. M. Abdellatif (✉) · S. A. F. Saleh · H. N. Hamed
Department of Accounting, Faculty of Commerce, Mansoura University, Mansoura 35516, Egypt
e-mail: elham@std.mans.edu.eg

S. A. F. Saleh
e-mail: prof_samir@mans.edu.eg

H. N. Hamed
e-mail: profhamed2007@mans.edu.eg

D. Magdi et al. (eds.), *Green Sustainability: Towards Innovative Digital Transformation*, Lecture Notes in Networks and Systems 753,
https://doi.org/10.1007/978-981-99-4764-5_3

1 Introduction

Corporate performance represents the basic concept in the current business environment, which is affected by rapid changes, fierce competition, and globalization, and includes all categories of stakeholders. It is also a long-term control tool for implementing the company's strategies, as well as a key indicator of a firm's health and efficiency [1].

The investor resorts to diversifying investments in some companies to avoid risks, and this requires an accurate prediction of the companies' performance in the future [2]. The company's performance is affected by a group of influencing factors that reflect several dimensions, including organizational aspects such as the size of the company, the history and structure of the company, environmental and social aspects, and others related to intellectual capital in addition to financial ratios [3]. Digital technology is changing places and workflows in the accounting profession, creating new opportunities and requiring different and advanced technological skills from employees. Changes based on new technologies ERP and AIS are not new and are part of an ongoing process in accounting. However, these current developments are particularly rapid and drastic due to the various market and regulatory forces driving this change at the same time. New techniques, especially those based on artificial intelligence, will have a significant impact on the overall structure and operations of accounting and thus significantly transform existing professional jobs and task profiles within a very short time [4]. Mullainathan and Spiess indicated that machine learning methods are particularly suitable for the basic prediction task in financial planning and analysis because of their focus on predictive performance [5]. These methods were able to identify generalizable patterns that worked well on new data. Thus, the important role of machine learning appears through its ability to analyze data, learn from it, and discover important patterns in addition to its predictive role in financial problems, as it proved its superiority in the results of studies that it is used in various problems, especially financial performance prediction.

2 Financial Performance Prediction

2.1 Importance of Financial Performance Prediction

For many stakeholders, predicting future corporate performance is critical. It is a necessary procedure for the profitable activity of banks, investors, and supply chain management. It is also important from a financial policy perspective. For example, modern banking regulations (e.g., Basel) require banks to build their internal model to assess the creditworthiness of client companies, which reflects their future performance estimates [6].

Mousa et al. pointed out that corporate annual reports are considered one of the most powerful financial communication channels. Annual reports contain text parts and financial statements. Text parts include the report of the board of directors, the auditor's report, and other textual information. They indicated that research has shown that financial statements lack relevance and that there is an increasing interest in non-financial information, the language and style of writing it, and that the textual contents contain information twice as valuable as the basic financial statements contain, which enables it to improve the quality and accuracy of predictive models for corporate performance. Various approaches have been used to address the classification problem, ranging from analysis of experience to complex decision support systems. In addition, statistical methods have been the traditional method for building such predictive models. However, with the development of information technology and the use of modern methods of data analysis, artificial intelligence techniques have been used, especially machine learning, as it has proven its superiority in predictive models [7].

2.2 Types of Financial Performance Prediction Models

A review of studies that dealt with predicting corporate financial performance included two different methodologies; the first is based on statistical methods, while the second methodology relies on artificial intelligence techniques, which is a recent trend resulting from the development of human behavior simulation models. The following are corporate financial performance prediction models [8]:

(1) **Statistical Models**. These models rely mainly on the quantitative data of the financial statements and focus on warnings of financial difficulties. Moreover, these models can be grouped according to the elements of financial statements used in the ratio calculations and the type of information they provide, as follows:

- **Classical (traditional) models for corporate performance prediction**. These models are created using mainly accounting data from the balance sheet and income statement (known as accrual ratios).
- **Cash flow models for corporate performance prediction**. These models contain information about current and estimated cash flows, as well as accounting data from the balance sheet and income statement.

(2) **Artificially Intelligent Experience Systems (AIES) models**. New information techniques are used in model building, where computers are "taught" how to solve problems based on past experience (e.g., neural networks, genetic algorithms, case-based reasoning models). These models usually combine quantitative and qualitative information to more accurately simulate human behavior, so their background is also based on statistical methodology. In addition to statistical models, artificial intelligence (AI) models also focus on the symptoms (warnings) of financial difficulties.

(3) **Theoretical models** depend on certain theoretical assumptions; they are built from qualitative information that must satisfy the theoretical assumptions that will be included in the model. These models usually use statistical methodology to quantitatively verify theoretical assumptions. The theoretical models, unlike previous ones, focused on the reasons for financial difficulties rather than their symptoms.

3 An Overview of Artificial Intelligence

3.1 Digital Transformation

Digitalization is one of the biggest changes in society in recent times, and it is already affecting many areas of our lives. Robots and artificial intelligence programs are everywhere, changing the way we work and live with increasing speed. The "digitalization" term has been replaced by the "digital transformation" term, which is being used increasingly recently as it reflects a social dimension more than a technological dimension [9]. Digital transformation represents an important opportunity for the financial prediction and analysis function, especially when appropriate tools are used to analyze large volumes of data, and here the important role played by machine learning techniques to support financial prediction and analysis appears by extracting important patterns and information from the data [10].

3.2 The Concept of Artificial Intelligence and Its Role in Financial Prediction

Artificial intelligence (AI) can be defined as systems with different levels based on a machine that achieve a certain set of human-defined goals by making predictions, recommendations, or decisions. Artificial intelligence techniques are increasingly relying on massive amounts of data sources and data analytics to feed data to machine learning models, which then use that data to learn and improve prediction and performance automatically through experience without being programmed by humans [11].

Financial prediction is an important part of analyzing financial statements, and financial prediction relies on the use of past and current financial information to achieve the best prediction of the future financial situation to avoid situations involving high risks and to increase benefits. These predictions are of interest to anyone who wants to know the state of potential financial resources in the future, including investors and decision-makers. AI affects the global business environment significantly and has many applications in many different fields, whereas it is difficult to analyze the financial statements because they contain a great deal of uncertainty because they are nonlinear, unstable, and characterized by a wide range of fluctuations. Therefore, AI can be used to classify financial data for analysis, allowing data to be examined and analyzed more quickly, helping to make more accurate decisions, significantly reducing human error, achieving better returns for clients, making more accurate predictions of potential outcomes, and facilitating risk control [12].

3.3 The Benefits of Using Artificial Intelligence in Business [13]

(1) Improving corporate efficiency by reducing costs and enhancing productivity, thereby increasing profitability (e.g., enhanced decision-making processes, automated execution, and gains from improvements in risk management, regulatory compliance, and other process improvements).
(2) Improving the quality of services provided as well as products, such as by introducing new products and offering high customization of products and services, will increase the competitive advantage.
(3) Assistance in making the right decisions is provided through recommendations and accurate predictions that are reached through the use of various techniques of artificial intelligence.
(4) AI techniques are well suited to complex (unstructured) decisions in which it is not possible to know a priori the relationships between variables.
(5) It can be used under conditions of uncertainty and ambiguity surrounding a problem, during which it is difficult to use traditional methods of data analysis, for example, fuzzy logic.

Based on the foregoing, researchers believe in the important role of artificial intelligence technologies by employing the nature of each of them with the nature of different decisions in addition to taking into account the nature of the available data to reach accurate results, whether in the form of predictions, such as predicting corporate performance, share prices, profits, etc., or classifying the case of a problem, such as classifying companies in terms of financial defaults into defaulting and not defaulting in addition to other problems. The advantages of these techniques are beneficial to internal and external decision-makers, in supporting decision-making.

3.4 Machine Learning

The "artificial intelligence" term is used broadly and generally to refer to any type of machine learning software [14]. Although there is no standard definition of "machine learning", it can be described as a set of methods that automatically make predictions from complex data. Machine learning determines the predictive relationship between inputs and outputs [10].

Machine learning has gained great attention in the social sciences because it enriches the field of data analysis, where machine learning techniques discover complex patterns in this data and choose the best variables to make accurate predictions through the provision of new methods for discovering patterns in accounting numbers. In addition, they also help the organizers of the profession monitor financial reporting practices [15]. Machine learning algorithms have made the decision-making process more effective and profitable in the field of business [16]. Thus, the important role of machine learning is due to its ability to analyze data, learn from it, and discover important patterns in addition to its predictive role in financial problems, as it proved its superiority in the results of studies that it used in various problems, especially predicting financial performance.

4 Predicting Corporate Financial Performance Using Artificial Intelligence Techniques in a Digital Transformation Environment

Modern prediction methods enable businesses to think about the future, i.e., to use the data gathered to improve prediction and understand what happened and the latent causes. Traditional decision support tools and business intelligence have been replaced by innovations and technical developments in line with digital transformation, where artificial intelligence, machine learning, and other modern techniques depend on big data analysis techniques. A researcher conducted a global survey in cooperation with the IBM Institute on the value of companies, and the results showed that the extensive use of information and its interpretation using modern data analysis techniques increases the company's performance. The company predicts and improves its expectations based on extracting important patterns of information from the data, such as analyzing consumer behavior, relationships with suppliers and competitors, and predicting future sales, and this allows companies to follow changes and take alternative measures when necessary, as well as avoid errors through the reduction of uncertainty [17].

Many studies have used machine learning to predict financial performance, with good results and even proving their superiority, as Lee et al. indicated that, as data processing technology advanced, studies began to use machine learning techniques to predict corporate performance and that many studies used the predicted corporate performance to make investment decisions. They used a model based on financial and technical information that reflected the company's technological competition. The study found good prediction results for performance, and one type of artificial neural network outperformed the other tools used. It also distinguished itself from others by significantly lowering prediction error [18].

The importance of narrative information in evaluating the company's performance and its value in the mandatory disclosure of information were also noted. However, machine learning and text mining have rarely been exploited to explore important information, which may have purposes similar to those of available financial data about the company's condition, such as disclosing data about the company's performance and maintaining market efficiency and integrity. The data mining techniques in the mandatory disclosures of annual reports are of value to investors when they are assessing the potential change in the company's financial performance in the long term [19].

Miyakawa et al. aimed to use machine learning methods to predict different measures of corporate performance and compare them with the predictive power of the scale designed by credit agencies by applying data on more than one million Japanese companies based on factors such as corporate characteristics and supply chain information over the years 2006–2011. Machine learning outperformed the credit measure. In addition, sales prediction and profit growth were greatly improved as measures used to predict financial performance, and in addition to that, the prediction model inputs that were selected by machine learning differed according to the performance measure being predicted [6].

Sinthupundaja et al. also used three methods of prediction, which are logistic regression as a traditional method, artificial neural networks, and support vector machines as modern methods (machine learning), to classify the company's performance into two groups; performance above or below the average performance in the next year based on the return on assets as a measure of the corporate financial performance. The developed models considered the independent variables of the company's internal and external factors. The results of this study include contributions from two sides; academic and administrative aspects. For the academic side, the results confirm that machine learning achieves better accuracy than logistic regression for rating prediction models. For the administrative side, these prediction models help managers and decision-makers track performance a year in advance and identify important business trends [20].

5 Conclusion

The advancement of data processing technology and machine learning algorithms has led to increased efforts to develop sustainable quantitative prediction models and update new training data. For example, the administration and economic sectors in the United States of America use machine learning algorithms such as artificial neural networks and decision trees to produce quantitative predictions. Machine learning techniques can be used to support corporate financial performance prediction, especially under the digital transformation environment that led us to use modern analytical techniques to obtain accurate results that are superior to traditional analysis techniques, taking into account that the proposed model of corporate financial performance prediction includes financial information such as financial ratios and non-financial information such as corporate governance. Also, new variables can be included, such as the social and environmental dimensions and other non-financial variables that can affect corporate financial performance.

References

1. Vintila G, Nenu EA (2015) An analysis of determinants of corporate financial performance: evidence from the bucharest stock exchange listed companies. Int J Econ Financ Issues 5(3):732–739
2. Bamiatzi V, Bozos K, Nikolopoulos K (2010) On the predictability of firm performance via simple time-series and econometric models: evidence from UK SMEs. Appl Econ Lett 17(3):279–282
3. Kourtzidis S, Tzeremes NG (2019) Investigating the determinants of firm performance: a qualitative comparative analysis of insurance companies. Euro J Manage Bus Econ
4. Leitner-Hanetseder S, Lehner OM, Eisl C, Forstenlechner C (2021) A profession in transition: actors, tasks and roles in AI-based accounting. J Appl Account Res
5. Mullainathan S, Spiess J (2017) Machine learning: an applied econometric approach. J Econ Perspect 31(2):87–106
6. Miyakawa D, Miyauchi Y, Perez C (2017) Forecasting firm performance with machine learning: evidence from Japanese firm-level data. RIETI
7. Mousa GA, Elamir EA, Hussainey K (2022) Using machine learning methods to predict financial performance: Does disclosure tone matter? Int J Discl Gov 19(1):93–112
8. Arinovic-Barac Z (2011) Predicting sustainable financial performance using cash flow ratios: a comparison between LDA and ANN method. Sarajevo Bus Econ Rev 31(1):33
9. Cong Y, Du H, Vasarhelyi MA (2018) Technological disruption in accounting and auditing. J Emerg Technol Account 15(2):1–10
10. Wasserbacher H, Spindler M (2021) Machine learning for financial forecasting, planning and analysis: recent developments and pitfalls. Digit Finan 1–26
11. Goodell JW, Kumar S, Lim WM, Pattnaik D (2021) Artificial intelligence and machine learning in finance: Identifying foundations, themes, and research clusters from bibliometric analysis. J Behav Exp Financ 32:100577
12. Lin SL, Huang HW (2020) Improving deep learning for forecasting accuracy in financial data. Discrete Dyn Nat Soc
13. OECD AI (2021) Machine learning and big data in finance: opportunities. Challenges, Implications Policy Makers
14. Wehle HD (2017) Machine learning, deep learning and AI: what's the difference. Data Sci Innov Day 2–5
15. Bertomeu J, Cheynel E, Floyd E, Pan W (2021) Using machine learning to detect misstatements. Rev Acc Stud 26(2):468–519
16. Mahoto NA, Iftikhar R, Shaikh A, Asiri Y, Alghamdi A, Rajab K (2021) An intelligent business model for product price prediction using machine learning approach. Intell Autom Soft Comput 29(3):147–159
17. Wroblewski J (2018) Digitalization and firm performance: are digitally mature firms outperforming their peers?.
18. Lee J, Jang D, Park S (2017) Deep learning-based corporate performance prediction model considering technical capability. Sustainability 9(6):899
19. Qiu XY, Srinivasan P, Hu Y (2014) Supervised learning models to predict firm performance with annual reports: an empirical study. J Am Soc Inf Sci 65(2):400–413
20. Sinthupundaja J, Chiadamrong N, Suppakitjarak N (2017) Financial predictions models from internal and external firm factors based on companies listed on the stock exchange of Thailand. Suranaree J Sci Technol 24(1)

Strategic Measures and Alternatives to Ensure the Financial Security of Machine-Building Enterprises

Galyna Azarenkova and Kateryna Oriekhova

Abstract The scientific article proposes analytical and applied support to analysis, estimating and forecasting threats of external and internal environment on formation of strategic measures financial security of machine-building enterprises. It is based on formation system of internal and integral financial indexes and estimation importance of external and internal environment threats. The scientific and practical approach to estimate effectiveness of strategic alternatives financial security management of machine-building enterprises is presented. It is based on choosing multi-criteria method for strategic alternatives financial security of machine-building enterprises. It is based on fuzzy sets, utility theory and simulation model for developing scenarios on ensuring financial security. It corresponds with current situations, taking into account forecast states for implementation of corrective actions in order to limit internal and external environment threats.

Keywords Capital movement · Financial interests · Financial security estimating · Financial security increasing · Minimization of financial security threats · Maximization of financial security

1 Introduction

Mechanical engineering is one of the system forming sectors of the national economy in Ukraine in terms of production and sales. This is due to the fact that mechanical engineering provides by means of production by other sectors of the economy (fuel, agro industrial, construction), thereby contributing to the renewal and accumulation of capital. In modern economic conditions, mechanical engineering plays an important role in accelerating scientific and technological progress.

G. Azarenkova (✉) · K. Oriekhova
V. N. Karazin, Kharkiv National University, Kharkiv 61022, Ukraine
e-mail: azarenkova.g.m@karazin.ua

K. Oriekhova
e-mail: oriekhova@karazin.ua

D. Magdi et al. (eds.), *Green Sustainability: Towards Innovative Digital Transformation*, Lecture Notes in Networks and Systems 753,
https://doi.org/10.1007/978-981-99-4764-5_4

The machine-building complex in Ukraine includes more than 20 industries, 58 sub-sectors, which employs 11,267 enterprises (146 large, 1834 secondary, 9287 small), with about 1.5 million employees. Despite the significant contribution of mechanical engineering to the development of the national economy in Ukraine, it should be noted that its share in the structure of industry for 2017–2021 decreased by 2.2% (2017—8.6%, 2018—7.1%, 2019—6.5%, in 2020—6.1%, in 2021—6.4%). In addition, in 2021, 22.6% of machine-building enterprises worked unprofitable, reaching a loss of 10 152.1 million UAH.

Problems of economic activity of machine-building enterprises in Ukraine are exacerbated through the use of outdated equipment and equipment; insufficiency of long-term sources of financing of current assets; low level of innovation activity and investment attractiveness; low competitiveness of products and insufficient number of skilled workers.

2 Literature Review

A significant number of domestic and foreign scientists such as: Alsayah Abedalsttar [1], Tarek Ghazouani [6], Imbrescu [10], Viktoriia Koilo [14], Tetiana Konieva [15], Viktor Koziuk [16], Tsaurai Kunofiwa [20], Tetiana Lepeyko [21], Morar [25], Oleksiv [28], Shmygol [37], Šlefendorfas [38], Timchev [40].

However, in their scientific research papers, the problem of providing financial security, which is extremely relevant for machine-building enterprises in Ukraine in the context of economic integration into the European Union, has not been appropriately reflected. This is primarily justified by such provisions. First, the financial security of machine-building enterprises depends on their ability to update equipment and technology, to form a reserve of production capacity, inventories of inventory. That is, financial security is able to ensure the viability of machine-building enterprises, their competitiveness and the possibility of development in the short and long term. Second, financially safe machine-building enterprises have advantages over other enterprises in the same industry in attracting investments, in the choice of suppliers, in the selection of skilled personnel. Third, financially safe machine-building enterprises do not come into conflict with the state to transfer taxes and non-tax payments and with society on payment of wages, dividends.

3 Materials and Methods

A set of methods of scientific knowledge that provided the conceptual unity of research is used. The methodological basis is the resource and functional approaches to the study of economic and social processes and phenomena, as well as the fundamental provisions of the theory of finance and probability theory and mathematical statistics.

The following methods were used to determine strategic measures and alternatives to ensure the financial security of machine-building enterprises in Ukraine: multi-criteria selection—to produce management decisions that correspond to a certain state of financial security of the enterprise; utility theories—for the choice and quantitative substantiation of the most effective management decisions.

The information base of the study is the annual financial reporting of machine-building enterprises in Ukraine; information obtained through sociological surveys of the collectives of these enterprises; scientific research papers of Ukrainian and overseas scientists on ensuring financial security.

4 Results

The aggressiveness of the influence of environmental threats and the unpredictability of the factors of the internal environment creates the need to develop a complex of effective tools for ensuring financial security, both the entire industry as whole and individual machine-building enterprises.

Further development of the script approach is the situational-resource approach proposed in the scientific research paper [2]. The basis of the implementation of the scientific approach is a general situation that characterizes possible resources of the enterprise.

The development of this direction in choosing a strategy is a functional resource adaptive approach, which was developed by Lisnichenko and Yermak [22]. Based on the provisions that describe situational-resource and functional adaptive approaches to the choice and development of an industrial enterprise strategy, taking into account the methodology of evaluation of financial security and its components [32], we offer a model of choosing a financial security strategy based on quantitative characteristics of complex evaluation of economic activity machine-building enterprises, their internal and external environment.

The introduction of a methodological approach to the choice and implementation of the strategy of ensuring the financial security of machine-building enterprises should be done through consistent use of appropriate means, the formation of the necessary information and the development of strategy options based on the use of matrix positioning.

As a means of implementing the financial security strategy of machine-building enterprises are:

First, the characteristics of financial security;

Second, the justification and selection of threats to the internal and external environment that influence it.

At the stage of formation of initial information, the current state of ensuring the financial security of machine-building enterprises by the following components is analyzed:

Actual and forecast level of financial security of machine-building enterprises;

Influence of threats to the external and internal environment on the financial safety of machine-building enterprises on fact and forecast.

The integral indicators of financial security, calculated by the method of complete reduction of factor space (method of taxonomic evaluation) allowed to obtain a bagotor quantitative and qualitative evaluation of economic activity of machine-building enterprises in the form of a complex integral indicator, which synthesizes the diverse influence of threats.

The values of the integral financial security indicator change in the range [0–1]: the greater the value approaches 1, the higher the level of financial security of machine-building enterprises.

The classification of financial security levels based on integrated ranges is highlighted, namely:

Financial security can acquire four levels: crisis, minor, normal, sufficient.

The influence of threats to the external and internal environment is classified by the degree of their influence: aggressive, moderate, weak and neutral. Four variants of financial security strategy are identified with the limits of the zones of each strategy.

The implementation of options for choosing a provision strategy is based on a matrix approach that has allowed identifying the areas of integral evaluation and the impact of financial security threats in four types [8, 9, 12, 13, 18, 25, 27, 28, 30, 42]. In particular, it's:

Growth (general nature of the offensive strategy);
Stabilization (the general nature of the waiting defence strategy);
Survival (the general nature of the forming defence strategy);
Diversification (the general nature of the offensive waiting strategy).

On the basis of the matrix approach, comparative financial security matrices and threats to the external and internal environment for each of the machine-building enterprises, taking into account the predictive data, were provided.

Using the information received and the matrix approach to choosing a strategy, the option of ensuring the financial security strategy of machine-building enterprises was selected (Table 1).

Thus, taking into account the current situation for PJSC "Kharkiv Machine-Building Plant "Shakhtar" and PJSC "Kharkiv Machine-Building Plant "Plinfa" for the forecast period, the most reasonable is the diversification strategy (Table 2), and for PJSC "HEAZ", PJSC "Kupyansky Machine-Building Plant", PJSC "Drohobych Machine-Building Plant", PJSC "Druzhkiv Machine-Building Plant" and PJSC "Odessa Machine-Building Plant"—survival strategy. The implementation and implementation of the strategies obtained will reduce the negative impact of internal and external threats on the financial safety of machine-building enterprises and to improve their effective activities.

In addition, the analysis showed that a significant level of financial security can be achieved even in the conditions of aggressive influence of threats (for example, PJSC "Kharkiv Machine-Building Plant "Shakhtar" in 2021), or, conversely, crisis—in the conditions of poor influence of threats (PJSC "Kharkiv Machine-Building Plant "Plinfa", 2018).

Table 1 Comparative matrix of financial security and threats to the external and internal environment of machine-building enterprises in Ukraine

Influence of threats to the external and internal environment	Aggressive	–	PJSC "Kharkiv Machine-Building Plant" "Shakhtar"	–	PJSC "Kharkiv Machine-Building Plant" "Plinf"
	Moderate	PJSC "HEAZ", PJSC "Kupyansk Machine-Building Plant"	PJSC "Drohobych Machine-Building Plant", PJSC "Druzhkiv Machine-Building Plant"	PJSC "Odessa Machine-Building Plant"	–
	Weak	–	–	–	–
	Neutral	–	–	–	–
		Significant	Normal	Insignificant	Crisis
		The total level of financial security of the enterprise			

Source According to financial reports of machine-building enterprises in Ukraine

Table 2 Financial security strategies machine-building enterprises in Ukraine

Enterprise	Option financial security strategy
PJSC "Kharkiv machine-building plant" "Shakhtar"	Diversification
PJSC "Kharkiv machine-building plant" "Plinf"	Diversification
PJSC "HEAZ"	Survival
PJSC "Kupyansk machine-building plant"	Survival
PJSC "Drohobych machine-building plant"	Survival
PJSC "Druzhkiv machine-building plant"	Survival
PJSC "Odessa machine-building plant"	Survival

Source According to financial reports of machine-building enterprises in Ukraine

This situation is due to the fact that internal and external threats do not always have a dominant effect on the overall financial security indicator, or cause a high level of its individual components. Therefore, when forming a financial security strategy in the face of threats, it is advisable to analyze their impact on financial security components. The procedure for determining the impact of internal and external threats on the financial safety of machine-building enterprises on the basis of the use of single-factor dispersion analysis is as follows:

As a factor used a four-level qualitative assessment of the impact of threats obtained scientific research papers [3, 11, 23, 39], on the basis of the obtained values of factorial variance, which characterizes the influence of the threats of the internal and external environment on local integral indicators of financial security, corresponding conclusions are formed: statistically reliable, the highest threat influence only on the use of capital of financial security (almost 0.70) and indicators of formation and distribution (0.54).

Thus, when producing and making management decisions within the framework of the above variants of strategies, it is advisable to set priorities for the formation, distribution and use of capital. The allocation of data components is due to the need to enhance the anti-digital orientation of the mechanism of management and development of a complex of preventive management decisions.

The development of management decisions within each variant of the strategy was carried out taking into account the results of cluster analysis [31, 35, 36, 42]. The establishment of clusters' profiles made it possible to orient the areas of implementation of financial security measures. Data corresponding to a high-level cluster of each component of financial security provides information to form management decisions for growth strategy, with average—for stabilization and low—for survival and diversification.

Table 3 was developed and according to it a list of management measures should be considered as a systematic list of typical management decisions.

The development of a set of reasonable management measures to ensure the financial security of machine-building enterprises, depending on a certain variant of the strategy is the main purpose of the decision-making mechanism for the formation and choice of a complex of preventive measures to ensure proper financial security.

Table 3 Complexes of typical management measures depending on the variant of the strategy for ensuring the financial security of machine-building enterprises

Growth	*Stabilization*
Capital formation Improvement of the profit distribution mechanism Change in the structure and growth of the value of shares Increase in equity due to additional issue of shares	Capital formation Improvement of the profit distribution mechanism Change in the structure and growth of the value of shares Increase in equity due to additional issue of shares
Capital allocation Repayment of the entire amount of receivables Lack of attraction of long-term and short-term capital in the form of a bank loan Increase in the amount of reinvested profit	Capital allocation Maximum possible repayment of receivables Reduction of attraction of long-term and short-term capital in the form of a bank loan Reduction of liabilities by reducing fixed and conditionally variable costs, reducing the terms of accounts payable for commodity transactions
Use of capital Capital investment in the form of long-term types of financial instruments (securities) Increase in innovative investments Use of currency risk hedging (forward/futures hedge, currency option) Creation of internal or external investment funds Development of the investment budgeting system (determination of forms of investment activity, sources of financing, structure of income and expenses)	Use of capital Cooperation with current investors, introduction of a system of benefits for various investors, attraction of foreign investments Insurance and investment guarantees for domestic and foreign investors Use of currency risk hedging (forward/futures hedge, currency option)
Survival	*Diversification*
Capital formation Increase in the authorized capital at the expense of the owners	Capital formation Increase in the authorized capital at the expense of the owners
Capital allocation Refinancing of receivables (factoring, forfeiting, bill accounting) Repayment of receivables by non-current assets Use of a targeted bank loan Restructuring of the company's debts Search and mobilization of reserves for cost savings for activities	Capital allocation Refinancing of receivables (factoring, forfeiting, bill accounting) Repayment of receivables by non-current assets Obtaining rehabilitation loans Covering losses at the expense of the company's own capital Use of various forms of debt restructuring Transfer of debt to another legal entity (guarantor, guarantor)
Use of capital Sale of part of unfinished construction objects Lease of part of fixed assets; Implementation of part of fixed assets Partial sale of own shares, debt obligations	Use of capital Sale of objects of unfinished construction Lease of the main part of fixed assets Implementation of the main part of fixed assets Partial/full sale of own shares, debt obligations

Source According to financial reports of machine-building enterprises in Ukraine

For this purpose, strategic alternatives were raised by multi-criteria selection based on additive rolls for each strategy and each of the directions of economic activity [8, 18, 30]. Thus, the proposed scientific and methodological approach to the choice of strategic financial security measures is a tool for improving the quality of strategic alternatives for development on the basis of the non-random choice of the most significant management measures that are adequate to the current situation and taking into account the forecast levels for the future (Table 4).

The purpose of the fourth stage is to formulate a list of the most appropriate strategic alternatives in terms of the effectiveness of their implementation using the use of utility theory [13, 36, 42]. The evaluation of the efficiency of the strategic decision is determined by the fuzzy expected utility of each alternative (possible measure) within a certain financial security strategy, which allows you to choose the most appropriate management decision, taking into account its usefulness of the adaptable for the specific mechanical engineering enterprises [34]. This approach has made specific recommendations for prompt decision-making in the form of a set of the most appropriate and analytically sound strategic alternatives based on the synthesis of national practice of forming a financial security strategy, taking into account the internal capabilities and potential of the enterprise in the areas of financial and economic activity. The proposed methodological tools for the choice of ways and means of strategic development and ensuring the proper level of financial security of mechanical engineering enterprises are an effective tool for improving the quality of strategic management in general.

The purpose of the fifth stage of the scientific and methodological approach is to evaluate the effectiveness of strategic alternatives to financial security management based on simulation experiments [4, 7]. The simulation modelling methodology allows to carry out simulation experiments under certain conditions and restrictions that reflect the results of specific management decisions made by the enterprise according to the financial security class, taking into account the dynamics of development and influence of threats to the external and internal environment, which is formed at the initial moment of the forecast period. In order to ensure effective practical implementation of specific management decisions, the methods of simulation will allow the management of mechanical engineering enterprises to introduce measures to control the purposefulness of the decisions made, as they allow to predict a certain result in the future.

The adequate situation of the external and internal environment is one of the central problems of strategic management. Its complexity is due, on the one hand, to significant uncertainty, so when choosing strategic alternatives of development, it is always necessary to predict the future, and on the other hand, the presence of a set of contradictory criteria, sometimes difficult to agree [11].

The task of choosing strategic alternatives to financial security may have a large number of productions, depending on the goals facing the enterprise, the available resource potential and the force of the influence of negative factors of the external and internal environment, the tendency of prompt decision-making to risk and possible forecasts for the future.

Table 4 Set of criteria for evaluating alternatives to management measures depending on the option of the strategy for ensuring the financial security of machine-building enterprises

Growth	*Stabilization*
Capital formation Improvement of profit sharing mechanism (a21) Change in structure and increase in share price (a22) Increase in equity due to additional issue of shares (a23)	Capital formation Improvement of profit sharing mechanism (a21) Change in structure and increase in share price (a22) Increase in equity due to additional issue of shares (a23)
Capital allocation Lack of attraction of long-term and short-term capital in the form of a bank loan (a11) Repayment of the entire amount of receivables (a12) Increase in the amount of reinvested profit (a13) Reduction of accounts payable for commodity transactions (a14) Search and mobilization of reserves for cost savings for activities (a15)	Capital allocation Maximum possible repayment of receivables (a11) Reduction of attraction of long-term and short-term capital in the form of a bank loan (a12) Reduction of liabilities by reducing fixed and conditionally variable costs (a13)
Use of capital Development of the investment budgeting system (determination of forms of investment activity, sources of financing, structure of income and expenses) (a31) Increase in innovative investments (a32) Use of currency risk hedging (forward/futures hedge, currency option) (a33) Creation of internal or external investment funds (a34) Capital investment in the form of long-term types of financial instruments (securities) (a35)	Use of capital Collaboration with current investors, implementation of a system of benefits for different investors, attracting foreign investments (a31) Insurance and investment guarantees for internal and external investors (a32) Using currency risk hedge (forward/futures hedge, currency option) (a33)
Survival	*Diversification*
Capital formation Increase in authorized capital at the expense of owners (a21) Issue of bonds and securities under the guarantee of third parties (a22)	Capital formation Increase in authorized capital at the expense of owners (a21) Sale of part of the stock on the stock exchange (a22)
Capital distribution Refinance of receivables (factoring, forfeiting, accounting of bills) (a11) Redemption of receivables by non-current assets (a12) Using a targeted bank loan (a13) Enterprise debt restructuring (a14) Search and mobilization of cost savings reserves (a15)	Capital distribution Refinance of receivables (factoring, forfeiting, accounting of bills) (a11) Redemption of receivables by non-current assets (a12) Obtaining remedial loans (a13) Cover losses at the expense of the enterprise's equity (a14) Using different forms of debt restructuring (a15) Transfer of debt to another legal entity (guarantor, guarantor) (a16)

(continued)

Table 4 (continued)

The use of capital Sale of part of unfinished construction objects (a31) Lease part of fixed assets (a32) Implementation of part of fixed assets (a33) Partial sale of own shares, debt (a34)	The use of capital Sale of unfinished construction (a31) Transfer of temporary management of investors or their authorized persons (a32) Leasing the main part of fixed assets (a33) Implementation of the main part of fixed assets (a34) Partial/full sale of own shares, debt liabilities (a35)

Source According to financial reports of machine-building enterprises in Ukraine

It is envisaged that the main purpose of prompt decision-making is to choose an effective strategic decision to ensure financial security under certain other conditions, with limited financial resources of machine-building enterprises.

On the basis of the analysis of scientific research papers [5, 17, 19, 24, 33, 41] and expert analysis, a set of criteria for evaluating alternatives is determined; the results of the criteria, depending on the option of financial security management strategy are presented in Table 5.

The resulting indicator of the model is the level of financial security of machine-building enterprises, the growth and decline of which is formed under the influence of factors:

Threats to the external environment at the current and previous time, formed by a certain set of factors of external influence;

Threats to the internal environment at the current and previous moments, formed by a certain set of factors of internal influence;

The dynamics of the indicator of financial security in the previous moments of time;

Management decision;

Scenarios for the development of situations.

5 Summary and Conclusion

As a result of the analysis of the developed rules, three classes of situations were formed: favourable; neutral and unfavourable. Simulation experiments have been conducted that reflect the results of management decisions made by the enterprise managers according to the class of the situation, which is drawn up at the initial moment and taking into account the forecast level of financial security for generating six possible scenarios:

Table 5 Set of criteria for evaluating alternatives to management decisions depending on the option of the strategy for ensuring the financial security of machine-building enterprises

Growth	*Stabilization*
Capital formation c1—risk from losses; c2—business activity; c3—management costs; c4—the capital structure of the enterprise	Capital formation c1—share price; c2—management costs; c3—business activity; c4—the risk of losses
Capital allocation c1—the cost of preparing documentation; c2—the cost of implementing projects; c3—the risk from losses that the company may incur; c4—interest rate for using a bank loan; c5 is the time of implementation of strategic development projects	Capital allocation c1—the cost of preparing documentation; c2—time of project implementation; c3—interest payments for the loan; c4—the risk of losses
Use of capital c1—the risk from losses that the company may incur; c2—the cost of preparing documentation; c3—product profitability; c5—time of project implementation; c_6—quality of products (goods, works, services)	Use of capital c1—management costs; c2—profitability of production; c3—the cost of assessing sales markets; c4—the risk of losses
Survival	*Diversification*
Capital formation c1—the cost of paying for brokerage services; c2—share price; c3—risk from losses; c4—management costs; c5—social significance of the enterprise	Capital formation c1—the cost of paying for brokerage services; c2—management costs; c3—risk from losses; c4—is the share price
Capital allocation c1—management costs; c2—the cost of preparing documentation; c3—time of project implementation; c4—availability of highly liquid assets	Capital allocation c1—the cost of preparing documentation; c2—availability of highly liquid assets; c3—interest payments for the loan; c4—risk from losses; c5—the price of the bond; c6—social significance of the enterprise
Use of capital c1—risk of loss; c2—business activity; c3—management costs; c4—product profitability	Use of capital c1—risk of loss; c2—business activity; c3—management costs; c4—product profitability

Source According to financial reports of machine-building enterprises in Ukraine

basic scenario of changing the integral indicator of financial security with certain initial and ultimate levels without taking into account the impact of threats and the lack of certain management actions (prompt decision-making does not make any decision);
a scenario that corresponds to the aggressive level of influence of threats to the external and internal environment within the specified intervals on the level of general financial security indicator and the lack of certain management actions;
a scenario that corresponds to the poor level of influence of threats to the external and internal environment within the defined intervals on the level of general financial security indicator and the lack of certain management actions;
a scenario that corresponds to the neutral level of influence of threats to the external and internal environment within the defined intervals on the level of general financial security indicator and the lack of certain management actions;
a scenario that corresponds to the aggressive level of influence of threats to the external and internal environment within the specified intervals on the level of general financial security indicator and the implementation of certain management actions for a given situation in priority;
a scenario that corresponds to the neutral level of impact of threats to the external and internal environment within certain intervals on the level of general financial security indicator and the implementation of certain management actions that are most effective for this situation and
a scenario that corresponds to a poor level of impact of threats to the external and internal environment within certain intervals on the level of general financial security and the implementation of certain management actions, a progressive reaction (making effective risky management decisions adapted to the situation).

Implementation of the developed approach through the use of methods of multi-criteria choice and utility theory, is a tool for improving the quality accepted for the implementation of development alternatives on the basis of non-random choice of complex measures for the appropriate level of financial security, taking into account their utility for mechanical engineering enterprises in the conditions of limited financial resources. In addition, it is an effective tool for improving the quality of strategic management for Ukrainian enterprises.

References

1. Alsayah A (2022) Strategic alignment and its impact on creating an organization's reputation and image. Prob Perspect Manage 20(1):501–513. http://dx.doi.org/https://doi.org/10.21511/ppm.20(1).2022.40
11. Alzate I, Manotas E, Boada A, Burbano C (2022) Meta-analysis of organizational and supply chain dynamic capabilities: a theoretical-conceptual relationship. Prob Perspect Manage 20(3):335–349. http://dx.doi.org/https://doi.org/10.21511/ppm.20(3).2022.27
3. Bielykh A, Pysarenko S, Dong Meng R, Kubatko O (2021) Market expectation shifts in option-implied volatilities in the US and UK stock markets during the Brexit vote. Invest Manage Financ Innov 18(4):366–379. http://dx.doi.org/https://doi.org/10.21511/imfi.18(4).2021.30

4. Babenko V, Demyanenko O, Lyba V, Feoktystova O (2021) Assessment the cost-effectiveness of information support for the business processes of a virtual machine-building enterprise in the framework of industry 4.0. Int J Eng Trans A: Basics 34(1):171–176. http://dx.doi.org/https://doi.org/10.5829/IJE.2021.34.01A.19
5. Dobrowolski Z, Sułkowski Ł, Panait M (2022) Using the business model canvas to improve audit processes. Prob Perspect Manage 20(1):142–152. http://dx.doi.org/https://doi.org/10.21511/ppm.20(3).2022.12
6. Ghazouani T (2021) Investigating the effects of environmental taxes on economic growth: Evidence from empirical analysis in European countrie. Environ Econ 12(1):1–13. http://dx.doi.org/https://doi.org/10.21511/ee.12(1).2021.01
7. Guryanova L, Yatsenko R, Dubrovina N, Babenko V (2020) Machine learning methods and models, predictive analytics and applications. In: CEUR Workshop Proceedings, 2020, 2649, pp 1–5. http://ceur-ws.org/Vol-2649/
12. Horvathova J, Mokrisova M, Petruška I (2022) Indebtedness and profitability—a threshold model approach. Invest Manage Financ Innov 19(3):13–27. http://dx.doi.org/https://doi.org/10.21511/imfi.19(3).2022.02
8. Hossam HS (2022) The impact of ownership structure on external audit quality: a comparative study between Egypt and Saudi Arabia. Invest Manage Financ Innov 19(2):81–94. http://dx.doi.org/https://doi.org/10.21511/imfi.19(2).2022.07
9. Hsu HW (2022) CEO educational backgrounds and non-GAAP earnings disclosures. Invest Manage Financ Innov 19(3):167–175. http://dx.doi.org/https://doi.org/10.21511/imfi.19(3).2022.14
10. Imbrescu CM (2016) Insolvency of the group of companies. Agricult Manage Lucrari Stiintifice Seria I, Manage Agricol 18(2):223–226
2. Iyoha AOI, Ohiokha G, Umoru D, Akhor SO, Igele GA (2022) Target capital structure for managerial decision making: dynamics and determinants. Invest Manage Financ Innov 19(3):322–334. http://dx.doi.org/https://doi.org/10.21511/imfi.19(3).2022.27
30. Kayode OS, Sibanda M, Olarewaju OM (2022) Analyzing the determinants of financial management behavior of administrators in Nigerian state-owned enterprises. Invest Manage Financ Innov 19(3):278–290. http://dx.doi.org/https://doi.org/10.21511/imfi.19(3).2022.23
14. Koilo V (2020) A methodology to analyze sustainable development index: evidence from emerging markets and developed economies. Environ Econ 11(1):14–29. http://dx.doi.org/https://doi.org/10.21511/ee.11(1).2020.02
15. Konieva T (2021) The impact of financing policy on the cost of debt. Invest Manage Financ Innov 18(4):177–189. http://dx.doi.org/https://doi.org/10.21511/imfi.18(4).2021.16
16. Koziuk V (2022) What do cross-country Bitcoin holdings tell us? monetary and institutional discontent vs financial development. Invest Manage Financ Innov 19(1):168–185. http://dx.doi.org/https://doi.org/10.21511/imfi.19(1).2022.13
17. Kozlovskyi S, Petrunenko I, Mazur H, Butenko V, Ivanyuta N (2022) Assessing the probability of bankruptcy when investing in cryptocurrency. Invest Manage Financ Innov 19(3):312–321. http://dx.doi.org/https://doi.org/10.21511/imfi.19(3).2022.26
18. Kozmenko S, Korneyev M (2017) Formalization of the impact of imbalances in the movement of financial resources on economic growth of countries in Central and Eastern Europe. Account Financ Control 1(1):48–58. http://dx.doi.org/https://doi.org/10.21511/afc.01(1).2017.06
19. Kravchenko M, Solntsev S, Babenko V, Zhygalkevych Z (2020) Applying sustainable innovations for the development of Ukrainian machine-building enterprises. Int J Technol Manage Sustain Develop 19(3):279–296. https://doi.org/10.1386/tmsd_00027_1
21. Lepeyko T (2020) On the waves of economic fluctuations. Environ Econ 11(1):65–66. http://dx.doi.org/https://doi.org/10.21511/ee.11(1).2020.06
22. Lisnichenko O, Yermak S (2016) Development of the business model of the enterprise in the context of forming value for stakeholders. Econ Space 07:189–198. http://nbuv.gov.ua/UJRN/ecpros_2016_107_20
23. Mansouri A, Nazari A, Ramazani M (2016) A comparison of artificial neural network model and logistics regression in prediction of companies' bankruptcy (a case study of Tehran stock exchange). Int J Adv Comput Res 6(24):81–92

25. Modern determinants of fiscal policy: local and international dimension: monograph (2016). In: Krysovatyy A, Gospodarowicz A (eds) Wroclaw 282
26. Morar GC (2016) Factors influencing insolvency at the level of companies. Manage Intercult 18(2):135–142
27. Moskalenko N, Romanenko O, Oliinyk T (2015) Approaches to enterprises' financial and economic security management. Econ J XXI. 7–8(1):54–57. http://nbuv.gov.ua/UJRN/ecchado_2015_7-8(1)__14
35. Nagarakatte SG, Natchimuthu N (2022) Impact of Brexit on bond yields and volatility spillover across France, Germany, UK, USA, and India's debt markets. Invest Manage Financ Innov 19(3):189–202. http://dx.doi.org/https://doi.org/10.21511/imfi.19(3).2022.16
28. Odintsova T, Nevmerzhytska O, Chaus O (2013) CVP-analysis as a financial results management tool. Financ Space (1):96–97. http://nbuv.gov.ua/UJRN/Fin_pr_2013_1_21
29. Oleksiv I (2014) Theoretical and practical approaches to identification of stakeholder interests of the company. Econ, Entrepreneurship, Manage 1(1):31–36. http://nbuv.gov.ua/UJRN/EEM_2014_1_1_7
31. Pasko O, Zhang L, Bezverkhyi K, Nikytenko D, Khromushyna L (2021) Does external assurance on CSR reporting contribute to its higher quality? Empirical evidence from China. Invest Manage Financ Innov 18(4):309–325. http://dx.doi.org/https://doi.org/10.21511/imfi.18(4).2021.26
32. Puspitaningtyas Z, Toha A, Prakoso A (2018) Understanding the concept of profit as an economic information instrument: disclosure of semantic meanings. Account Financ Control 2(1):27–36. http://dx.doi.org/https://doi.org/10.21511/afc.02(1).2018.03
33. Ramazanov S, Babenko V, Honcharenko O, Moisieieva N, Dykan V (2020) Integrated intelligent information and analytical system of management of a life cycle of products of transport companies. J Inf Technol Manage 12(3):26–33. https://doi.org/10.22059/jitm.2020.76291
24. Sari MP, Budiarto A, Raharja S, Utaminingsih NS, Risanda A (2022) The determinant of transfer pricing in Indonesian multinational companies: moderation effect of tax expenses. Invest Manage Financ Innov 19(3):267–277. http://dx.doi.org/https://doi.org/10.21511/imfi.19(3).2022.22
36. Shapoval Y (2021) Relationship between financial innovation, financial depth, and economic growth. Invest Manage Financ Innov 18(4):203–212. http://dx.doi.org/https://doi.org/10.21511/imfi.18(4).2021.18
37. Shmygol N (2014) Retrospective analysis of the United States experience in taxation, financial support to enterprises and its application for the formation of strategic objectives. Bull Zaporizhzhia Nat Univ Econ Sci (1):135–143. http://nbuv.gov.ua/UJRN/Vznu_eco_2014_1_18
38. Šlefendorfas G (2016) Bbankruptcy prediction model for private limited companies of Lithuania. Ekonomika Econ 95(1):134–152
13. Tchapchet-Tchouto JE, Koné N, Njoya L (2022) Investigating the effects of environmental taxes on economic growth: evidence from empirical analysis in European countrie. Environ Econ 13(1):1–15. http://dx.doi.org/https://doi.org/10.21511/ee.13(1).2022.01
40. Timchev M (2016) Accounting and balanced business analysis of the insolvency (bankruptcy) risk and of the company competitiveness in time of global economic crisis. Inst Account Control Anal Globalization Circumstances. (2):116–124
20. Tsaurai K (2022) Role of financial sector development in foreign direct investment inflows in BRICS. Invest Manage Financ Innov 19(3):215–228. http://dx.doi.org/https://doi.org/10.21511/imfi.19(3).2022.18
39. Vuong THG, Dao TH, Le TTH, Nguyen HM (2022) Debts and corporate cash holdings: Evidence from ASEAN-5. Invest Manage Financ Innov 19(1):186–200. http://dx.doi.org/https://doi.org/10.21511/imfi.19(1).2022.14
34. Wats S, Sikdar C (2016) Dynamics in futures and spot markets: a panel study of advanced and emerging economies of Asia. Invest Manage Financ Innov 19(1):64–76. http://dx.doi.org/https://doi.org/10.21511/imfi.19(1).2022.05

41. Yeletskyh SY (2014) The elements of innovative management in a financially sustainable development of industrial enterprises. Market Innov Manage. (4):84–90. http://nbuv.gov.ua/UJRN/sonkp_2014_4_11
42. Zhuravlyova IV, Lelyuk SV (2014) Management of enterprise financial security and its intellectual component based on creating multiagent decision support system. Actual Prob Econ. (7):163–170. http://nbuv.gov.ua/UJRN/ape_2014_7_21

Digital Transformation as a Tool for Implementation of the "Green Deal" Concept in the National Economy of Ukraine

Victor Zamlynskyi, Irina Kryukova, Olena Chukurna, and Oleksii Diachenko

Abstract Ensuring decent living conditions both present and future generations in an ecologically clean environment, restoring biological resources and preserving resource potential, and achieving equality and justice in society are ambitious and strategic tasks for the world's population. The solution of these vital tasks within the European space is achieving due to the introduction into practice of tools of the Green Deal concept, which takes into account the most relevant issues for the population of Europe and the world. The purpose of the study is to analyze the state of transformational transformations within the framework of this strategic program in Ukraine and to identify the impact of digitalization on the fulfillment of its main tasks. The results of the research showed that Ukraine is actively integrating into the European space and has adopted the key priorities and goals of the Green Deal as strategic steps for social and economic development for the period up to 2030. The key sector for green transformations in the economy of Ukraine was identified as the agrarian sector, as well as industry and transport. It was determined that digitalization is one of the main tools that will be used to implement the Green Deal tasks in the near future. The digital economy creates for business and society the basic advantages necessary for green transformations: information availability, transparency, partnership opportunities and global presence, and cybersecurity. At the same time, the level of development of digital processes in the economies of countries in the world today differs significantly. The leaders of digital transformations are the USA, China, Hong Kong, Sweden, and Denmark. Today, the digital economy of the

V. Zamlynskyi (✉) · O. Chukurna
State University of Intelligent Technologies and Telecommunications, Odessa 65023, Ukraine
e-mail: zam.agrariy@gmail.com

O. Chukurna
e-mail: elenchukurna@ukr.net

I. Kryukova · O. Diachenko
Odessa State Agrarian University, Odessa 65012, Ukraine
e-mail: kryukovaia1@rambler.ru

O. Diachenko
e-mail: oleksa1983.03@gmail.com

D. Magdi et al. (eds.), *Green Sustainability: Towards Innovative Digital Transformation*, Lecture Notes in Networks and Systems 753,
https://doi.org/10.1007/978-981-99-4764-5_5

EU countries is rapidly overcoming the lag behind the USA and uses the powerful capabilities of digital tools to solve the tasks of the Green Deal, improve the quality of human development level, ensure gender equality, and prevent climate change. In Ukraine, activation of the processes of digitization of the national economy will allow for additional growth of the national GDP by 4% per year.

Keywords Digital economy · European Green Deal · Digital transition · Architecture of sustainable development

1 Introduction

The active development of innovative technologies of social production, high rates of intensification of national economies, the desire to increase the level of competitiveness, and profitability of business over the past decades have led to the need to develop joint measures to ensure the preservation of bioresource potential and the solution of environmental and social problems of society not only within the national scale, but also at the level of the international economy. The set of views, decisions, and actions of the world community in the direction of solving the most urgent problems of the development of the economy and society led to the emergence of the concept of sustainable development, whose tenets are now prevalent in the world's leading nations.

2 Literature Review

Aspects of ensuring sustainable food ecosystems, protecting the environment, preserving natural and biological resources, preventing climate change, promoting sustainable urban and community growth, addressing poverty and inequality, ensuring equal access to quality education, and promoting gender equality, as well as fostering socially responsible production and consumption (Fritsche et al. [1]). Searching for such anthropogenic influences on the potential of natural resource part of modern methods to sustainable development will ensure the possibility of full restoration of resources for current and future generations Freitag et al. [2]. In an inextricable relationship, the environmental and production-economic components are considered with the social factor and without a high level of knowledge and responsibility, and it is impossible to solve the task of sustainable development. So, sustainable development should benefit natural resources and the environment and only secondarily satisfy the needs of society. Global trends in sustainable development and the implementation of interstate and national programs for their implementation in practice have not brought the expected results, which indicates an unfinished global concept of sustainable development until 2030, adopted at the UN summit in 2015. The universal principle of the concept of "sustainable development" is to provide for the needs of the present generation without risks and threats for future generations to provide for their own. The main emphasis of the concept of sustainable development of the national economies of the world is the orientation toward meeting the needs of humanity (under the conditions of improving the quality of life of the population) under the conditions of preservation and restoration of bio- and ecosystems.

Today, within the framework of the joint policy of the EU countries, the content of further sustainable development within the single economic space is interpreted within the framework of the "Green Deal" concept—the strategic program of the "Green Economy", which the EU launched in 2019. The European Green Deal was adopted for the existing hazards and threats to the generations and nations of Europe's future as the first project in the world to transform Europe into a climate-neutral continent and strengthen its unity [3].

The strategy of sustainable development for the next eight years should ensure: a neutral or positive impact on the environment and natural resources, mitigation of the consequences of climate change, prevent further loss of biodiversity, and ensure food security and public health through safe and ecologically clean food (from the farm to the table) [4]. As noted by Ionescu et al. [5], in addition to environmental bonuses, the Green Deal master plan should provide additional benefits and socio-economic effects in the form of a reduction in unemployment, raising the level of education of the population of European countries, strengthening the level of convergence and unity of nations, taking into account cultural and logical aspects, improving basic conditions, and an acceptable quality of life for the population of the regions.

3 Materials and Methods

The research used abstract, analytical and empirical methods, methods of economic and statistical research, methods of graphical display of the results of research, and critical analysis of indicators of the level of digital development of the economies of the world. Scientific literature was used, which contains analytical studies by scientists regarding the content and effectiveness of the implementation of the strategic European Green Deal course, assessment and opportunities of digital transformations in the development of society, the dynamics of digital processes in the world, and their advantages and potential benefits for the national economies of the world and society. The theoretical and methodological base and statistical data showing the importance of the development and carrying out the duties for the achievement of the global goals of sustainable human development and the implementation of the tasks of the Green Deal course were investigated. On the basis of information from scientific sources, reports of the European Commission, the study of expert opinions of global analysts, the trends that take place in the modern processes of digitalization development, their possible impact on the probability of achieving the indicators, and indicators provided for by the adopted Green Deal agreement were identified. Based on the systematization of the key trends of digitalization of social life, the views and opinions of scientists and experts, a critical view of the practical basis of the advantages of the digital economy over the traditional one was formed. Using the abstract logical method and the method of statistical observations, an assessment of the most developed digital economies of the world was carried out.

4 Results

The strategic guidelines for the implementation of the Green Deal concept have many common directions with strategic goals for the development of the economy of Ukraine, in particular: the National Economic Development Strategy for the period until 2030, the Food Security Strategy—2030, the projects of the Strategy for the Development of the Agro Sector until 2030, and Sustainable Development Strategy of Ukraine until 2030. The implementation of the key tasks of the strategic initiative of the European Union Green Deal is to build a perspective, resource-efficient, and competitive economy based on the reduction of carbon dioxide emissions; ensuring economic growth that is not determined purely by resource factors and formation of a socially just and equal society. The strategic priorities of the Green Agreement for European countries are: preservation of clean resources and biodiversity; increasing energy efficiency; healthy and affordable food; transition to new, larger clean energy sources, and active implementation of innovative technologies; transformational processes in the direction of the development and spread of a closed economy; creation of new jobs; and increasing the level of competitiveness of industries and national economies. European Environment Agency [6] for the period up to 2030, the key areas of implementation of transformational changes of the Green Deal are: climate; affordable, safe, and sustainable energy; environment and oceans; agriculture, wild nature, transport, industry; research and innovation; finance and regional development; and the new European Bauhaus. The main principles of the implementation of the Green Deal concept in the European space are defined as: sustainability, inclusiveness, equality, and security. Today, the economy of Ukraine and Ukrainian society aspires to European values and strategic priorities. Current national programs and plans for the development of industries and activity types and transformations of regions and society all incorporate the core tenets of the European Green Deal effort. At the level of a methodical approach to the implementation of the Green Deal concept for the national economy (for all types of socio-economic activity), three main elements are dominant: (1) stability in the environment; (2) economic stability; and (3) social stability. All three elements ensure a logical relationship and the implementation of common goals and objectives in the context of the Green Agreement: natural resources are the basis of the organization of the production process in the agricultural sector, production results from the prerequisites for economic growth, and the provision of individual and public social development (Fig. 1).

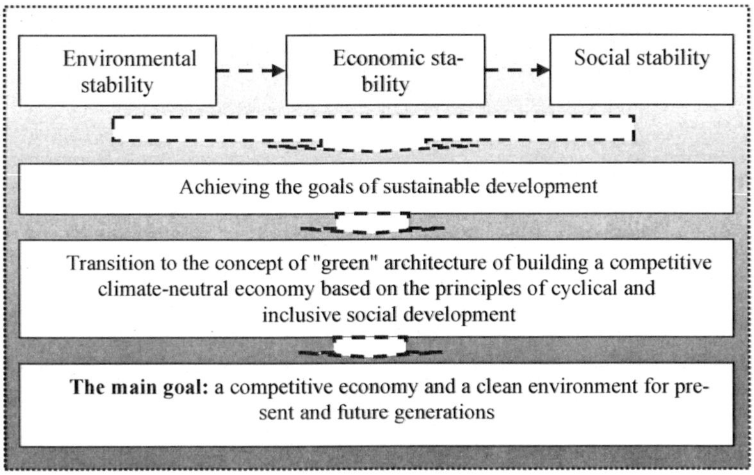

Fig. 1 Architecture of sustainable development of the national economy of Ukraine. *Source* Compiled by the authors

The economy of Ukraine now is mostly agrarian, the agriculture of which strategically determines the key priorities of development, forms the export potential, up to a quarter of foreign exchange earnings, and the basis of food security of the population. The agricultural sector is fundamental for solving the tasks of the Green Deal both for Ukraine and for the EU countries. Practical analyses of the existing condition of Ukraine's national economy's transformations in the context of completing the duties under the Green Deal allowed it to be determined that the main measures of the joint implementation of the Green Course should be: (1) ensuring the quality and safety of food products and agricultural raw materials; (2) promotion of products on world markets; (3) financial support for ecologically clean agricultural production, stimulation of the introduction of eco-technologies, and measures of gradual transition to the "green economy"; (4) financial and institutional support for the development of rural areas; (5) increasing the level of innovative activity of agricultural production on the basis of eco- and biotechnologies; (6) solving the problems of low quality of life of the population of rural areas; and (7) preservation and revival of the natural resource potential of the agricultural sector for future generations (Fig. 2).

In addition to the agrarian sphere, the following are extremely important areas of involvement of the national economy of Ukraine in the implementation of tasks together with European practice in the near future (until 2030) of the Green Deal: (1) improvement of the management of sustainable development of territories and types of activities; (2) social protection of the rural population, improvement of living conditions, ensuring gender equality of the population; (3) development of transport infrastructure; (4) availability of quality education; (5) providing the population of Ukraine with quality medical care; (6) increasing the level of employment and

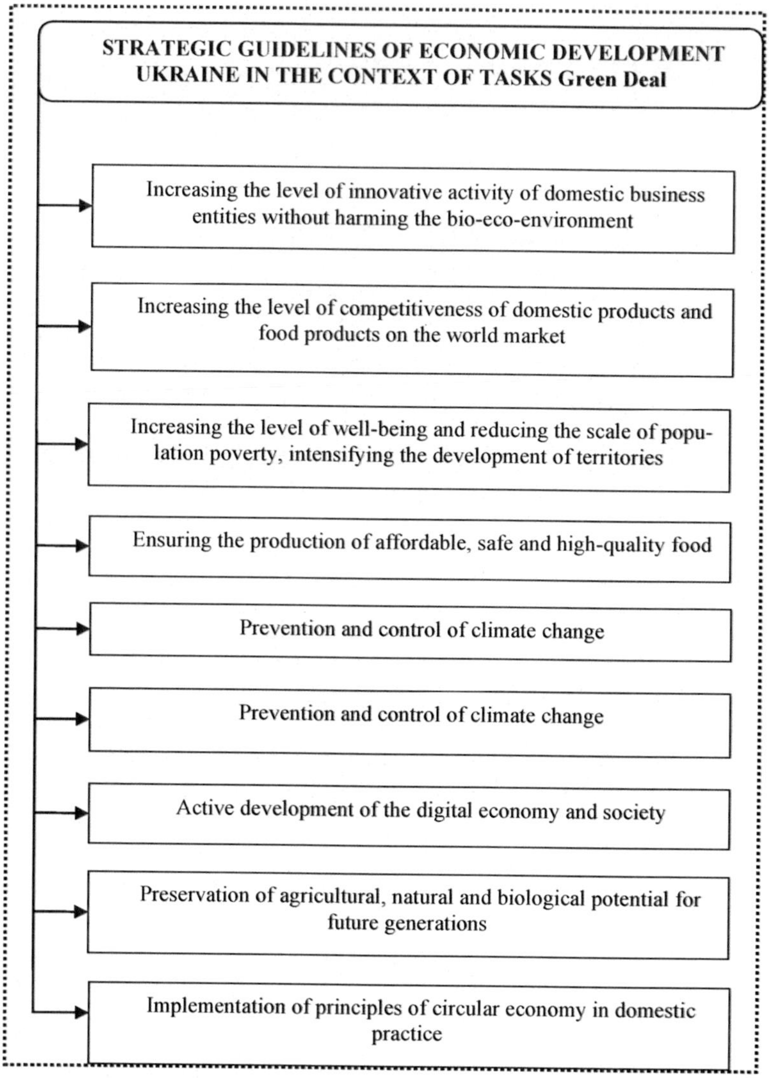

Fig. 2 Strategic guidelines for the sustainable development of the national economy of Ukraine in the context of European priorities

reducing unemployment; (7) diversification and increasing the level of innovativeness of the national economy; (8) inclusive development of entrepreneurship and small businesses; (10) rational use of natural resource potential and preservation of biodiversity; (11) man-made and ecological safety of regions and territories; and (12) financial, logistical, and innovative support for the sustainable development of regions.

Today, Ukraine takes an active part in the implementation of the global initiative of transition to the principles of sustainable development of the agricultural sector of the national economy. Together with the World Bank, studies on climate change modeling and assessment of their consequences for agriculture and the economy as a whole have been completed in the country. The results of this study should help the agro-entities of Ukraine to adapt their economic activities to the requirements of world practice and the European strategy of sustainable development for the period 2023–2030. To accomplish these objectives, the IFS Company is initiating a new financial support project aimed at ensuring the stable and strategic sustainability of Ukraine's agricultural sector. The project is set to provide funding amounting to 11 billion USD in the USA [7]. The primary focus of the project's implementation will be the introduction of standards and technologies for assessing greenhouse gas emissions (CO_2).

Support for Ukraine's policy in the context of the Green Deal initiative is also provided by other international organizations, in particular FAO. Thus, within the framework of the cooperation agreement, the FAO provides assistance in the implementation of projects in the following key areas: (1) development of agro-food production and assistance to small farms in accessing markets; (2) strengthening of producer relationships within the parameters of technology connections; (3) consulting on sustainable use of natural resources; (4) provision of technical support in matters of formation of the agricultural land market; (5) enhancing management of healthy food, safety, and hygiene; and (6) provision of assistance in the conservation of resources within territories with degraded landscapes. The key tasks within the framework of the intergovernmental program in the realm of enduring stability are: overcoming rural poverty, ensuring the sustainability of natural resources, food security of the country's population, and joint fight against decline in the climate-related factors in Ukraine [8].

Change to the inclusive agro-economy tracks occurs through the achievement of specific goals and indicators, including: a decrease in the amount of used (sold) pesticides, mineral fertilizers, and other chemical substances (as a result of the transition of farmers to eco-agro technologies); reducing CO_2, increasing the area of crops under organic production; increasing the share of organic agro-products on the market; agricultural products; the volume of exports of food for people across the world, increasing the gross agro-added value of the sector in the Ukrainian economy. For example, the joint agro-policy of the EU adopted strategic indicators until 2030 as the basis for further indicators: a 50% reduction in the use of harmful chemicals in agricultural production; the share of agricultural land under organic production is at least 25%; reduction of antimicrobial sales by 50%; reduction of the use of mineral fertilizers by 20%; and the share of agricultural land with a diverse landscape is at least 10%. The strategic plan of the EU SAP, which will come into effect in 2023, envisages a significant impact of the "green" course on the sustainable development of agriculture and rural areas [9]. The strategic development programs of EU must demonstrate imperatives in the field of environmental protection and the consistency with which a shared strategy is carried out aimed at preventing climate change [10]. The priority of financial support (in the amount of up to 35% of the common agro-budget of the

EU countries) will be ecological schemes for ensuring and supporting safe technologies of agro-production, preserving the biodiversity of the animal and plant world, protecting landscapes, supporting the climate, and developing the bio-economy of rural regions. Achieving the set tasks has a clear mechanism of control by the EU institutional management bodies, requirements for company reporting, specific indicators, and indicators that will be key for discussing issues of further cooperation and participation of Ukraine in international projects, programs, and processes of further European integration. Key tools for achieving the goals of the Green Deal concept should be science, innovation, modern technologies, as well as sufficient financial support for strategic and current development programs. Scientists today agree that digitalization can be one of the most effective tools for achieving the goals of the Green Deal [11]. In the coming years, digital tools and technologies will rely on digital technologies and tools. One of the promising tools is already the Internet of Things, which can expand and enhance the nutritional value of humans with safe and high-quality food products [12]. Today, the Web of Things is used in the formation of supply chains and the sale of goods of the circular economy and forms platforms and methods of extending the life cycle of goods, creating markets for used goods and returning them to the production cycle. Artificial intelligence is used as a means of optimizing energy production and water environment clean-up [13]. Digital smart technologies help in solving challenges with inclusivity in the industry equal access assets, markets, and jobs, thereby contributing to increasing the level of happiness for all facets of humankind. Use of internet information network tools is unrestricted which are defined in the spectrum of combating climate change and biodiversity loss. Digitization provides free access to large amounts of information and provides opportunities for building a sustainable society of the future. Digitization is the integration of modern online information resources into everyday life, thanks to which new opportunities are formed and the quality of life increases [14]. The exclusive property of digitalization of Appio et al. [15] considers the ability to provide information necessary to justify decisions for effective utilization of resources, finances, and assets to have a significant impact on achieving the goals of the Green Deal. Digitization, as indicated by Habibi and Zabardas [16], today appears to be an effective tool for achieving the goals of reducing inequality and poverty of the all people that fall under the heading of national people thanks to the provision of online access to quality education. Digital technologies expand the possibilities of agrarian management, that is, inclusive and contribute to land evaluation, soil condition monitoring, control of plant and animal, food quality, the development of inclusive agricultural relationships, and the circularity principles in agricultural ecosystems [17]. In the opinion of scientists, agro-food systems should be the main area of digital transformations, which will form the prerequisites for solving the global problem of food security [18]. In general, digitalization in the near future is the technological basis on which industrial industry 4.0 is based, where the dominant elements are not only automation, but also the use and management of smart objects and technologies [19]. Smart management, smart production, smart technologies, and smart products and services will form the basis of socio-economic relations of the future.

A novel system of social and industrial interactions, the economy of online information data can help all of its players create likely preferences. Fundamentally new methods and equipment that are connected to fresh database formats and applications appear at the core of such a system. The internet serves as the primary system foundation in this instance. A key prerequisite for the activation of the date economy is the network sector of internet life, which is a digital software distributor that become tools for a strengthening of social, personal, and business ties. The agency WEA (Bureau of Economic Analysis, USA) determines the presence of signs of the digital economy by three key-type variables: (1) the use of computers and the availability of online data, primarily ICT products and services; (2) efficiency and effectiveness of online commerce; and (3) volume, cost, speed of services, and satisfaction with them for consumers [20].

Experts' extensive assessment work's findings demonstrate that the digital economy forms four basic long-term rewards for potential business: information availability, transparency, proximity (of suppliers and buyers), global presence, and cybersecurity. The International Centre for Strategic Studies adds to the set of these elements: development of personnel skills, flexible regulatory tools, customer protection, and financial literacy [21].

The systematization of the advantages of the processes of further digitalization made it possible to determine their list in the following: prompt and convenient access to the necessary information required (for both personal and professional use); providing access points for sources of resources, capital, and data for stakeholders; minimization of barriers in the international relations that are becoming more intense; increasing the level of efficiency and effectiveness of production and individual choices; reducing costs, increasing competitiveness; and optimization of supply chains due to optimization (reduction) of the number of market intermediaries.

In the line with organizational mechanism of its actions, the digital economy forms the prerequisites for the development of an inclusive economy. Fair chances and universal access to network resources can serve as the cornerstones of an inclusive socio-economic system. Profitability, brand recognition, and other aspects of the participants are also irrelevant. At the same time, the use of digital tools creates not only an economic, but also a powerful social effect, which is manifested in the improvement of the quality of life of the population, the expansion of employment opportunities, self-realization, additional personalization, and the improvement of communications.

At the same time, the degree of intensification of digital processes in the world is characterized by significant differences, and the processes of digital transformations are characterized by specific signs, in particular: the presence of an uneven distribution of access to digital potential; cybercrimes; imperfection of data security mechanisms; violation of the privacy boundaries of the participants; emergence of digital monopolies; and the emergence of addiction in society (as a psychological factor in the use of the application of information solutions).

The inevitability of the transformational processes determines the further active development of digitalization with different levels of its intensities and advantages within national economies. The main indices characterizing the digitization of the world include: Digital Evolution Index, DiGiX, Networked Readiness Index, IMD World Competitiveness, DESI, Bloomberg Innovation Index, Digital IMD Index, and ICT Development Index.

The top five nations in the world ranking of the European Center for Digital Competitiveness have been showed leading countries in the digital economy in 2021: USA, China, Hong Kong, Sweden, and Denmark. In the block of G20 countries, China has been the undisputed leader for the past 3 years, and Saudi Arabia and Brazil showed the largest increase in the integrated digital index for 2021. The USA, along with Germany, Japan, and Great Britain, showed a decrease in the level of digital activity. Ukraine took the 64th position among 64 countries in the world [22, 23]. Key indicators of the state of development of the digital economy in the world today are the volumes of the digital economy and its growth rates. The USA has established itself as the undisputed leader in the internet sector and offers modern digital services as of 2022. China has high rates of urbanization as well, accounting for about 36.8% of GDP (Fig. 3). The analysis' findings indicate on the state of development and the impact of the digital economy on economic growth in the USA, and the CSIS agency outlined the directions for the further development of digital processes for the future: inclusive identification systems, order improvement and intensification of the digital payment system, further improvement of functioning tools of the digital bank infrastructure (bridging inclusive gaps in access to data networks), and improving the security of data protection [24].

According to the McKinsey Center, China currently has the largest reserves and possibilities for the further intensification of informational phenomena and processes and real prospects to become a leader in terms of investments in digital projects and the formation of digital infrastructure with a significant impact on the world economy. Already in the past years, the value of e-commerce transactions in China today is many times greater than in France, Germany, Japan, Great Britain, and other countries together with the USA [25].

Fig. 3 Value of the world's largest digital economies, 2022. *Source* Author's construction according to [26]

The drivers of the intensification of China's digital economy are: financial technologies, business tools of virtual reality, educational innovations and mechanisms, 3D printing, use of Big-Date data, and robotics [27]. The high pace of digital development in China is due to the action of favorable factors that allow for the rapid commercialization of digital business models and to maximize the size of business and industrial relations. The information economy of the EU countries is smaller than that of the US in terms of volume, but the existing gap is closing rapidly. In the USA, only the digital services sector provides the same contribution as the digital economy of the euro area. For most countries of the European Union, the annual growth of the digital economy is 0.1% [28]. The information sphere and means of future non-contact communication are included in most global indices of the competitiveness of economies (IMD index, GCI), the WDCR, Digital Evolution Index, DESI, IST Development Index, EIBIS, Bloomberg Innovation Index, Network Readiness Index, ICT Index, etc. In the European practice of analysis and economic research, the integrated DESI index is widely used to assess the level of level of use of digital services, which combines more than 30 indicators in the key areas of digitization: (1) internet connectivity; (2) human capital and digital skills; (3) use of the internet; (4) degree of integration of digital technologies; and (5) digital public services. According to conclusions by the European Investment Bank, in recent years EU countries have been more active in implementing digital technologies compared to Japan and South Korea. Today, Finland and Norway are world leaders in the field of digital public services. Unconditional leadership in the field of access to the internet will belong to South Korea, which is also the leader in terms of human capital in the digital plane. The most common digital technologies in the EU in recent years have been online platforms and advanced robotics and the Internet of Things. The European Union identified the need to provide leadership in the direction of green digitalization as a strategic course of digitalization development. To ensure the realization of this goal, the EU adopted the "Digital Europe" target program, the main task of which is to invest in digital infrastructure to gain competitive advantages based on the transition to the "green economy" and technological sovereignty. For this, it is planned to provide financial support in the amount of 7.6 billion euros in the near future, including 2.2 billion euros—for the purchase of supercomputers, 2.1 billion euros—for the development and implementation of artificial intelligence, 1.6 billion euros—for ensuring cybersecurity measures, and 1.1 billion euros—for the intensification of the distribution of modern information tools in the life of EU countries [29]. Ukraine, according to Digital Riser in the Eurasian space, has lost 66 points in the overall digital rating over the last year and is at the bottom of the overall list. According to the WDCR index in 2021 Ukraine improved its rating position by 4 points, which was the result of leveling off economic shocks from the COVID-19 pandemic. Analyzing the results of the conducted research, it is possible to determine the main problematic aspects of the development of the informative Ukrainian economy at the current stage: (1) a low level of technical perfection and limited access to the internet in large areas; (2) low level of innovative activity and lagging behind the pace of innovative development compared to developed countries; (3) imperfection of the hard and soft infrastructure of digital ties; (4) insufficient amount and low effectiveness of information mechanisms for the development of socio-economic processes; (5) limited number of technological products and management tools; (6) the presence of inclusive gaps in access to tools of public communication and broadband internet traffic by income level and place of residence of the population of Ukraine; and (7) insufficient amounts of funding for projects related to the digitalization of the national economy.

To solve these systemic tasks for the economy of Ukraine, the government of the country approved the strategy for the implementation of digital development, digital changes, and digitalization in the system of public finance management until 2025. There are projects of digital transformation strategies for various types of economic activity. However, today, following the example of the world's leading countries in the field of digital technology development, a comprehensive strategy is necessary that will outline the strategic direction and orientations of the digital transformation of Ukrainian economy as a whole and will become part of the mechanism for implementing the goals and objectives of the National Economic Strategy 2030 [19]. Taking into account the priority and importance of the further transformation of digitization processes to ensure competitiveness and the improvement of the institutional management mechanism in Ukraine, about 60 tasks for increasing the level of digital activity are also indicated in the State Strategy for Regional Development until 2017.

According to the specialized state management bodies, the activation of digitalization processes of the national economy will allow for additional growth of the national GDP by 4% per year. The priority strategic areas of digitalization of the economy of Ukraine today are: improvement of digital infrastructure mechanisms, development of digital skills, development of the ICT sector, digitalization of all spheres of life, and types of economic activity [30].

Today, promising digital economy tools that shape the potential for active growth of Ukrainian economy are: cross-border digital flows of information, online platforms that unite market participants, 3D printing tool set, e-commerce, and other forms of economic relations in virtual reality, artificial intelligence, use of Big-Date data, cloud accounting, and expert analytics.

As Ukraine integrates into the global digital landscape, efforts to establish a unified cryptographic market with EU countries are ongoing. This initiative involves collaboration with leading global companies such as Microsoft, Rakuten, Apple, Amazon, Google, IBM, Palantir, Mastercard, and Visa. Together, they are working on projects under the 'digital lend lease' framework, aimed at enhancing financial institutions to facilitate modern business operations, achieving complete digitization of public services, and advancing digital education, customs, and the judicial system, among others. Additionally, an approved project is in place to establish a digital hub for Asian and European digital traffic in Ukraine [31].

Promising areas of further digitalization of the Ukrainian business and state regulation include the financial, accounting, statistical and control spheres, e-commerce, provision of educational and scientific services in digital format, digitalization of public services, business administration and control, analysis, implementation and ensuring profitability of startups, and stimulation of innovative activities based on the creation of digital online platforms. The key tools of digital transformations should be drivers, among which quantum information flows and related services, integrated circuits and blockchain, 5G and 6G technologies, the creation and provision of a mechanism for institutional support of digital platforms for entrepreneurship, not profitable sustainable development projects, the activation, and use of innovative Web 3 applications occupy an important place digital literacy of the population and cyber protection. The plane of financial digital technologies, which can become a financial and investment basis for finding donors and attracting investment flows to the national economy, deserves separate detailed attention.

5 Summary and Conclusion

The conducted studies showed the determining priority of digitization processes to ensure the implementation of the strategic European Green Deal initiative and the further development of world economies and global society. In Ukraine today, it is possible to ascertain the presence of a powerful potential of prospects for the further development of digital trends, which can form the prerequisites and the necessary basis for increasing the level of competitiveness of the national economy and the quality of life of the country's population. The priority directions of further transformations should be the continuation of the formation of digital infrastructure (support and service), increasing the level of digital literacy of the population and the use of digital financial technologies, which are able to form an investment base for achieving all the goals of the development of the national economy, in particular, in its digital plane.

For Ukraine, the European approach to solving global environmental, social, and economic problems, which is built on the foundations of a democratic society, care for the environment, the quality of life of the population, and the needs of future generations, becomes the most useful. Thanks to its fair tools and accessibility, digitalization is able to form the prerequisites for society's transition to equality, justice, responsible production, and consumption. In accordance with the global initiatives of sustainable development and the Green Deal, digitalization should contribute to the ecological transition from a traditional society to an ecologically and socially oriented one, provide opportunities to radically rethink traditional economic models, improve coordination and cooperation of business partnerships of society, and form the prerequisites for solving the tasks of inclusiveness of social development and growth welfare of the population.

The digital economy is growing much faster than the traditional one, which increases and destroys the contradictions and competitive positions of the leaders and outsiders of the economic system, strengthens the technological dominance and global influence of a few leading countries, which in the conditions of sustainable development is an essential risk when creating modern growth potential and a policy of equal opportunities for billions of people and their descendants. Among the obstacles to digital transformation are information security risks (increased likelihood of technological failures and man-made disasters, threats to the national security of countries due to an increase in the likelihood of criminal interventions of various kinds using digital technologies, the threat of cyber attacks) in all areas of the economy and public administration, weakening control over virtual economic ties and transactions, the growth of temporarily unemployed workers, the accumulation of non-critical violations of the law, a decrease in the quality of life, which ultimately leads to permanent degradation, which is characterized by the lack of sustainable development of society and the expansion of the horizons of people's consciousness and entails the need for total digital control. Further digitalization processes should be aimed at implementing the principles of sustainable development in the world economy, supporting the process of socio-technical transformations that reduce socio-economic and gender inequality, reduce the load on resources and the biosphere, improve working conditions, facilitate the solution of the global problem of food security, and stimulate and simplify the introduction of innovations into social life. For this, it is necessary to create the basic infrastructural conditions for adapting technical and organizational solutions to the key tasks of the Green Course, continue to develop digital inclusiveness, support the development of digital ecosystems, and create new management models for both the business and the state.

References

1. Fritsche U, Brunori G, Chiaramonti D, Galanakis CM, Hellweg S, Matthews R, Panoutsou C (2020) Future transitions for the bioeconomy towards sustainable development and a climate-neutral economy-knowledge synthesis final report. Publications Office of the European Union, Luxembourg
2. Freitag C, Berners-Lee M, Widdicks K, Knowles B, Blair G, Friday A (2020) The climate impact of ICT: a review of estimates, trends and regulations. https://arxiv.org/ftp/arxiv/papers/2102/2102.02622.pdf
3. Von der Leyen U (2019) A Union that strives for more: my agenda for Europe, European Commission. https://ec.europa.eu/commission/sites/beta-political/files/political-guidelines-next-commission_en.pdf
4. European Green Deal (2020) https://ec.europa.eu/food/horizontal-topics/farm-fork-strategy_en
5. Ionescu GH, Firoiu D, Pîrvu R, Enescu M, Rădoi M-I, Cojocaru TM (2020) The potential for innovation and entrepreneurship in EU countries in the context of sustainable development. Sustainability 12:7250
6. European Environment Agency (2020) EEA greenhouse gas: data viewer. https://www.eea.europa.eu/ds_resolveuid/f4269fac-662f-4ba0-a416-c25373823292
7. Szoke E (2022) Ukraine is ready to make a sustainable shift in its agriculture sector. https://ceenergynews.com/climate/ukraine-is-ready-to-make-a-sustainable-shift-in-its-agriculture-sector/
8. FAO (2013) Ukraine and FAO sustainable food systems for food security and nutrition. https://www.fao.org/3/cb0813en/CB0813EN.pdf
9. Babenko V (2020) Enterprise innovation management in industry 4.0: modeling aspects. In: Emerging extended reality technologies for industry 4.0: early experiences with conception, design, implementation, evaluation and deployment, pp 141–163. https://doi.org/10.1002/9781119654674.ch9

10. Viktor Z, Anatolii L, Olha Z, Svetlana M (2022) The digital agricultural revolution: innovations and challenges in agriculture through technology/Roheet Bhatnagar, Nitin Kumar Tripathi, Nitu Bhatnagar, Chandan Kumar Panda. https://doi.org/10.1002/9781119823469.ch11
11. Ionescu RV, Zlati ML, Antohi VM, Virlanuta FO (2020) Digital transformation in the context of European Union's Green Deal quantifying the digitalisation impact on the EU economy. https://www.researchgate.net/publication/358816255_Digital_transformation_in_the_context_of_European_Union's_Green_Deal_QUANTIFYING_THE_DIGITALISATION_IMPACT_ON_THE_EU_ECONOMY_CASE_STUDY_GERMANY_AND_SWEDEN_VS_ROMANIA_AND_GREECE
12. Zamlynskyi V, Kryukova I, Zamlynska O, Skrypnyk N, Reznik N, Camara BM (2022) Coaching as a tool for adaptive personnel management of modern companies 2022. https://doi.org/10.1007/978-3-031-08954-1_26
13. Babenko VA (2013) Formation of economic-mathematical model for process dynamics of innovative technologies management at agroindustrial enterprises. Actual Probl Econ 1(1):182–186
14. Mondejar ME, Avtar R, Diaz HL, Dubey RK, Esteban J, Gómez-Morales A, Hallam B, Mbungu NT, Okolo CC, Prasad KA, She Q (2021) Digitalization to achieve sustainable development goals: steps towards a smart green planet. Sci Total Environ 794:148539
15. Appio FP, Frattini F, Petruzzelli AM, Neirotti P (2021) Digital transformation and innovation management: a synthesis of existing research and an agenda for future studies. J Prod Innov Manag 38:4–20
16. Habibi F, Zabardast MA (2020) Digitalization, education and economic growth: a comparative analysis of Middle East and OECD countries. Technol Soc 63:101370
17. Dubey RK, Dubey PK, Abhilash PC (2019) Sustainable soil amendments for improving the soil quality, yield and nutrient content of Brassica juncea grown in different agroecological zones of eastern Uttar Pradesh. India. Soil Tillage Res 195:104418
18. Bu F, Wang X (2019) A smart agriculture IoT system based on deep reinforcement learning. Fut Gener Comput Syst 99:500–507
19. Lasi H, Fettke P, Kemper HG, Feld T, Hoffmann M (2014) Industry 4.0 bus. Inf Syst Eng 6:239–242
20. Digital Economy (2022) Official site BEA. https://www.bea.gov/data/special-topics/digital-economy
21. Firoiu D, Pirvu R, Jianu E, Cismas LM, Tudor S, Lat G (2022) Digital performance in EU member states in the context of the transition to a climate neutral economy. Sustainability 14:3343
22. Digital riser report 2021 (2021) European center for digital competitiveness. https://digital-competitiveness.eu/digitalriser/
23. European Commission, DG Connect (2022) Digital economy and society index. https://digital-agenda-data.eu/datasets/desi/visualizations. Accessed 10 Jan 2022
24. Becchetti L, Piscitelli P, Distante A, Miani A, Uricchio AF (2021) European Green Deal as social vaccine to overcome COVID-19 health and economic crisis. Lancet Reg Health Eur 2:1478
25. McKinsey Global Institute (2017) Digital globalization: the new era of global flows. https://www.mckinsey.com/~/media/mckinsey/business%20functions/mckinsey%20digital/our%20insights/digital%20globalization%20the%20new%20era%20of%20global%20flows/mgi-digital
26. Digital Economy (2022) Top 10 countries with a digital economy in 2022. https://ankingroyals.com/top-10-countries-in-digital-economy-2021
27. European Central Bank (2008) The digital economy and the euro area. European Central Bank. https://www.ecb.europa.eu/pub/economic-bulletin/articles/2021/html/ecb.ebart202008_03~da0f5f792a.en.html
28. European Investment Bank (2021) Digitalization in Europe 2021–2022. European Investment Bank. https://www.eib.org/attachments/publications/digitalisation_in_europe_2021_2022_en.pdf

29. EU Policy (2021) Digital transformation: importance, benefits and EU policy. https://www.europarl.europa.eu/news/en/headlines/society/20210414STO02010/digital-transformation-importance-benefits-and-eu-policy
30. Fedorov M (2020) Digitization of the economy will allow to achieve at least 4% additional GDP growth per year. https://thedigital.gov.ua/news/mihajlo-fedorov-cifrovizaciya-ekonomiki-dozvolit-dosyagti-minimum-4-dodatkovogo-zrostannya-vvp-na-rik
31. Overview of the digital transformation of Ukraine's economy (2022) National Institute of Strategic Studies. https://niss.gov.ua/news/komentari-ekspertiv/ohlyad-tsyfrovoyi-transformatsiyi-ekonomiky-ukrayiny

Digital Transformation Cloud Computing and Mobility

Intelligent Mechanism for Virtual Machine Migration in Cloud Computing

Karam M. Hassan, Fatma El-Zahraa A. El-Gamal, and Mohammed Elmogy

Abstract Cloud computing has risen in its importance and is now hosted in massive data centers based on the virtualization technology that in turn allows creating multiple virtualized environments and several virtual machines (VMs) to provide multiple services on a single physical host. Despite its advantages, the virtualization technology might fail at any time or be updated or loaded. Accordingly, the VM must be transferred from the utilized host to another. This movement has now become a significant factor in saving the available resources, reducing energy consumption, increasing resource utilization, maintaining the quality of service in cloud data centers, increasing reliability, and achieving load balancing. Multiple methods for moving VMs have been developed for best utilization of the resources. Among these methods, pre-copy migration is considered as a common approach where it migrates the state of the VM's memory from the original host to the intended host through a number of iterations before the shutting down of the VM on the original physical host with an amount of time called downtime. The problem with this approach is that it might cause a little disruption to the services operating in the VM. Therefore, various research attempts focused on studying and selecting proper destination hosts with the available resources to the future usage on the VM. Thus, this paper aims to highlight the current scientific work that target the aforementioned goal. Then, the paper tries to illustrate the current challenges and possible future directions in this research area.

Keywords Cloud computing · Virtualization · Migration · Resource utilization · Downtime

---

K. M. Hassan (✉) · F. E.-Z. A. El-Gamal · M. Elmogy
Faculty of Computers and Information, Mansoura University, Mansoura, Egypt
e-mail: karammohamed@mans.edu.eg

F. E.-Z. A. El-Gamal
e-mail: fatma_zahraa@mans.edu.eg

M. Elmogy
e-mail: melmogy@mans.edu.eg

D. Magdi et al. (eds.), *Green Sustainability: Towards Innovative Digital Transformation*, Lecture Notes in Networks and Systems 753,
https://doi.org/10.1007/978-981-99-4764-5_6

1 Introduction

Recently, one of the famous and widely utilized technologies is the cloud computing. Comparing the cloud computing with the existing traditional information technology infrastructure shows the hallmark ability of the cloud computing to deliver different computing services (e.g., physical hosts, different kinds of storage, databases, and networking) in an online form. In turn, this online-based availability of the services, provided through the cloud providers to their customers, enables quicker development, more scalable resources, minimal congestion, good quality of service, and cost saving [1].

In cloud computing, users only pay for services they used instead of purchasing these computing services. In turn, this leads to an inexpensive maintenance of these services. According to this power of the cloud computing, it was described as "pay as you go" technology. As a result, if the companies need to save their money, they put their information and computer centers in the cloud where this will improve the companies' services to their customers [2]. Additionally, the cloud services are generally accessible around-the-clock from devices such as personal computers or mobile phones and applications such as the chrome web browser. This gives everyone a good amount of freedom to decide how to carry out their work. Accordingly, the success of the organizations can depend on the services provided by highly reliable systems and the Internet [3].

By utilizing a cloud-based platform with built-in redundancy, a company can prevent data loss. Actually, data loss has negative effects on performance, income, and company reputation. Therefore, since the data is stored, available, and backed up in the cloud, the businesses can be recovered from catastrophic events and the operations can be resumed quickly [4].

Recently, the rapid advancement of virtualization technologies [5, 6] led to an increasing number of data centers that are turned to cloud computing to build these data centers [7, 8] due to its benefits on resource separation, server consolidation, and live migration. Consequently, in the virtualization technologies, the virtual machines (VMs) that are being migrated have always been accessible throughout the migration process [9–11].

2 Background of Cloud Computing

2.1 *Cloud Computing Types*

Choosing the kind of cloud on which to deploy the cloud services is the first step where the choice can be between public clouds, (on-premise) private clouds, hybrid clouds, or community clouds as shown in Fig. 1.

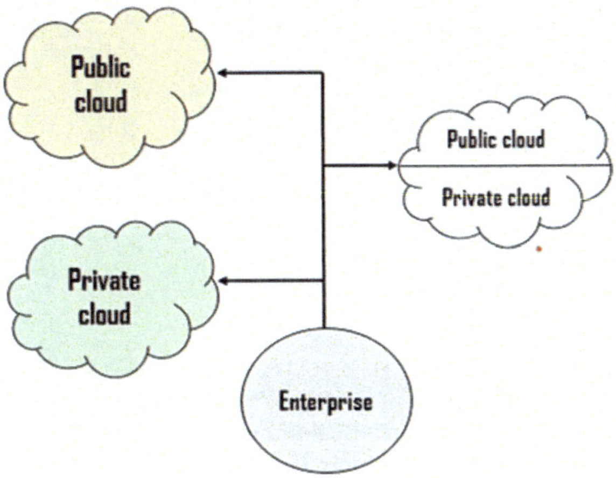

Fig. 1 Main cloud computing types: public clouds, private clouds, and hybrid clouds

2.1.1 Public Cloud

To make use of the services under this type of cloud computing and to allow the users to control their account, they will need a web browser (e.g., Mozilla Firefox, Chrome, etc.). There are different providers who present the cloud services under this type (e.g., Microsoft Azure, IBM cloud, and Oracle) where they own and operate public clouds that in turn offer all information technology infrastructures (i.e., hardware, software, and networking). Among the available service platforms, the platform with the most revenue is Microsoft Azure, followed by Google cloud and Amazon Web Services (AWS) where these three major public cloud providers achieved substantial revenue growth in 2019, 2020, and 2021 as shown in Fig. 2.

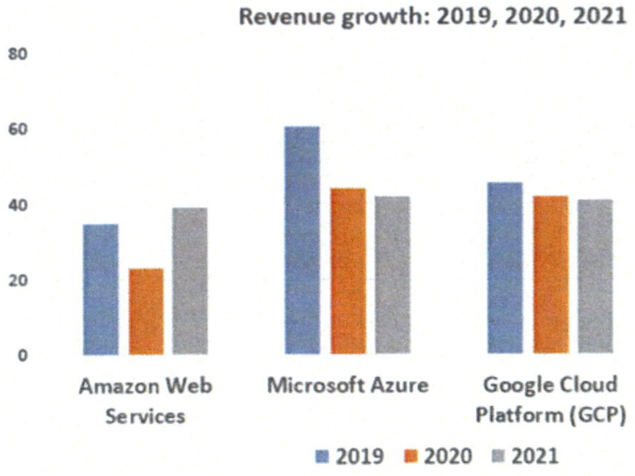

Fig. 2 Significant public cloud providers

2.1.2 Private Cloud

This type of cloud is called internal cloud computing and on-premises where the resources are dedicated to and accessed by a single customer or organization. The private cloud can be hosted on the premises at a data center of the company. Despite its implementation benefits, private cloud computing has a few drawbacks such as the remote access to the data [12].

2.1.3 Hybrid Cloud

By enabling data and applications to flow between private and public clouds, hybrid clouds can combine the best features of both. Companies choose the hybrid cloud when the infrastructure of the companies is insufficient and the company needs to develop its infrastructure to satisfy its business needs.

2.2 Services of Cloud Computing

The three well-known services offered by cloud service providers are infrastructure as a service (IaaS), platform as a service (PaaS), and software as a service (SaaS) Fig. 3, where they can be compared from several perspectives to better understand their origination. These service models' characteristics distinguish cloud computing from other technologies.

2.2.1 Infrastructure as a Service (IaaS)

Platforms for virtualization, which are an expansion of virtualized server solutions that have existed for years, are often offered by cloud infrastructure providers such as Microsoft Azure. Customers purchase resources (e.g., servers, software, or data center space), rather than having to build these infrastructures, and pay-per-use. They control and maintain their own software that they deploy on VMs. Major

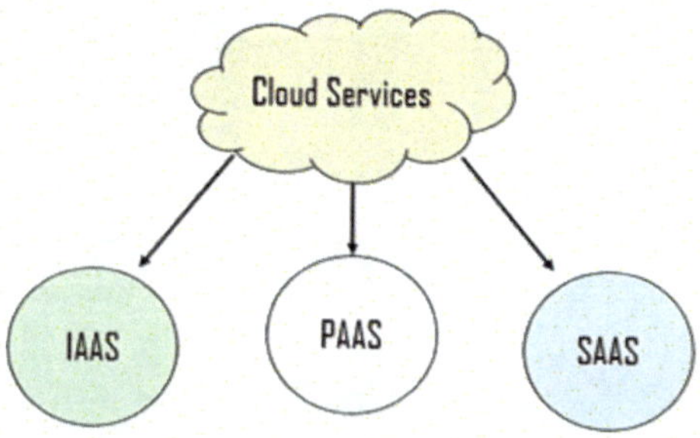

Fig. 3 Services of cloud computing

virtualization services here include the Rackspace Cloud and ServePath's GoGrid. Also, there are IBM Smart Business cloud solutions, Nimbus, RightScale, Oracle Cloud Computing, and GigaSpaces.

2.2.2 Platform as a Service (PaaS)

A monitored relatively high software infrastructure is offered by PaaS providers. PaaS enables clients to use the provider's tools, programming languages, and integrated development environment to create and implement particular kinds of services and applications software. Because it is hidden behind the platform, customers have minimal control over the underlying software and hardware that include hosts, storage, networks, and operating systems [12].

2.2.3 Software as a Service (SaaS)

SaaS is a method of on-demand software distribution through the Internet which is well-known as web-based e-mail. Most cloud computing software services are web-based programs that may be accessed using a client interface such as Chrome web browser that can be accessed from different client devices. Clients of these services are neither managing nor controlling the actual infrastructure or application platform. Google Apps is a popular software solution that is also meant for personal usage. Also, calendar, contacts, chat capabilities, and Google documents component represent examples of this solution as well where users may view and distribute documents, presentations, and spreadsheets. There is also Box.net that is a different document-sharing and backup solution. Additionally, there is a video and photo-sharing service called SmugMug that uses Amazon S3.

2.3 Challenges of Cloud Computing

When employing cloud computing, data security is one of the biggest challenges for many organizations. Due to the fact that the majority of their clients share one physical server, security is normally handled by the cloud service provider. Despite the cloud service providers' assurances regarding data integrity, data leakage, data stolen, and many other security issues are still common. It is usually a very challenging process and necessitates reconfiguration when a business employs cloud services from a certain provider for a long period of time and wishes to transit to another cloud-based service provider. Actually, a loss of flexibility can result when attempting to switch between clouds. There is also the performance that represents one of the key issues since if cloud performance is inadequate, the customers may choose another service provider. Additionally, the fault tolerance, when operations continue as necessary even when one or more components fail, represents another source of challenges.

3 Virtual Machine Migration

The method of virtualization enables the physical host to operate many VMs, or reserved resources from a particular physical host. Also, running multiple operating systems simultaneously in a single physical host is referred to as virtualization. It provides load balancing between data centers which means lowering the physical servers needed to handle a particular workload. Accordingly, the migration of VMs seeks to transfer them from the original host to the intended host Fig. 4.

3.1 Types of Virtual Machine Migration

Mainly, there are two types of VMs migration that are: (i) cold (non-live migration) where the VM moves from the origin host to the intended host which occurs when the VM is shut down; and (ii) hot (live migration) where the VMs are transferred from the source to the intended host when it is running where there are two techniques for transferring, pre-copy memory migration and post-copy memory migration.

In pre-copy live memory migration Fig. 5, there is the warm-up phase followed by a stop-and-copy phase. In the warm-up phase, while the VM is still running on the origin host, the hypervisor moves all memory pages from the original physical host to the intended host. If memory pages are updated, then they become "dirty pages" and accordingly they will be copied again up until the point where the amount of copied pages outpaces the amount of dirtying pages [13]. After this phase, the stop-and-copy phase takes place where the VM suspended its operations from the work on the original host. The intended host will receive copies of the remaining dirty pages. Then, the VM restarted and resumed its work on the recipient host [9]. Time of suspending the VM from the work on the original host and resuming it on the intended host is known as "down-time" that should be kept short as possible to avoid

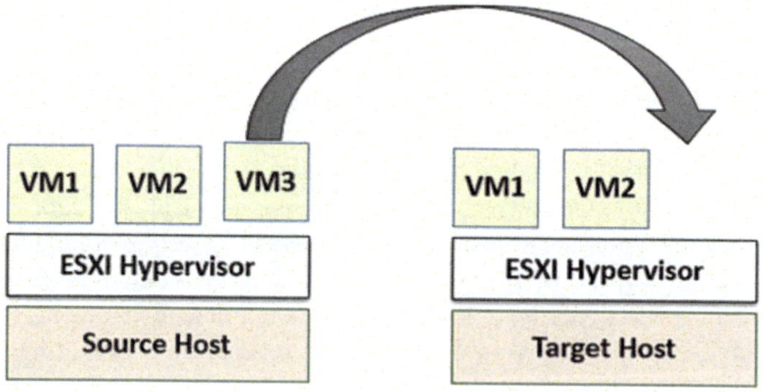

Fig. 4 Migration of virtual machine from source host to target host

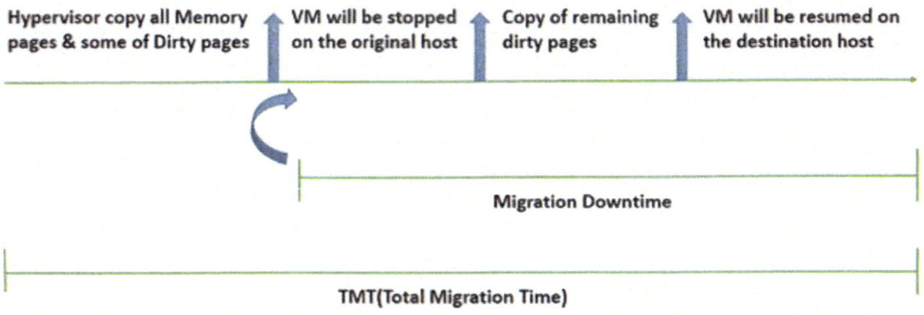

Fig. 5 Pre-copy live migration

disrupting the VM's services. Downtime can be milliseconds to seconds depending on the VM's memory capacity and the applications it is running [14].

On the other hand, when the origin VM in post-copy Fig. 6 is suspended from its work, then a small part of its execution state, including its registers and central processing unit (CPU) information, is migrated to the intended host [15]. After that, the VM is restarted on the intended host. Additionally, the original host pushes the remaining pages to the intended host. When the VM accesses a page that hasn't yet been pushed to the intended host, a page fault occurs. When a network error occurs, the intended host notifies the origin host, which then sends back the faulty page. It is important to note here that the performance of programs that run inside the VM may suffer from an excessive number of network errors. Actually, pre-copy is better than post-copy in situations where there are a lot of dirty pages. It can cut down on overall migration time and downtime during live migration in many applications when compared with post-copy [9].

Finally and returning to the hot and cold types of the VM migration, the choice between these migrations depends on the degree of the risk the one is prepared to take and the available time. Hot migrations allow you to migrate workloads in real time while maintaining full access to them. As a result, VMs will be vulnerable to unauthorized access corruption. Also, hot migrations aren't suitable for some

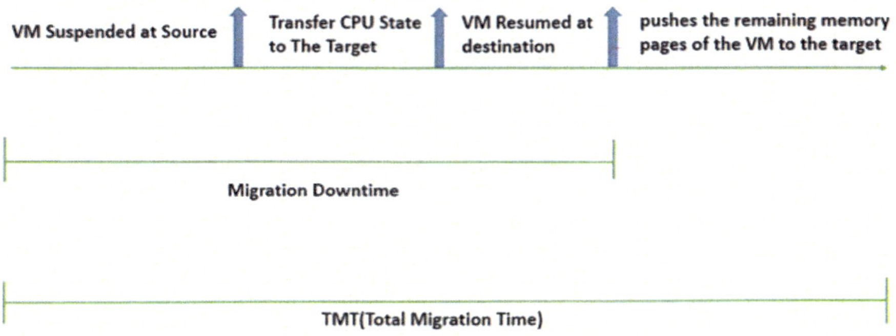

Fig. 6 Post-copy live migration

physical machines, such as active directory controllers. On the other hand, cold migrations allow you to turn off workloads and move them between hosts without losing data but often result in a long downtime.

4 Virtual Machine Live Migration Metrics

Actually, the period of time between terminating a VM on its original host and restarting it on the intended host when the VM's service is unavailable, the downtime, as well as the migration time are two significant considerations in each VM migration, Fig. 7. Since different types of VM migrations result in varying levels of downtime and migration duration, these aspects should be addressed when choosing the migration method.

Additionally, there are common factors that can be addressed through the VM migration techniques during the migration process as depicted in Fig. 8. The VM migration techniques may consider factors like the preparation time, downtime, resumed time, transferred data, time of the migration, and degradation of application. The preparation time states that during the interval between starting the process of the migration and sending the state of the VM's CPU to the intended host, the VM can be still active and use memory. Then, there is the downtime where the execution of the migrated VM is halted at this time. At the absolute least, the transfer of processor state is included in this. Any remaining dirty pages for pre-copy are also transferred. There is also the period of time where resuming the VM's execution at the destination, till the migration is finished, may affect the performance of the migration. Besides that, there is the total transmitted amount between the origin and intended hosts which is described as the amount of information moved through the entire process of the migration based on the amount of bandwidth. Additionally, there is the application degradation which refers to the total time, from the start to end. Actually, this metric is significant since it influences the resource's allocation on multiple hosts and the amount of time that the migration takes to slow down the VM's applications.

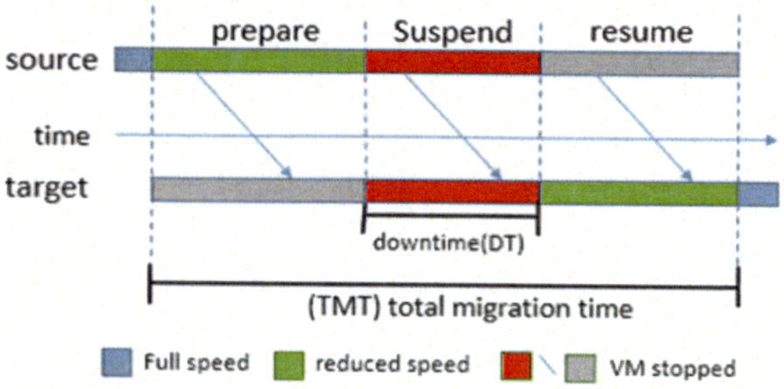

Fig. 7 Downtime and migration time

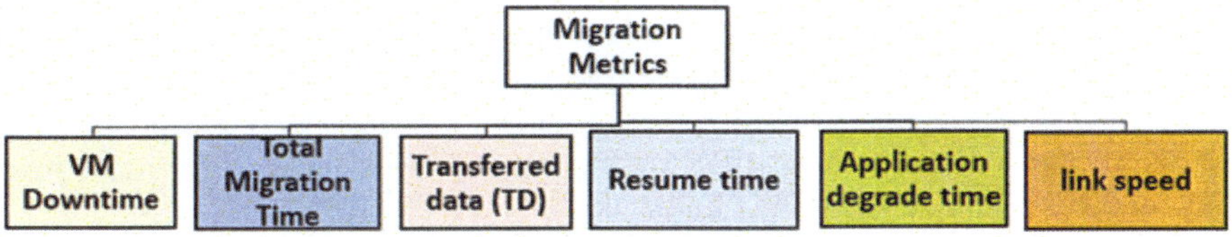

Fig. 8 Performance metrics of the process migration of the VM

On the same context, there are several directions that can be used to deal with the dirty pages. This includes using CPU scheduling to reduce the number of dirty pages, using machine learning techniques to predict dirty pages, and using compression algorithm to minimize the size of the transferred data from the source to the destination.

5 Scientific Efforts on Live Migration Cloud Computing

Throughout the time, different scientific contributions have been proposed to address the live migration cloud computing. For example, Sharma and Chawla [16] proposed an optimization method that operates in three phases: (i) reducing memory page transfers, (ii) eliminating duplicate page transfers by identifying often and infrequently pages that are updated, and by applying a straight forward compression strategy, (iii) the final iteration of migration can send the least amount of information possible. Each of the proposed phases reduced the total amount of pages sent as a result of downtime and migration time. In a Xen virtualized environment, the suggested approach was assessed using several typical workloads. The strategy was found to decrease the overall migration time by 70%, the total pages transmitted by 71%, and the downtime by 3% for greater loads. Also, it seemed to have no significant overhead when compared with the standard pre-copy method.

Additionally, Arif et al. [17] explored a machine learning technique that minimized the downtime during live migration across a large area network where it was an intelligent live migration model based on predictive mechanisms for common workloads. In this study, the machine learning methods were utilized to decrease the downtime. Throughout the migration process, the downtime noticed to be improved by up to 15%.

Wang et al. [18] suggested a live VM migration technique for memory cutting based on introspection where it was both quick and effective. At the beginning, they divided the memory pages into five groups based on how they are used by the operating system: cache pages, kernel, anonymous, and free. Then, during migration, they removed inconsequential free pages and redundant pages. Accordingly, a considerable amount of useless data was avoided, and the migration time was cut in half. Using introspection, the memory pages could be categorized into certain groups. In addition to the cache pages, this technique eliminated the unused pages, in contrast to most other cases, where the free pages that the operating system (OS) designated it as zero pages were removed, where this resulted in achieving 72% of the total migration time on the kernel-based virtual machine (KVM) hypervisor. Despite the obtained results and due to the processing of incomplete cache pages and page introspection, the proposed technique had a time limit.

Cui et al. [19] proposed a pre-copy migration approach that in turn was comprised of two stages (i.e., the warm-up, and stop-and-copy stages). Before the VM shuts down due to a problem, the original server's pages of memory were all moved to the intended server. This scheme proposed a VM migration paradigm based on dynamical creation of flexible topologies depending on the requirements of the VMs to minimize the cost of VM communication and migration. This technique showed that the issue of the traffic-aware VM migration in a dynamic topology is non-deterministic polynomial-time hardness (NP-hard) problem. The proposed technique migrates the VM with a validated approximation ratio for repeated traffic. So this approach achieves high-velocity flows but the migration time becomes long.

Alrajeh et al. [20] proposed three machine learning techniques that estimated the live migration for several testing and training sets obtained from the prediction performance. The findings demonstrated that while some VMs migrate quickly, others migrate slowly, and some VMs cannot migrate while the workload is being executed.

Patel et al. [21] explored the integrated model to anticipate dirty pages during a migration. The results indicated that in the case of an environment with many dirty pages, the model can forecast 93% of the time. In comparison with Xen's pre-copy process, the combined solution could save 10.76% overall migration time and 19.16% downtime.

Duggan et al. [22] suggested a model that estimated the CPU and network bandwidth requirements for the live migration. In this study, a recurrent neural network (RNN), a sequence prediction system, was compared with linear and nonlinear forecasting methods. Experiments showed that a multi-time ahead prediction system minimized the bandwidth consumption during key times, while also improving a data center's overall efficiency. The tests showed that an RNN's memory retention and sequence prediction provided the highest accurate prediction for resource utilization and bandwidth. The RNN had the ability to cut bandwidth utilization during peak times, and to enhance a cloud data center's overall efficiency.

El-Moursy et al. [23] proposed regression host consumption techniques that used the Euclidean distance and absolute summing to integrate memory, CPU, and bandwidth utilization results of the energy and service-level agreement violations. Then, the authors specifically consider real-world workloads. The advantage of these suggested algorithms over existing techniques was further demonstrated by the extensive simulation analysis. Comparatively speaking to the most recent multiple regression techniques, the energy usage, taking into account the amount of service-level failures, VM migrations, and energy consumption, were improved by at least 80% and at least 12%.

Sui et al. [24] suggested a VM technique using a machine learning algorithm to achieve a load balancing (i.e., to solve the load imbalance issue in cloud data centers) through proposing a load forecasting method based on k-means clustering and an adaptive differential evolution algorithm. The experimental findings demonstrated that, compared with other classical algorithms, the suggested VM scheduling approach minimized the number of VM migrations by 94.5% and the energy usage of cloud data centers by 49.13%.

Al-Said Ahmad and Andras [25] proposed a method that assessed the system's overall scalability. Even when identical experiments were run in different clouds, the response time differed for services provided by cloud technical scalability measurements based on elasticity metrics. It used two cloud-based systems in order to show the value of measurements and contrast their scalability performance on Microsoft Azure and Amazon EC2. The comparative analysis of the experimental data compared three sets: a cloud-based software solution that uses two separate auto-scaling strategies but exists on the same cloud platform, the deployment of the same cloud-based software solution across two different public cloud infrastructures, and lastly, two alternative auto-scaling strategies for the same services of cloud that is hosted on the same cloud platform.

Rajapackiyam et al. [26] proposed a way for reducing the total migration time of typical memory migrations by mirroring VMs at the destination host. It primarily focused on the entire migration time and the maximum profit margin for suppliers. The suggested method prevented repetitive memory page movement between the origin host and the intended host. When compared with the pre-copy migration technique, the performance analysis provided evidence that this suggested algorithm could cut down the data transfer throughout the moving process by 36% and the migration time by 26%.

Elsaid et al. [27] explored artificial intelligence techniques to choose the ideal time to carry out a live migration request. The machine learning was used to forecast live migration costs and data center network utilization in an efficient timing technique. Once a live migration request had been submitted, data center administrators could be notified of this ideal scheduling recommendation. According to the testing results, the live migration time was reduced by 50% and cut down 32% of VM migration time for memory-heavy applications. For network-intensive applications, the suggested method could reduce the migration time by as much as 27%, on average saving of 21%.

Moghaddam et al. [28] suggested an intelligent VM migration method that was intended to minimize the number of migrations and to reduce the power energy. This was accomplished by employing a placement strategy that prioritized the efficiency and deferring migration by anticipated future resource needs. The strategy minimized the number of migrations in two ways: (i) utilizing a modified cellular learning automata-based evolutionary computing (CLA-EC) to determine where on physical servers to install VMs and replace them, and (ii) preventing unnecessary migrations by predicting the future resource used in hosts using a neuro-fuzzy algorithm. The suggested approach could lower the data center's average migration number, energy use, and service-level agreements (SLA) violations by 59.05, 8.5, and 70.76%, respectively.

Rajabzadeh et al. [29] proposed a resource monitoring mechanism that followed a specific sequence and efficiently used the memory and bandwidth. The sequence began with a test of the critical condition of physical devices, followed by a definition of the VM to migrate and the application of the VM migration policy. This approach generated an overhead by continuously tracking load among the node servers where the interaction between the users and the application was handled by the node server.

Motaki et al. [30] predicted six crucial parameters for each live migration procedure using an ensemble-learning approach that made use of linear and nonparametric regression methods which in turn were supplied by the user or the operator. To migrate a VM, the proposed approach allowed considering the optimum algorithm-metric combination. The results demonstrated that the suggested model cut down the rate of service-level agreement violation by 31–60%, while also reducing the overall CPU time required for the prediction process.

Surya and Rajam [31] explored techniques to forecast the hosts that will compete for resources and to reduce the frequency of VM migrations in order to maximize CPU utilization. Based on the forecasts, the VMs were transferred from overloaded hosts to potentials or regularly loaded hosts and the intended host were prevented from going into overload after the migration of the VM. The suggested technique used a number of VM movements as criteria that assessed how well the suggested work was performed. Due to migration overhead, moving a VM from one system to another results in latency and a reduction in CPU usage. Results demonstrated that by decreasing the frequency of VM movement, the proposed techniques could increase the performance.

Vatsal and Agarwal [32] addressed the NP-Hard problem when there are restrictions on the network capacity. Heuristic techniques could be used to discover the optimal VM migration solution to address this problem. VM migration was supplemented by various additional approaches used in data centers to achieve green cloud computing solutions. According to the study, utilizing suitable virtualization techniques for resource allocation and VM migration could reduce the operational expenses and power consumption.

Gupta and Namasudra [33] proposed basic algorithms. The target of these algorithms was to help improving the cloud computing environments' performance by reducing the time of the migration process through Host Selection Migration Time (HSMT) which all user queries are initially received by this algorithm which then forwards these queries to the development of VMs to allow continuous tracking of all nodes and VM Reallocation Migration Time (VMRMT) When Algorithm 1 is finished, Algorithm 2 begins a search operation for the slave host with the highest capacity among the current slave hosts and designates it as the intended host. Finally, VM Reallocation Bandwidth Usage (VMRBU) VM moves from the overloaded host to the intended host once Algorithm 2 is finished. It initially loads the memory pages and the user demand onto the intended host. The response to the customer's request that has been accepted is then saved in a variable. Once the problem has been resolved and the memory page and workload have been transferred from the faulty node to the intended host, the response to that customer's request must be provided to the user who made it, when a produced result fails even before it reaches the user.

Mason et al. [34] proposed a machine learning technology that used the neural networks to predict the host CPU utilization. Evolutionary optimization and several cutting-edge swarm approaches (i.e., the differential evolution, particle swarm optimization, and covariance matrix adaptation evolutionary strategy) were employed to train the neural networks to forecast host utilization. The results demonstrated that no further training was necessary because the trained networks maintained their accuracy when used to analyze the CPU utilization data from many hosts.

Talwani et al. [35] proposed a method based on machine learning for dynamically integrating VMs based on adaptive estimates of the utilization thresholds to satisfy service-level agreement requirements. Dynamic data was produced in runtime to compare the effectiveness of the suggested method with those of other machine learning methods. The effectiveness of the categorization systems was evaluated using the false negative rate (FNR) and the true positive rate (TPR). The classifier result was taken into consideration to be TPR if the VM migration was correctly classified into the class called "migrated" due to high usage. Additionally, if the classifier result was deemed FNR, VM migration from the current server was rated as "not migrate".

Tuli et al. [36] suggested utilizing an enhanced artificial bee colony (PEA) to facilitate a better dynamic migration and deployment of various VMs and physical computers. This process consisted of two distinct steps. Limiting the number of VM migrations after first selecting a location for the physical machine PM with a delay in access to the site where it needs to be transferred. Also, the service-level agreement violation (SLA-V), energy use, number of hosts that were shut down, and resource utilization of potential approaches were compared. The findings showed that the SLA-V had decreased by 20 and 31% over time, the migrations had increased by 16 and 25%, the resource's usage had increased by 8%, and the energy consumption had increased by 13%.

Vatsal and Agarwal [37] investigated methods to rethink the concept of green cloud computing. The number of VM migrations might be cut down, while also using less energy through turning off the idle hosts after a smart dynamic VM migration and reallocation to an appropriate destination host. Here, the target host was given a VM in line with the actual and anticipated resource requirements. The algorithmic strategy demonstrated its efficacy by reducing the number of migrations, energy consumption, and SLA breaches.

Finally, Toutov et al. [38] moved the VM using a multicriteria strategy. Overload and overheating are the two main factors that the study took into account. It indicated that there are more tasks assigned to the server or that the CPU is being used more actively when a VM is transferred to another physical server due to the overload. To allocate VMs that were being transferred to real servers, the Hungarian method was employed. A summary of different scientific work in this context can be found in Table 1.

Table 1 Summary of the substantial efforts on live migration of VM

Authors	Contribution	Advantages	Disadvantages
Sharma and Chawla [16]	It operates in three phases: (i) reduce memory page transfers (ii) eliminate duplicate page transfers	Reduced migration time by 70%, pages transferred by 71%, and downtime by 3%	0.05% space overhead, 0.05% CPU processing overhead
Wang et al. [18]	Avoid moving memory pages from the cache and free space by removing inconsequential free pages and cache pages that are redundant	72% total migration time on KVM hypervisor	Due to the processing of missing cache pages and page introspection, it has a time limit
Cui et al. [19]	Pre-copy migration process is divided into two steps: warm-up first, then pause and copy, which decreases the cost of VM movement	Reducing the time it takes for a flow completion while lowering the cost of VM migration	On the host server, bad memory pages are put back, migration time is lengthy
Patel et al. [21]	Pre-copy approach based on statistical prediction and compression model is used	10.76% migration time, and 19.16% downtime	Higher migration time
Ahmad and Andras [25]	Cloud platforms used are Amazon elastic compute cloud and Microsoft Azure, to improve system performance in the event of any fault tolerance	The performance is high	Significant variations in response time in different clouds
Rajapackiyam et al. [26]	Prevents repetitive memory pages from moving between the original host to the intended host	36% of data communicated, 26% of migration time	Long migration time and high VM downtime
Rajabzadeh et al. [29]	Employ memory and bandwidth for virtual machine migration, resource monitoring approach is provided	Save bandwidth throughout the VM transfer procedure	Approach causes overhead
Gupta and Namasudra [33]	Suggested three algorithms: host selection migration time, virtual machine reallocation migration time, and virtual machine reallocation bandwidth to improve migration process	Downtime is reduced by 70–80%, the number of CPU cores decreased by 60–70%, time spent migrating is cut by 40–50% and 40–50% less data is sent per second	Don't take into account the deallocation of the VM

6 Future Work Directions

According to the aforementioned scientific efforts, as well as the disadvantages presented in table, the suggested future directions to construct powerful migration models can be as follows:

- The filtration and reduction techniques can be used to distribute the data properly throughout the all locations, allowing the workload to be shared equally.
- Since another issue that can be with the live migration is the lack of bandwidth; a better utilization of the network capacity can be obtained by dynamically assigning it.
- Memory and CPU transfer must be further improved to lessen the complexity of the post-copy procedure.
- In the post-copy migration strategy, the page failure detection can be accomplished by hiding the actual page and creating a virtual page which can be used in the cloud environment to manage the target node failure.
- It is possible to migrate VMs using migration tools where the pre-copying and post-copy migration strategies can be used together as a mixed technique. If the memory reuse idea is employed during VM migration, the best outcomes can be obtained.
- More focusing on the machine learning approaches, including the multi-classes ones, can help improving the performance results.

7 Conclusion

Recently, VM migration shows a great impact in different perspectives such as achieving load balancing to avoid system failure, improving system performance which transfers the load from overused to underused hosts, and providing efficient as well as low latency service excitation when migrating from a host to another in the case of the system maintenance. Therefore and due to its importance, this article focuses of addressing the scientific work in this research area. The majority of the studied works in this survey were found to employ CPU utilization as a metric for energy optimization. Consideration of many parameters, such as memory, network bandwidth, and application was encouraged in some work. According to the analyzed results of the related research efforts, the advancement of the energy optimization in VMs showed a lot of potential. Numerous similar studies employed CPU utilization as an objective function, and other factors such as VM bandwidth consumed during migration and RAM used for VM allocation can also be used. Most of scientific efforts in this area of VM migration directed to in the future to use machine learning approaches for improving the process of migration and the other directed to utilizing resources to achieve load balancing in the environment of cloud.

References

1. Ibrahim S, He B, Jin H (2011) Towards pay-as-you-consume cloud computing. In: Proceedings of the 2011 IEEE international conference on services computing. IEEE, New York, pp. 370–377
2. Saini H, Upadhyaya A, Khandelwal MK (2019) Benefits of cloud computing for business enterprises: a review. In: Proceedings of international conference on advancements in computing and management (ICACM)
3. Abdulkader SJ, Abualkishik AM (2013) Cloud computing and e-commerce in small and medium enterprises (SME's): the benefits, challenges. Int J Sci Res 2(12):285–288
4. Motahari-Nezhad HR, Stephenson B, Singhal S (2009) Outsourcing business to cloud computing services: opportunities and challenges. IEEE Internet Comput 10(4):1–17
5. Barham P, Dragovic B, Fraser K, Hand S, Harris T, Ho A, Neugebauer R, Pratt I, Warfield A (2003) Xen and the art of virtualization. ACM SIGOPS Oper Syst Rev 37(5):164–177
6. Waldspurger CA (2002) Memory resource management in VM ware ESX server. ACM SIGOPS Oper Syst Rev 36(SI):181–194
7. Nurmi D, Wolski R, Grzegorczyk C, Obertelli G, Soman S, Youseff L, Zagorodnov D (2009) The eucalyptus open-source cloud-computing system. In: Proceedings of the 2009 9th IEEE/ACM international symposium on cluster computing and the grid. IEEE, New York, pp 124–131
8. Sotomayor B, Montero RS, Llorente IM, Foster I (2009) Virtual infrastructure management in private and hybrid clouds. IEEE Internet Comput 13(5):14–22
9. Clark C, Fraser K, Hand S, Hansen JG, Jul E, Limpach C, Pratt I, Warfield A (2005) Live migration of virtual machines. In: Proceedings of the 2nd conference on symposium on networked systems design and implementation, vol 2, pp 273–286
10. Nelson M, Lim BH, Hutchins G et al (2005) Fast transparent migration for virtual machines. In: USENIX annual technical conference, general track, pp 391–394
11. Ye K, Jiang X, Ye D, Huang D (2010) Two optimization mechanisms to improve the isolation property of server consolidation in virtualized multi-core server. In: Proceedings of the 2010 IEEE 12th international conference on high performance computing and communications (HPCC). IEEE, pp 281–288
12. Mell P, Grance T et al (2011) The NIST definition of cloud computing
13. Hacking S, Hudzia B (2009) Improving the live migration process of large enterprise applications. In: Proceedings of the 3rd international workshop on virtualization technologies in distributed computing, pp 51–58
14. Moghaddam FF, Cheriet M (2010) Decreasing live virtual machine migration down-time using a memory page selection based on memory change pdf. In: Proceedings of the 2010 international conference on networking, sensing and control (ICNSC). IEEE, New York, pp. 355–359
15. Hines MR, Deshpande U, Gopalan K (2009) Post-copy live migration of virtual machines. ACM SIGOPS Oper Syst Rev 43(3):14–26
16. Sharma S, Chawla M (2016) A three phase optimization method for precopy based VM live migration. Springerplus 5(1):1–24
17. Arif M, Kiani AK, Qadir J (2017) Machine learning based optimized live virtual machine migration over wan links. Telecommun Syst 64(2):245–257
18. Wang C, Hao Z, Cui L, Zhang X, Yun X (2017) Introspection-based memory pruning for live VM migration. Int J Parallel Program 45(6):1298–1309
19. Cui Y, Yang Z, Xiao S, Wang X, Yan S (2017) Traffic-aware virtual machine migration in topology-adaptive DCN. IEEE/ACM Trans Netw 25(6):3427–3440
20. Alrajeh O, Forshaw M, Thomas N (2017) Machine learning models for predicting timely virtual machine live migration. European workshop on performance engineering. Springer, New York, pp 169–183
21. Patel M, Chaudhary S, Garg S (2018) Improved pre-copy algorithm using statistical prediction and compression model for efficient live memory migration. Int J High Perform Comput Netw 11(1):55–65

22. Duggan M, Shaw R, Duggan J, Howley E, Barrett E (2019) A multitime-steps ahead prediction approach for scheduling live migration in cloud data centers. Softw Pract Exp 49(4):617–639
23. El-Moursy A, Abdelsamea A, Kamran R, Saad M (2019) Multi-dimensional regression host utilization algorithm (MDRHU) for host overload detection in cloud computing. J Cloud Comput 8(1):1–17
24. Sui X, Liu D, Li L, Wang H, Yang H (2019) Virtual machine scheduling strategy based on machine learning algorithms for load balancing. EURASIP J Wireless Commun Netw 2019(1):1–16
25. Al-Said Ahmad A, Andras P (2019) Scalability analysis comparisons of cloud-based software services. J Cloud Comput 8(1):1–17
26. Rajapackiyam E, Subramanian AV, Arumugam U (2020) Live migration of virtual machines using mirroring technique. J Comput Sci 16(4):543–550
27. Elsaid ME, Abbas HM, Meinel C (2020) Live migration timing optimization for vmware environments using machine learning techniques. In: CLOSER, pp 91–102
28. Moghaddam J, Esmaeilzadeh A, Ghavipour M, Zadeh AK (2020) Minimizing virtual machine migration probability in cloud computing environments. Clust Comput 23(4):3029–3038
29. Rajabzadeh M, Haghighat T, Rahmani AM (2020) New comprehensive model based on virtual clusters and absorbing Markov chains for energy-efficient virtual machine management in cloud computing. J Supercomput 76(9):7438–7457
30. Motaki SE, Yahyaouy A, Gualous H (2021) A prediction-based model for virtual machine live migration monitoring in a cloud datacenter. Computing 103(11):2711–2735
31. Surya K, Rajam V (2021) Prediction of resource contention in cloud using second order Markov model. Computing 103(10):2339–2360
32. Vatsal S, Agarwal S (2021) Energy-efficient virtual machine migration approach for optimization of cloud data centres. In: Proceedings of the 2021 2nd international conference for emerging technology (INCET). IEEE, pp 1–7
33. Gupta A, Namasudra S (2022) A novel technique for accelerating live migration in cloud computing. Autom Softw Eng 29(1):1–21
34. Mason K, Duggan M, Barrett E, Duggan J, Howley E (2018) Predicting host CPU utilization in the cloud using evolutionary neural networks. Fut Gener Comput Syst 86:162–173
35. Talwani S, Singla J, Mathur G, Malik N, Jhanjhi N, Masud M, Aljahdali S (2022) Machine-learning-based approach for virtual machine allocation and migration. Electronics 11(19):3249
36. Tuli K, Kaur A, Malhotra M (2023) Efficient virtual machine migration algorithms for data centers in cloud computing. International conference on innovative computing and communications. Springer, New York, pp 239–250
37. Vatsal S, Agarwal S (2023) Safeguarding cloud services sustainability by dynamic virtual machine migration with re-allocation oriented algorithmic approach. Smart trends in computing and communications. Springer, New York, pp 425–435
38. Toutov A, Toutova N, Vorozhtsov A, Andreev I (2021) Multicriteria optimization of virtual machine placement in cloud data centers. In: Proceedings of the 2021 28th conference of open innovations association (FRUCT). IEEE, pp 482–487

Blockchain Technology to Enhance Performance of Drugs Supply Chain in the Era of Digital Transformation in Egypt

Aya Mohammed A. Moussa, Fatma El-Zahraa A. El-Gamal, and Ahmed Saleh A. El Fetouh

Abstract Medicine is one of the products that must be delivered to the patient safely, undamaged, and during its validity period. However, the risks of the drug supply chain have become increasingly apparent. To address these risks and due to the rapid advances in the technology field, various digital transformation trials around the globe start to be considered in order to replace the traditional drug supply chain until its delivery to the patient. Therefore, the aim of this paper is to contribute in the digital transformation goal of the drug supply chain in Egypt. Thus, this paper proposes a trusted automatic system, based on the blockchain and Internet of Things (IoT) technologies to enhance the performance of the drug supply chain and accordingly make the drug manufacturing and distribution system in Egypt more efficient and reliable.

Keywords Drug supply chain · Damaged medicines · Internet of Things · Blockchain

A. M. A. Moussa (✉)
Business Information Technology Program, Faculty of Computer and Information, Mansoura University, Mansoura, Egypt
e-mail: ayamhmd91.ym@gmail.com

F. E.-Z. A. El-Gamal
Information Technology Department, Faculty of Computer and Information, Mansoura University, Mansoura, Egypt
e-mail: fatma_zahraa@mans.edu.eg

A. S. A. El Fetouh
Delta Higher Institute for Management and Accounting Information Systems, Mansoura, Egypt
e-mail: elfetouh@mans.edu.eg

D. Magdi et al. (eds.), *Green Sustainability: Towards Innovative Digital Transformation*, Lecture Notes in Networks and Systems 753,
https://doi.org/10.1007/978-981-99-4764-5_7

1 Introduction

Pharmaceutical supply materials and the network through which medicines are distributed, starting from raw materials, and going through all the operational and organizational activities needed to manufacture the medicine until it is delivered to the patient and the basic components of this network (i.e., manufacturers, distributors, wholesalers, retailers, pharmacies, hospitals, and health offices). The pharmaceutical supply chain includes many complex steps that in turn might count different sources of unwanted events which might harm the medicine, and thus pose a danger to the patient. These events include achieving a higher percentage than the permissible during transportation or storage such as: high temperature and humidity, or direct exposure to sunlight. Actually, according to the Food and Drug Administration (FDA), it is possible to take some medicines after their expiration date within a year or two if they are at the right temperature and humidity [1]. The problem is when any of these errors occur, it might be difficult to track it because there is no reliable system between the parties of the chain.

The emergence of the Internet of Things (IoT) technology along with its sensors presents huge roles in addressing this vital problem. This is due to the IoT sensors' capability to: (i) track the medicine at all the supply chain stages, (ii) monitor the natural factors (i.e., temperature, humidity, and sunlight) during the transportation and storage of medicines, and (iii) offer immediate report of any abnormality in the environment in which the medicine is kept.

Besides the powerful capabilities of the IoT, a strong and secure system where the information and transactions are exchanged between supply chain parties is highly needed. For this purpose, the blockchain power could be utilized where the nature of blockchain based on distributed ledger technology and a decentralized database can perfectly address the security target. It will bring transparency and trust between the parties to the chain and, due to the use of a consensus mechanism, will ensure that data is not manipulated to make it immutable.

2 Related Work

Numerous studies tried to address different challenges of the drug supply chain. Pharmaceutical distribution and supply chains in developing and middle-income nations might be thought of as a challenging and complicated procedure. Also, drugs that are not kept under standardized settings, such as at the proper temperature, waste a lot of resources since these unsanitary storage conditions have negative impacts on the drugs' efficacy.

Numerous literary reviews as well as the World Health Organization reports uncovered that improper storage factors harm the health system and have a negative impact on it, in addition to being extremely costly to the economy. Also, it was discovered that one of the difficulties confronting supply chains is the absence of standards for proper storage practices. In addition, there aren't any reliable ways to assess and keep tracking of the storage of the data throughout the supply and distribution phases [2–6].

Lingayat et al. [7], Panda and Satapathy [8] focused on providing supply chain system equipped with blockchain technology to make it more transparent by ensuring the tracking and tracing of medication. Also, it helped preventing the delivery of the counterfeit drugs but only the arrival of the original product to the patient.

Musamih et al. [9], Uddin [10], Ahmadi et al. [11] depended on the decentralization of blockchain technology and the basics of encryption to achieve non-tamperable records of transactions within the supply chain and to ensure security from electronic attacks and piracy. Also, they focused on mandating that all entities participating in the chain must be authenticated using digital certificates before adding or accessing data and the system records. Additionally, in their work, all the transactions and data are not subject to change.

Badhotiya et al. [12] presented a solution based on blockchain to reduce redundancy and X-I that exists within the supply chain. The research contribution was based on a given that the medicine industry is unregulated and contains many sellers and distributors, and blockchain technology will help upgrade network management. Also, the study relied on smart contracts to reduce costs.

Ouf [13] relied on the semantic web to improve the representation capacity of supply chains and blockchain components by explaining them in semantically rich languages. The system allowed the use of these proposed technologies to provide immutable distributed storage as well as a mechanism to track and detect any suspicious activities.

Beside the aforementioned research efforts, many other solutions have been proposed where they utilized blockchain technology as a solution to combat drug counterfeiting and to protect events and transactions within the supply chain between the parties from the manufacturing to the consumer, Zhu et al. [14], Soundarya et al. [15], Kumar and Tripathi [16], Tseng et al. [17], Uddin et al. [18], Pandey and Litoriya [19], Liu et al. [20].

Despite the former efforts that provided solutions to improve drug supply chains and make them more transparent and reliable between the parties, there are other elements that must be taken into consideration for the integration of previous solutions. These elements include for example: (i) preserving the drug from damage during its storage or transportation and not exposing it to undesirable external conditions (e.g., high temperatures and humidity or direct exposure to sunlight for long periods); and (ii) alerting the decision-maker in the event of any defect that is not suitable for the storage and transportation conditions of the drug. Therefore, the aim of this paper is to build an integrated system of the blockchain and IoT technologies to achieve a reliable pharmaceutical supply chain system. Additionally, the interplanetary file system (IPFS) was utilized to achieve tamper-proof storage, which is required to avoid unauthorized modifications. Accordingly, the system can provide a huge assistance to the EDA to track and ensure the safety of the transmitted drugs.

3 Research Methodology

To construct this research work, a search for academic studies that discussed the challenges and risks of supply chains, particularly in developing countries, has been conducted. This search was due to the challenges of the pharmaceutical supply chain that have recently emerged clearly, and particularly during COVID-19 pandemic (e.g., the shortage of medicines, the emergence of many counterfeit medicines, and the large number of expired medicines). After a critical review of the types, purposes, and methods of the scientific and academic research, the qualitative methods, which are the best in this research area, have been utilized. Accordingly, this article starts with determining the risks of poor storage of medicines that can negatively affect the health system. Then, the article proposes a scientific strategy that attempts to solve these risks. Also, to complement the previous efforts, the proposed work presents a state-guided solution that integrates blockchain and IoT to make the supply chain stronger, safer, and more reliable for all parties involved and for the end user.

3.1 Technical Background

The aim of this subsection is to present a brief background about IoT as well as the blockchain technologies as they present the base of the proposed supply chain solution. Actually, these technologies allow securing the supply chain, make it more transparent, reliable, and allow the trust between the parties by providing records that cannot be changed.

Starting with the IoT, this technology represents a network of physical devices, which consists of networks of sensors to provide tracking of drug status such as temperature, humidity, and sunlight during shipment and storage, as well as sending notifications in the event of any defect in the measurements to reduce damage and address real-time problems. Actually, this characteristic can be utilized in increasing the overall performance of the pharmaceutical supply chain.

Despite its advantages, the IoT devices can store data on a central system where this can consequently affect the data privacy and security. Accordingly, the blockchain has been utilized in order to address these challenges due to the decentralization of the data on the blockchain where this can make the data more secure. It is more transparent and not subject to tampering, so this article focuses on building a system based on the blockchain that connects all parties to the pharmaceutical supply chain as well as their transactions with each other, using records that cannot be changed and data that can be accessed from authorized parties only.

3.2 *Proposed IoT and Blockchain-Based Drug Supply Chain (IBDSC) Architecture*

According to the aforementioned power of the IoT and blockchain technologies, they were utilized to construct the proposed architecture that in turn aims to assist the Egyptian drug authority (EDA) in tracking and ensuring the safety of the transmitted drugs. Therefore, the aim of the following text is to present the details of the proposed architecture including the role of these technologies in the processes sequence (Fig. 1).

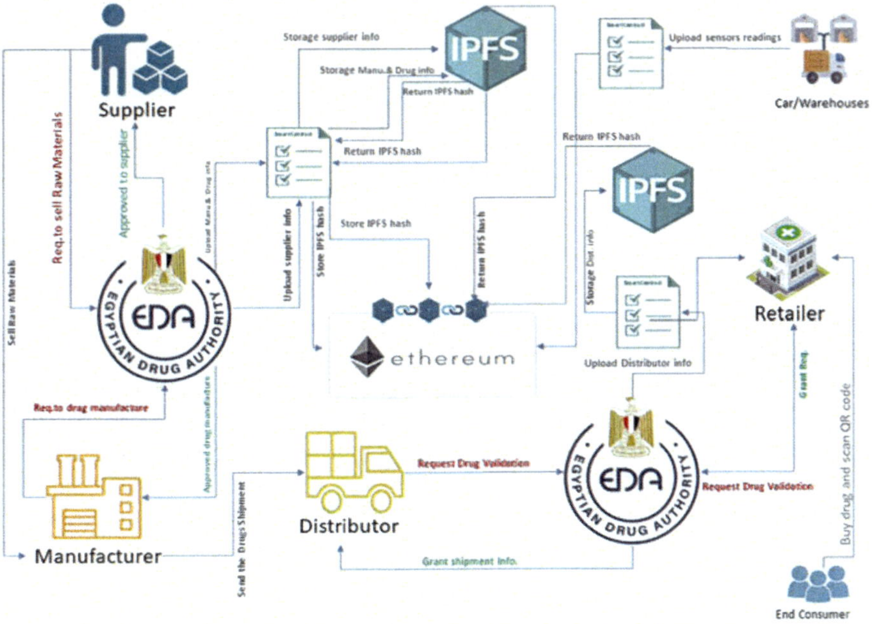

Fig. 1 Proposed IoT and blockchain-based drug supply chain architecture (IBDSC)

The EDA: Is the authority that is responsible for implementing the system through which all pharmaceutical factories, distribution companies, and pharmacies are registered in its system and ensuring that their warehouses and transport vehicles are equipped with IoT sensors to measure different factors (i.e., temperature, humidity, light). Once these registered entities meet the specifications, they will have the right to manufacture, sell, and distribute the drugs through the system.

Supplier: Is the person or company that is specializing in selling raw materials (i.e., the chemical compositions used in drug manufacturing). The suppliers are registering in the system by adding raw material data where the most important data is the expiration date of the raw material.

Manufacturer: The manufacturer requests uploading the drug's data to the system which will then be accepted or rejected by the drug authority. The manufacturer will be allowed if he is registered, and if he is not registered, he makes a request to register his data and checks his stores and transporting vehicles. After the acceptance, the drug data will be uploaded, including the QR code of the drug, where all of these data will be recorded in the rest of the stages.

Distributor: Before receiving the drug shipment, the distributor must enter the system through his data to make sure that he is an authorized person, then he must confirm that he received the shipment to be able next to start transporting it.

Transport vehicles and warehouses: These entities contain IoT sensors to record temperature, humidity, and light readings throughout the transportation and storage period, and send this data directly to be stored in the blockchain. The obtained readings can be retrieved at any time by the supply chain parties, and can ultimately inform the patient whether the drug is effective or not.

Hospitals and pharmacies: These entities need to log into the system first, then they can receive the shipment, confirm its receipt, and check the previous stages by reading the QR code to ensure the safety of the medicine.

End user: All the end user must do before buying the medicine is to scan the QR code on it to make sure if it is valid or damaged. This validity information is based on monitoring the factors that affect the drug throughout its storage and transportation period, as well as the maximum acceptable time of leaving the medicine in different conditions (e.g., high temperatures, humidity, light, etc.). For example, according to an FDA article entitled "Insulin loses some effectiveness when exposed to extreme temperatures", the longer the exposure to extreme temperatures, the less effective the insulin becomes. This can result in loss of blood glucose control over time. Despite that and only under emergency conditions, you might still need to use insulin that has been stored above 86 °F [21].

4 Implementation and Discussion

In this article, the Go Ethereum (Geth) client has been chosen to be worked with, along with Solidity language for smart contracts programming. The Ethereum protocol specifies an interface allowing people to interact with smart contracts and each other over a network. Also, the web3.py library has been used to communicate with Ethereum nodes and smart contracts in the EVM with an HTTP link to access up-to-date details regarding contract status, and new transactions. It is important to note here that the smart contract is a program that runs within a blockchain where it contains a set of rules that constitute an agreement made between two or more parties. When these rules are met, the digital contract executes the transaction. Additionally, with the IPFS, this article could address vast volumes of data and put immutable, permanent links in transactions, time stamping, and protecting content, without putting the data itself on-chain.

In the implementation of the proposed IBDSC architecture/system, each user of the system is given access to the client application's user interface to initiate a transaction after successful identity verification.

The following Fig. 2 shows the control screen of the EDA, where it can show all the warehouses and transport vehicles equipped to store and transport medicines.

Then, Fig. 3 shows the screen through which the drug details can be added (e.g., drug name, type, quantity, expiration date, and medication leaflet). These details are necessary to save allowed storage temperatures, humidity, and light of the drugs. Finally, the distributor name can be added followed by generating a QR code through the factory.

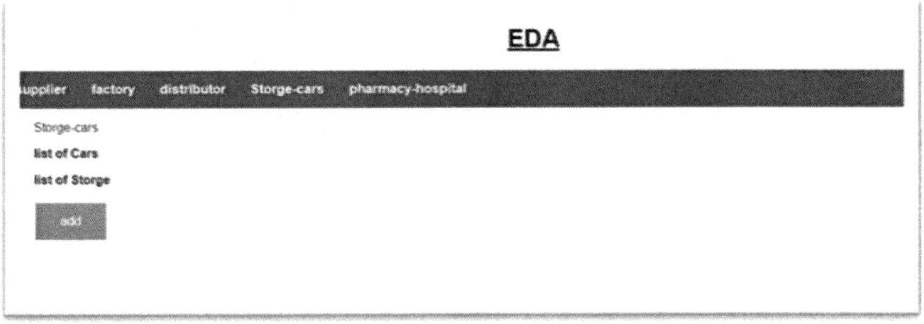

Fig. 2 EDA control screen of the warehouses and transport vehicles

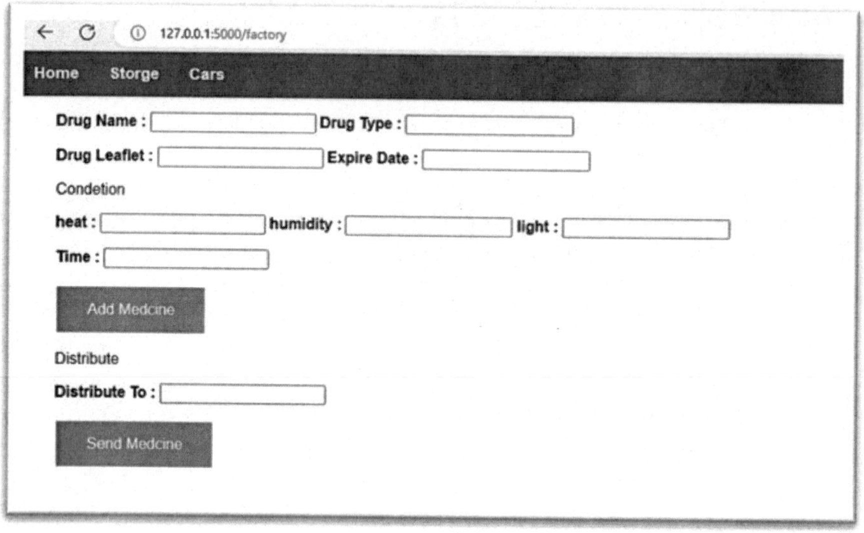

Fig. 3 Screen of creating and setting drug details

If the distributor was concerned about the quality of the medicine in the prior transaction, they would look at the information shown in the following Fig. 4. Accordingly, they can refuse the cargo since the medicine has been harmed by being exposing to a strong light or a rise in temperature.

Regarding the patients themselves, they can access a special interface that is designed for them, as shown in Fig. 5, where they can scan the drug's QR codes where there will be no hidden information about the medicine. The whole drug information sheet, including the medicine's name, kind, number of sales, distributors, and whether drug is valid or damaged, will be available through the patients' interface.

In addition to the aforementioned system details, a tool, blockchain network benchmark, to determine how many smart contracts function calls and transactions can be processed in a certain time in order to stress test the network has been built. The tool has been tested for 1, 2, and 3 min using 16 GB of RAM, an 8-core Core i7 CPU, and a 4 GB Nvidia GTX 960 M GPU where the obtained results are represented in Fig. 6. Finally, a comparison between the proposed solution and other related work is presented in Table 1.

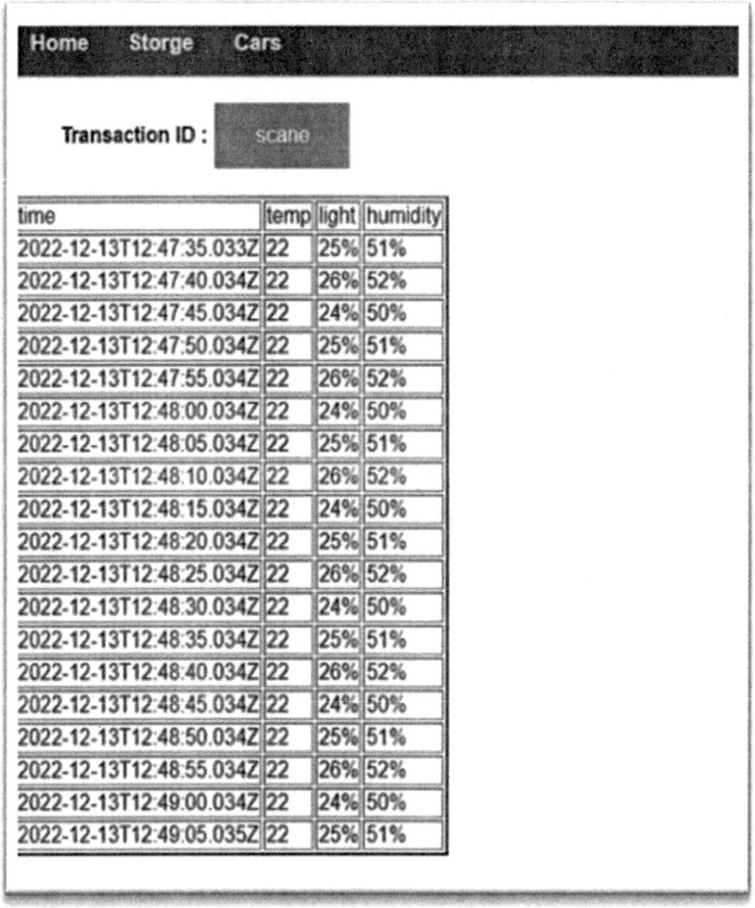

time	temp	light	humidity
2022-12-13T12:47:35.033Z	22	25%	51%
2022-12-13T12:47:40.034Z	22	26%	52%
2022-12-13T12:47:45.034Z	22	24%	50%
2022-12-13T12:47:50.034Z	22	25%	51%
2022-12-13T12:47:55.034Z	22	26%	52%
2022-12-13T12:48:00.034Z	22	24%	50%
2022-12-13T12:48:05.034Z	22	25%	51%
2022-12-13T12:48:10.034Z	22	26%	52%
2022-12-13T12:48:15.034Z	22	24%	50%
2022-12-13T12:48:20.034Z	22	25%	51%
2022-12-13T12:48:25.034Z	22	26%	52%
2022-12-13T12:48:30.034Z	22	24%	50%
2022-12-13T12:48:35.034Z	22	25%	51%
2022-12-13T12:48:40.034Z	22	26%	52%
2022-12-13T12:48:45.034Z	22	24%	50%
2022-12-13T12:48:50.034Z	22	25%	51%
2022-12-13T12:48:55.034Z	22	26%	52%
2022-12-13T12:49:00.034Z	22	24%	50%
2022-12-13T12:49:05.035Z	22	25%	51%

Fig. 4 Screen showing drug's transaction details

As shown above, and through Table 1, this article utilized the blockchain technology features from a business perspective where it was found that the Ethereum-based blockchain has more flexible network types and cryptocurrency choices. Unlike the proof of work, which necessitates high-resource hardware to achieve high verification speeds and a large number of transactions per second, proof of stake can promote decentralization while using much less energy. Furthermore, the validator node can run on a standard laptop. In terms of economic security, proof of stake has shown to have a superior blockchain implementation for business and proof of concept.

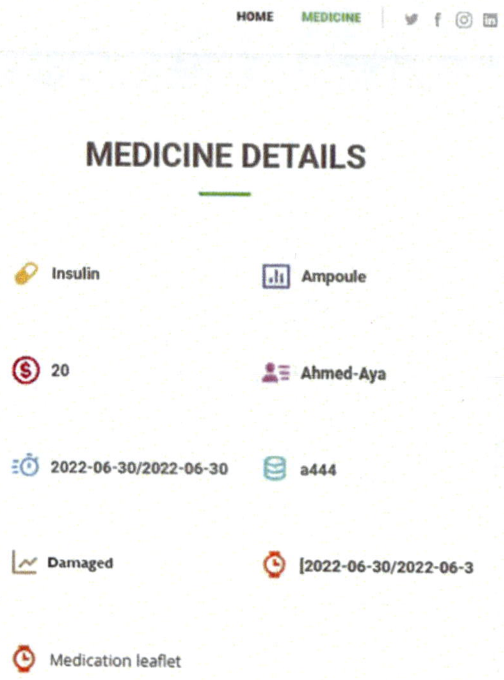

Fig. 5 End user (patient) interface

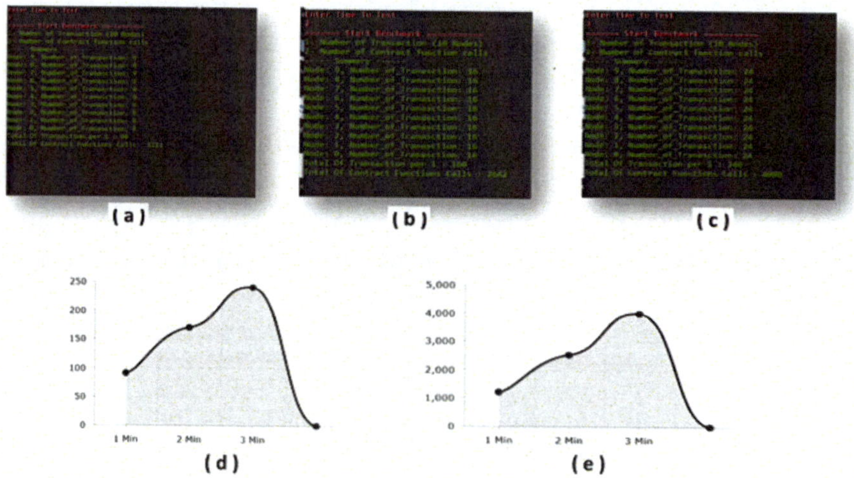

Fig. 6 Test of the blockchain network benchmark in: **a** 1 min, **b** 2 min, and **c** 3 min between 10 nodes, where it is found that **d** Total transactions are 90, 160, and 240 and **e** Total contract functions calls are 1211, 2642, and 4000

Table 1 Comparison between our proposed solution and other blockchain-based solutions

Features	Musamih et al. [9]	Kumar and Tripathi [16]	Tseng et al. [17]	Soundarya et al. [15]	SCAD [22]	The proposed solution
Algorithm consensus	(Proof of stack)	(Proof of work)	(Proof of work)	(Proof of work)	(Proof of work)	(Proof of stack)
Storage	Off-chain data storage	On-chain data storage	On-chain data storage	On-chain data storage	On-chain data storage	Off-chain data storage
Drug tracking (IoT implementation)	Limited	Limited	Limited	Limited	Limited	Unlimited

5 Conclusion

Data integration and sharing are huge challenges in the pharmaceutical sector. Text, dates and times, images, and other media are all valid data formats. It's possible that only a select group of stakeholders should have access to certain of these data, since they're very private and sensitive. Accordingly, it is important to think about the principles of "what sorts of data will be held on-chain, while the other is off-chain". To realize the on-chain and off-chain concepts, two options exist. Putting raw data off-chain and the hash of that data on-chain is one option. Alternatively, the encrypted data can be stored on-chain and the decryption key can be stored off-chain. The first choice offers various benefits, including reducing the block size, enhancing system speed, and protecting user privacy. The risk of a system's collapse due to a single weak spot is likewise increased. The second option may degrade system performance and compromise data privacy if the encryption can no longer be trusted to keep information private during storage and transfer. Keeping raw data off-chain and just keeping metadata, tiny important data, and hashes of the raw data on-chain is the method that was settled on this study. The study ensures the blockchain system's capabilities of data storage and computing, and conforms the confidentiality standards of the pharmaceutical industry. Accordingly, the proposed work provides a guideline for using on-chain and off-chain data in medication scenarios based on the stated principles.

References

1. FDA (2021) Don't be tempted to use expired medicines. https://www.fda.gov/drugs/special-features/dont-be-tempted-use-expired-medicines
2. Sumera AA, Savera AA, Nadir S (2016) Importance of storing medicines on required temperature in pharmacies and role of community pharmacies in rural areas: literature review. Imanagers J Nurs 6:32

3. Chinedu Obitte N, Chukwu A, Odimegwu D (2009) Survey of drug storage practice in homes, hospitals and patent medicine stores in Nsukka, Nigeria harnessing anti-respiratory syncytial agents from natural products view project harnessing basically researched biomaterials for pilot scale-up studies. View project
4. Jaberidoost M, Olfat L, Hosseini A et al (2015) Pharmaceutical supply chain risk assessment in Iran using analytic hierarchy process (AHP) and simple additive weighting (SAW) methods. J Pharm Policy Pract 8:1–10. https://doi.org/10.1186/s40545-015-0029-3
5. Yadav P, Lega H, Who T, Babaley GM (2011) The world medicines situation 2011 storage and supply chain management
6. Schöpperle A (2023) Analysis of challenges of medical supply chains in sub-Saharan Africa regarding inventory management and transport and distribution Project Thesis
7. Lingayat V, Pardikar I, Yewalekar S et al (2021) Securing pharmaceutical supply chain using blockchain technology. ITM Web Conf 37:01013. https://doi.org/10.1051/itmconf/20213701013
8. Panda SK, Satapathy SC (2021) Drug traceability and transparency in medical supply chain using blockchain for easing the process and creating trust between stakeholders and consumers. Pers Ubiquit Comput 21:1–17. https://doi.org/10.1007/s00779-021-01588-3
9. Musamih A, Salah K, Jayaraman R et al (2021) A blockchain-based approach for drug traceability in healthcare supply chain. IEEE Access 9:9728–9743. https://doi.org/10.1109/ACCESS.2021.3049920
10. Uddin M (2021) Blockchain medledger: hyperledger fabric enabled drug traceability system for counterfeit drugs in pharmaceutical industry. Int J Pharm 597:235. https://doi.org/10.1016/j.ijpharm.2021.120235
11. Ahmadi V, Benjelloun S, Kik ME, Sharma T, Chi H, Zhou W (2020) Drug governance: IoT-based blockchain implementation in the pharmaceutical supply chain. In: Proceedings of the 6th international conference on mobile secure services (MobiSecServ), pp 1–8
12. Badhotiya GK, Sharma VP, Prakash S et al (2021) Investigation and assessment of blockchain technology adoption in the pharmaceutical supply chain. Materials today: proceedings. Elsevier Ltd., Amsterdam, pp 10776–10780
13. Ouf S (2021) A proposed architecture for pharmaceutical supply chain based semantic blockchain. Int J Intell Eng Syst 14:31–42
14. Zhu P, Hu J, Zhang Y, Li X (2020) A blockchain based solution for medication anti-counterfeiting and traceability. IEEE Access 8:184256–184272. https://doi.org/10.1109/ACCESS.2020.3029196
15. Soundarya K, Pandey P, Dhanalakshmi R (2018) A counterfeit solution for pharma supply chain. EAI Endorsed Trans Cloud Syst 3:154550. https://doi.org/10.4108/eai.11-4-2018.154550
16. Kumar R, Tripathi R (2019) Traceability of counterfeit medicine supply chain through blockchain
17. Tseng JH, Liao YC, Chong B, Liao SW (2018) Governance on the drug supply chain via gcoin blockchain. Int J Environ Res Public Health 15:1055. https://doi.org/10.3390/ijerph15061055
18. Uddin M, Salah K, Jayaraman R et al (2021) Blockchain for drug traceability: architectures and open challenges. Health Inform J 27:228. https://doi.org/10.1177/14604582211011228
19. Pandey P, Litoriya R (2021) Securing E-health networks from counterfeit medicine penetration using blockchain. Wirel Pers Commun 117:7–25. https://doi.org/10.1007/s11277-020-07041-7
20. Liu X, Barenji AV, Li Z et al (2021) Blockchain-based smart tracking and tracing platform for drug supply chain. Comput Ind Eng 161:669. https://doi.org/10.1016/j.cie.2021.107669
21. FDA (2017) Information regarding insulin storage and switching between products in an emergency. https://www.fda.gov/drugs/emergency-preparedness-drugs/information-regarding-insulin-storage-and-switching-between-products-emergency
22. SCAD College of Engineering and Technology (2019) Institute of electrical and electronics engineers proceedings of the international conference on trends in electronics and informatics (ICOEI 2019), pp 23–25

Governance Model for Cloud Computing Service

Mohamed Gamal, Iman M. A. Helal, Sherif A. Mazen, and Sherif Elhennawy

Abstract Cloud computing services are one of the IT solutions whose popularity has grown across various industries. Big challenges face cloud computing services, like how to secure them and raise awareness of their usage. Furthermore, these concerns and obstacles are linked to cloud computing services, which may result in several critical risk areas. The adoption process of moving to cloud computing services is slowed by a lack of governance, as the role of governance is to assess the performance and analyze the adherence to agreed-upon goals and objectives and it is hard to apply in the cloud computing environment. This work aims to present a governance model which ensures that the required controls are implemented in Amazon Web Services (AWS) cloud services and analyzes the risk if the controls are not applied. In addition, the model will provide the user with various client-side controls as well as certain service level agreement (SLA) guiding parameters.

Keywords Cloud computing · Governance · Risk assessment · SLA · AWS

M. Gamal (✉) · I. M. A. Helal · S. A. Mazen · S. Elhennawy
Faculty of Computers and Artificial Intelligence, Cairo University, Giza, Egypt
e-mail: mohamed.gamal@acu.edu.eg

I. M. A. Helal
e-mail: i.helal@fci-cu.edu.eg

S. A. Mazen
e-mail: s.mazen@fci-cu.edu.eg

D. Magdi et al. (eds.), *Green Sustainability: Towards Innovative Digital Transformation*, Lecture Notes in Networks and Systems 753,
https://doi.org/10.1007/978-981-99-4764-5_8

1 Introduction

Several cloud computing models are defined in a variety of ways. Cloud computing is a model for enabling ubiquitous, suitable, on-demand access to a shared pool of various computing resources that can be rapidly provided and issued with minimal management effort or service provider interplay [1]. Cloud computing services have many benefits for the users such as scalability, as it enables the client to scale horizontally by adding or removing new stations easily [2]. As users of cloud services can access their data and services from anywhere if they utilize a cloud service that enables employees to be more adaptable in their work practices because of cloud computing allows for greater flexibility in their work routines. Business continuity is one of the most valuable benefits of cloud computing, as service providers mostly have a disaster recovery plan. Cloud service providers put data security protections in place in their data centers which make the data more secured.

For enterprise systems, cloud computing services have many benefits [3]. It helps in growing business for organizations. Cloud service providers (CSPs) relieve users from the burden of setting up and maintaining IT infrastructure and services. Therefore, businesses can concentrate on their core competencies. One of the most appealing features of cloud computing is its capacity to address a variety of issues that small and medium enterprises encounter.

There are many challenges faced when transferring to the use of cloud computing services like data security, data breaches, performance expectations, and cyber-attacks. The controls will be divided between the cloud service provider and the cloud client according to the service model. Thus, that will lead to issues in regulatory compliance for the cloud client. Cloud clients should have an SLA that ensures the service availability, security, compliance articles, and the articles needed to cover the challenges that may face them. Because of these reasons, many organizations have concerns about how to migrate their data and functionalities to the cloud.

In this paper, the authors intend to provide a model to help the auditor in testing the controls on CSP's side. This model can provide the auditor with the required controls that he must apply on the client-side. Finally, it provides the client with guidance parameters that should be found in the SLA.

2 Background

Cloud computing provide an on-demand service, in which the customer can contribute computer resources (e.g., server time, network, and storage) as required. Broad network access is another feature, as all resources are accessible over the network and may be accessed via regular protocols. Using a multi-tenant approach, the CSP's computing resources are pooled to support numerous Cloud Clients (CCs), with distinct physical and virtual resources constantly assigned and reassigned based on CC demand [4].

Cloud computing services have three service models and four deployment models as part of the cloud model [5]. The three main cloud delivery models are (1) Infrastructure as a Service (IaaS), (2) Platform as a Service (PaaS), and (3) Software as a Service (SaaS) [6]. The deployment models for cloud computing have been split into four categories [7]:

- **Private cloud**: In which the organization owns its cloud data center is considered the most secure deployment model.
- **Public cloud**: A public cloud is one in which a cloud service provider provides its cloud client with apps, storage, and a variety of other cloud services. This cloud model is designed to provide all organizations with boundless memory storage and extended data sharing across the Internet.
- **Community cloud**: Its infrastructure is monitored and then used by a range of organizations with the same core enterprise, projects, or shared specifications, such as software and hardware so that IT operating costs can be minimized.
- **Hybrid cloud**: Consists of public and private models. It helps in reducing the cost of spending benefits for its establishment as costs are spread among businesses.

Governance ensures that stakeholders' needs, conditions, and choices are considered to produce balanced and mutually agreed-upon corporate goals. It also guides by prioritizing and making decisions. Governance is responsible for assessing the performance of the agreed-upon goals and analyzing adherence to clear objectives [8]. Information systems' controls are a collection of policies and technological measures that an organization uses to protect the security and efficiency of its information systems.

Service level agreement (SLA) lays forth the minimal performance and quality standards that the provider promises to meet [9]. It advises that the supplier be subjected to remedial procedures and fines. An architecture was proposed that categorizes the SLA articles by performance (e.g., response time), customer satisfaction level (CSL), pricing, and security [10].

In the second quarter of 2021, AWS achieved a record $14.8 billion in net sales, accounting for little more than 13% of Amazon's total net sales. In terms of cloud computing platform market share, AWS has consistently outperformed competitor Microsoft Azure, rising to 30% in recent quarters. As illustrated in Fig. 1, AWS provides a wide range of cloud services to its customers [11].

Benchmarks are the only consensus-based, best-practice security configuration guides established and endorsed by the government, business, industry, and academia. The Center for Internet Security (CIS) offers prescriptive advice on selecting security choices for a subset of AWS, with a focus on basic, testable, and architectural settings [12]. The CIS benchmarks are created using a unique consensus-based method, including cyber-security professionals and subject matter experts from all around the world [13].

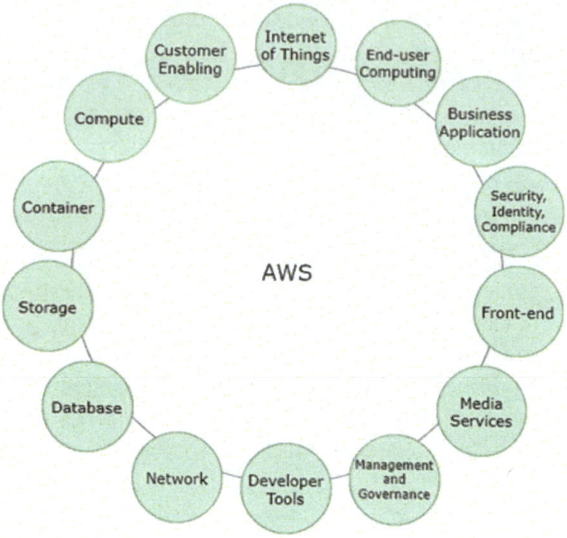

Fig. 1 AWS's services

3 Related Work

3.1 Concerns in Cloud Computing

There is a lack of a mechanism to protect the physical layer against unwanted access which is a major concern [14]. Authentication control is very important, which determines who has access to information and what information is available [10]. The majority of CSPs employ the software development life cycle (SDLC) rather than the secure software development life cycle (SSDLC), which might result in security vulnerabilities [15].

Data separation is a big issue because data infiltration can occur through client code injection or the use of any software from other clients [16]. Data location is a major problem since data is held in several locations, making auditing difficult and resulting in potentially fluctuating governance regulations. In addition, CC has several flaws, including a lack of security knowledge, inadequate recruiting methods, and personnel screening. Cloud computing can increase pricing [17]. Scaling back also saves money. CC has many worries regarding cloud computing services such as data breaches, which occur when protected or sensitive information, such as credit card numbers is leaked, viewed, stolen, or used online without permission [18].

In cloud computing, authentication and permission are crucial for data security. Once the contract is up for renewal, the level of data encryption and the process of removing data files are critical [15]. Malware (dangerous code or software) that may be remotely injected into the cloud and alter data poses a severe concern [10]. Malicious insider, who might be a current or former employee or a business partner who has been granted access to the information system and has willfully misused that access to compromise the information system's security and privacy [19].

3.2 Service Level Agreement Articles

Every type of cloud computing service has specific SLA parameters [20]. Scaling is a very important feature of the cloud and one of the best benefits of using cloud services. The first is autoscaling, which determines whether or not it can be completed automatically. The second parameter is the amount of time required to scale up, finally, the needed time to scale down. Clients think that the availability of cloud services is 100% but that may be wrong. So, the SLA must determine the availability of service percentage and it must be 99.5% or higher. Another important parameter is the response time which refers to the time needed for completing the process.

For PaaS, one of these articles is the number of developers who may use the service at the same time, among other things. Another crucial component is integration, which determines how other third-party developers must understand how to utilize the available APIs, necessitating the supply of suitable training resources. The licensing of the software that the CSP employs, as well as the fees of these licenses, are two of the most essential elements. Third-party developers must be able to utilize the APIs that are available, which involves the provision of appropriate training materials. Many APIs are provided by CSP, on the other hand, they do not provide enough documentation for these APIs [21].

The *geographic location* refers to the location (list of available countries) that which data will be stored in. Other parameters are related to security like *data transferring methodology*, *authorization*, and *authentication*. One of the most significant advantages of cloud computing is *recovery*. So, it is important to know the ability to recover and the time taken for recovery after a disaster. The place where the *backup* is stored is another important parameter too. *Documentation* is an underrated parameter even if it contains the list of the documents that the provider makes available, to demonstrate compliance with data protection requirements and obligations; also, *customizability*, as many types of users can use it easily via a small number of edits to enable users to deal with the interface smoothly. This can also refer to another parameter which is *usability* [22].

3.3 Governance in Cloud Computing

In IaaS, a taxonomy was provided, covering security challenges in cloud computing. It is set up to alert the CC to any unexpected behavior so that the controls may be examined [23]. A model for selecting CSP by reviewing and assessing all risk scenarios to make CSP selection easier [24]. An approach that assesses the risk in the cloud by defining the harms that may happen and their impact to check financial loss in case of risk appearance [25].

In 2017, Medhioub et al. [26] proposed a paradigm based on gathering baseline knowledge about the cloud environment, evaluating risks, and deciding whether preventive techniques are appropriate. The practice of monitoring and updating risk circumstances allows for a better knowledge of the changing risk scenario and improves reaction to it. Another developed framework focused on assessing the risks of the needed service model for the CC and checking if it can reduce these risks or mitigate them to accept this model [27].

A conceptual framework based on knowledge management (KM) has made it possible for corporations to cope with vast amounts of data and procedures. Also, internal communication is critical in implementing any technology, particularly when the work is highly interrelated and carries a high level of risk, especially when the task is highly interdependent and contains significant uncertainty. He also showed that users' problems should be reported to cloud computing experts as soon as possible, so that they may be resolved since the amount of training provided during the early stages of installation will influence how widely a system is used and how frequently problems [28].

A qualitative tool was developed to employ a variety of techniques, including a systematic literature review. In the first phase, experts gather data to assess which hazards are most likely to be encountered in cloud computing services. The tool performs an iterative risk assessment that continually reviews the controls applied to each risk [29].

Storage, software, network security, trust management, Internet and services, compliance, legal, and virtualization were identified as cloud computing security areas by [30]. These domains are then subjected to security policies to show that IT governance is successful while using the cloud.

There is a strategy for determining the critical success factors in cloud service implementation. It includes (1) organizational (defining responsibilities, business, and IT alignment), (2) technological (automation of data integration life cycle, having data metrics), (3) strategic points of control, (4) training and increasing data stakeholders' awareness, and (5) monitoring compliance [31].

OpenStack core services are used to map the three profiles (virtual, cloud, and end user) established by the CIS benchmark, incorporating the ideas of shared responsibility and cloud layers. Their assessment procedure is carried out by gathering evidence regarding the target system, such as by monitoring activity testing on a certain service. The evidence gathered allows you to check if the suggestions (standards) have been met or not. This benchmark evaluation is a score that represents the level of security in cloud computing. It focuses on the IaaS and provides continuous evaluation of the security of cloud computing [32].

4 Proposed Model

Our proposed model is based on the CIS benchmark for AWS to check the controls applied on the CSP's side. It checks the controls based on the used resources. The model gives the CC auditor a list of controls that are appropriately implemented, high-risk non-controlled resources, and medium-risk non-controlled resources. In addition, it gives some controls that must be applied on the CC's side according to the service model. Applying the needed controls on resources on both CSP and CC sides will lead to achieving governance. Finally, it guides the auditor with the main SLA parameters that are needed according to the service model, these parameters ensure compliance and alignment with regulations.

This model will be beneficial in identifying missing controls on the CSP's (AWS) side that might contribute to risk when the service is launched, as well as providing suggestions for all risk areas. So, it will assist the auditor in implementing missing controls and maximizing service use. Based on the service and SLA's major features, the model will also give the auditor the needed controls to decrease risk areas (IaaS, PaaS, and SaaS). The auditor will be able to accomplish governance by following the suggestions. To achieve governance, we have to apply relevant controls which in turn need to identify the resources and their risk areas that need to be controlled. As a result, we'll break down our model into three steps: (1) specify the used resources from AWS; (2) learn the possible risk areas that need controlling; and (3) check the controls of the used resource in a real-time setting.

4.1 Main Resources

Many categories need checking when using cloud computing services. As a result, the first step is to create a resource list. Computing, database, management, network, security, and physical storage are the most important resources.

1. Compute Resources
   (a) *Lambda* is a serverless compute service that automatically maintains CC's underlying computing resources and executes CC's code in response to events. Lambda may be used to add custom logic to other AWS services or to create back-end applications that take advantage of the size, performance, and security of AWS.
   (b) *Elastic Compute Cloud (EC2)* is used to create as many or as few virtual servers as you require, as well as handle security and networking and manage storage. The CC may scale up or down to accommodate shifting demands or popularity spikes using Amazon EC2, eliminating the need to forecast traffic.

   (c) *Elastic Load Balancing (ELB)* Incoming application traffic is automatically distributed to a variety of destinations, including Amazon EC2 instances, containers, IP addresses, Lambda functions, and virtual appliances.
   (d) *Elastic Load Balancer version 2 (Elbv2)* is used to disperse incoming traffic over several destinations such as the EC2 instances at CC. It allows CC to boost the application's availability.

2. Management Resources

   (a) *CloudFormation* is a service that simplifies the design and provisioning of a collection of linked AWS and third-party resources for developers and enterprises.
   (b) *CloudTrail* is used to gather logs, monitor, and record account activity associated with its AWS infrastructure.
   (c) *CloudWatch* is an AWS, hybrid, on-premises, and infrastructure resources monitoring and management solution.

3. Database Resources

   (a) *Relational Database Service (RDS)* is a web service that makes creating, running, and scaling a relational database simple. It performs basic database management duties while providing cost-effective, resizable capacity for an industry-standard relational database. It has many instances (i.e., DB.r6g, DB.m5d, DB.m5, DB.x2g), every one of them supports different types of database management systems (i.e., SQL, Oracle, MariaDB, PostgreSQL).
   (b) *Redshift* is a petabyte-scale data warehouse solution that allows analyzing all data fast and cost-effective utilizing the cloud client's existing business intelligence tools simple and cost-effective. It's made for datasets ranging from a few hundred gigabytes to a petabyte or more, and it costs < $1000 per terabyte per year or roughly a tenth of the price of most traditional data warehousing systems.

4. Network Resources

   (a) *Route 53* is in charge of DNS for both TCP and UDP traffic requests; the term Route could refer to routing, or it could refer to a popular highway naming convention.
   (b) *Virtual Private Cloud (VPC)* by using it CC may now deploy AWS resources into a previously specified virtual network. This virtual network is quite similar to a typical network that CC would run in its own data center, but with the added benefit of leveraging AWS's scalable infrastructure.

5. Security Resources

   (a) *AWS Certificate Manager (ACM)* makes creating, storing, and renewing public and private SSL/TLS X.509 certificates and keys for your AWS websites and apps a breeze. CC may either issue certificates directly using ACM or import third-party certificates into the ACM management system to offer certificates for the linked AWS services.

(b) *AWS Identity and Access Management (IAM)* allows CC to create and manage AWS users and groups, as well as grant and deny access to AWS resources, using permissions.
(c) *Key Management Service (KMS)* allows CC to have centralized control over the cryptographic keys that are used to safeguard his data.
(d) *Secret Manager* helps the CC to securely encrypt, store, and retrieve credentials for his databases and other services. Instead of hardcoding credentials while building his apps, the CC's user can make calls to the secrets manager to retrieve his credentials whenever needed.

6. Storage Resources

(a) *Simple Storage Service (S3)* is a web service that makes it easy to set up, run, and scale a relational database. It performs basic database management duties while providing cost-effective, resizable capacity for an industry-standard relational database. It has several instances (e.g., DB.r6g, DB.m5d, DB.m5, and DB.x2g), each of which supports a different type of database management system (i.e., SQL, Oracle, MariaDB, PostgreSQL).

4.2 Main Risks

Every resource has some concerns. To assure the accuracy of the cloud computing services in this part, we evaluate these factors. We will discuss samples of these issues for the aforementioned resources that we discussed in the previous section.

(1) Compute Risks

(a) EC2 risks

For computing resources, there are many risk areas, e.g., the AMI should not be public. Another danger involves checking if the ports are open, which might result in large-scale attacks on the resources. One of the critical risks is the usage of default security groups which might indicate that the principle of least privilege is not being consistently applied. To maintain correct power and access control, use customized security groups.

(b) ELB risks

Security group whitelists non-elastic IP addresses setup is a violation of company regulations. Also, the EBS volume must be encrypted. AWS IP ranges contain addresses that may be assigned to EC2 instances in any AWS account, as well as, services that can be used to connect to any AWS account which may lead to activating these IP ranges that might expose your AWS account to outside activities. ELB Access Logs are very important as they contain information such as the time the request was made, the client's IP address, latency, request routes, and server responses that can be used to investigate traffic patterns and security issues.

(c) ELBv2 risks

The load balancer should employ a secure protocol (HTTPS or SSL) that follows best practices for encrypted communication. With a load balancer that doesn't have a listener and utilizes an encrypted protocol, eavesdropping, and man-in-the-middle attacks are conceivable. Another issue is utilizing an older version of the SSL/TLS policy, since using the most recent version will resolve any difficulties discovered in older versions.

(2) Database Risks

(a) Redshift risks

One of the most important controls for the database is encrypting all data. Clusters that are publicly accessible pose a significant risk since they allow other AWS users to view your cluster and the data it contains.

(b) Relational Database Service risks

As a convenience control, CC must do an automated minor version upgrade. CC's database will be updated as soon as a new minor database engine version is published.

(3) Management Risks

(a) CloudFormation risks

Passing a role to CloudFormation stacks may result in privilege escalation for management since IAM users with rights inside the CloudFormation scope inherit the stack's role's permissions by default. Duplication of global service logging is also an issue since having too many log entries makes it more difficult to analyze potential problems.

(b) CloudTrail risks

It will be a major issue if the trail is not linked to CloudWatch since it will be difficult to track real-time and historical data, as well as set up alarms and notifications for unusual account activity, without a link to CloudWatch. You may monitor real-time and historical data, as well as set up alarms and notifications for odd account behavior if data events recording is set up as without a connection to CloudWatch.

(4) Network Risks

(a) VPC risks

The network is connected with several dangers. For instance, there are no flow logs to analyze illicit network traffic events like an attacker exfiltrating data or pivoting to other sites. Peering routing tables are critical because they lessen the impact of a breach by preventing the peered VPC from accessing resources outside of these routes.

(b) Route53 risks

To prevent someone from moving the user's domain to another registrar without their permission, the domain must be locked. Also, if the domain transfer locked is not supported by the top-level domain (TLD), it will lead to an unauthorized registrar. If the automatic renewal of the domain is set off it will lead to losing control over the client's domain names.

(5) Security Risks

(a) ACM risks

To maintain security on the CSP's side, all security certificates must not be expired. Another issue is if the ACM certificate's transparency logging is deactivated, which may cause browsers to refuse to trust your certificate as a result of the logging.

(b) IAM risks

Users should only be allowed to take on responsibilities that they are qualified for. Another essential consideration is the use of strong passwords to increase security. Multi-factor Authentication (MFA) should be set to require users to present their user name and password as well as an authentication code from their AWS MFA device, particularly for the root account.

(c) KMS risks

Customer Master Keys (CMKs) are not rotated and will cause risk as it will increase the probability of usage of compromised keys.

(6) Storage Risks

(a) S3 risks

Without versioning the buckets, CC cannot recover from both inadvertent user activity and application faults. Disabling bucket access logging is a concern since it will result in the loss of detailed logs of bucket requests. And these server access logs can assist the client in security and access audits, and understanding the Amazon S3 bill. Clients and S3 buckets can communicate via unencrypted HTTP if HTTPS is not imposed on the bucket policy. As a result, sensitive data might be sent across the network/Internet in plain text.

4.3 Model Components

1. **Test Engine:** Test engine is responsible for making the tests of compliance that are stored in resources to be checked and running the report generator. It checks the resources by recursively testing conditions for a path using the "utils" file, which runs the rules for each resource. To do this, it needs to evaluate all the "id" possibilities (Fig. 2).
2. **Rules:** It includes two parts. The first part is containing all the rules that will be stored as a result of testing resources in the last step; every rule is related to a specific resource. The second part is the base file which calls for the rules to be tested to eliminate one or more rules if they are not needed to be shown at present. Figure 3 illustrates the relation between rules and resources. Checking

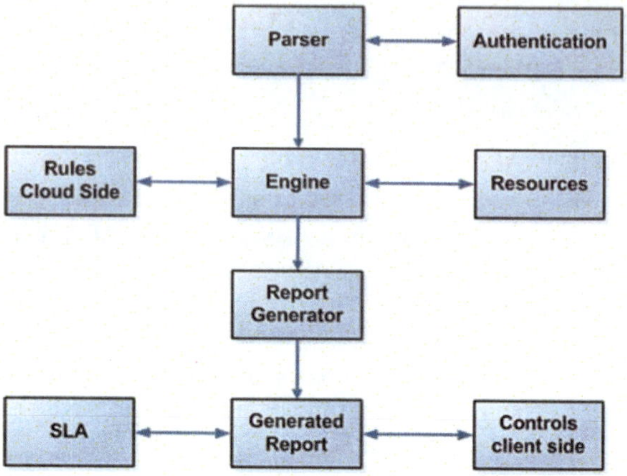

Fig. 2 Component diagram

whether the elastic block storage (EBS) is encrypted is an example of a rule. The description represents what will appear at the head of the report (EBS volume not encrypted). The rationale will show why this missing feature will cause a risk (when you enable encryption on EBS volumes, data is encrypted both at rest and in transit.). Finally, the condition that checks if the "Encrypted" parameter is "false" in EBS then will give a critical warning as shown in Listing 1.1. Listing 1.2 shows another control example that if the data event logs are configured or not to track the account behavior.

Listing 1.1: Rules example 1

```
{
" d e s c r i p t i o n " : "EBS Volume Not Encrypted " ,
" r a t i o n a l e " : "When you enable encryption on EBS volumes , data i s
encrypted bo " dashboard name " : " Volumes" ,
" path" : " ec 2 . r e g io n s . id . volumes . id " ,
" c o n d i t io n s " : [
"and" ,
[
```

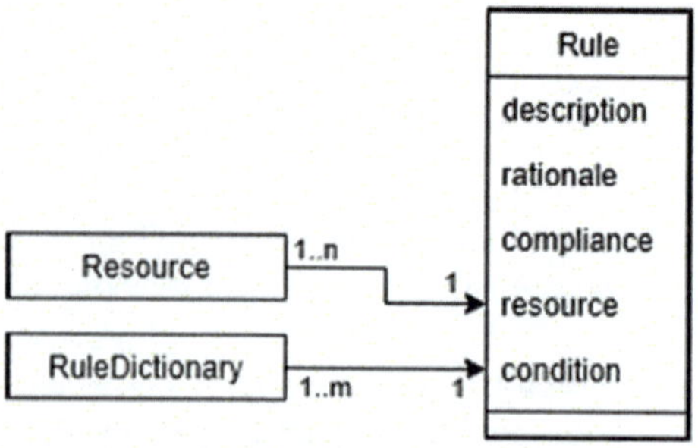

Fig. 3 Rules structure

```
" Encrypted " ,
"f a l s e",
""
]
]
}
```

Listing 1.2: Rules example 2

```
{
"d e s c r i p t i o n" : "Data Events Logging Not Configured " ,
"r a t i o n a l e" : " It ' s hard to track re a l =time and h i s t o r i c a l data , as
w e l l as s e
" dashboard name " : " Co n f igu ra t io n s " ,
"d i s p la y p a th" : "c l o u d t r a i l . r e g io n s . id . t r a i l s . id" ,
" path" : "c l o u d t r a i l . r e g io n s . id . t r a i l s . id" ,
"c o n d i t io n s" : [
"and" ,
[
"t h i s" ,
" withKey" ,
" DataEventsEnabled "
]
],
"i d_s u f f i x" : "c l o u d t r a i l =data=events=d i s a b l e d"
}
```

3. **Report Generator:** The report generator is responsible for reading from the result file and generates a web-based interface that includes all resources and the level of risk of each used resource in this account. This report will assist the auditor in determining the risk level of each resource and making recommendations for risk reduction based on the CIS benchmark. The report is divided into three parts. The first part is the result of tested resources (active directory, applications, network, security center, key-vault, storage, database, virtual machines) used on the cloud side. The second part will divide the risk areas according to the service type (IaaS, PaaS, SaaS, general) risks. The model will provide some of the needed controls that the CC's auditor must check if the CC's staff members apply it internally on the client-side (i.e., the activities associated with the CC) or not. Table 1 will give a sample of the controls that may be needed on the client-side. Many of the SLA criteria are listed in the report's last section, which the CC will need to review to see if the provider is using them. It increases the awareness of the auditor to know more details needed to check in the SLA (some non-functional requirements, security articles, and some legal articles). Some of these guiding SLA factors are given in Table 2 that shows some of these guiding SLA parameters. Figure 4 shows an example of the results of verifying the AWS account's utilized resources.

Table 1 Client-side controls example

Control	Concern
Each user has a unique identity	Identification and authentication
A list of users and their unique identities are maintained	Identification and authentication
The updated files must be kept an eye on	Secure data management
Information and data that is subject to personal data protection regulations. Special permission/licensing is required to disclose this material	Handling security incidents
The process of recovery	Access control and resource use
The applicability of users' accounts will be reviewed periodically	Logical access control
A structured method will be used by management to review user permissions at regular periods	Protection software (development and maintenance)
Determine all audit requirements by the existing legal and regulatory framework	Compliance
New applications must be supported by adequate documentation	Protection software (software changes)
The needs evaluation must include a risk assessment	Protection software (software changes)
Data monitoring for all systems with multi-user access	Compliance
To create apps, scientifically accepted guiding techniques should be applied	Protection software (development and maintenance)
Software development and testing systems must be kept distinct from the operating systems	Protection software (urgent changes)
Users must take precautions to ensure that their accounts are used safely	Access control and resource use
At least once a year, log files should be checked	Compliance
Permission from the data protection authorities is necessary	Access control and resource use

4.4 *Proposed Model Evaluation*

The proposed model is used to check the controls that are supposed to be tested while using some AWS services. There are main risks that are categorized, and every category contains some controls to be checked. The model is qualitative as it has constant levels of risk. There are many results for each control indicating the level of risk. The first level is "critical" which refers to high risk. The second level is "medium" which refers to medium risk. The third level "controlled" means that the needed controls are being applied well. The last result is "not used" which means that the related resources to the control are not used. Table 3 displays some of the output outcomes. This sample result covers the main resources we discussed in section four. Also, it shows the groups of each targeted resource. It represents all risk levels that

Table 2 SLA parameters example

Parameter	Description
Service availability	The uptime of service
Performance	Latency or the response time
Data locality	Location of the data
Access to the data	Data in a readable format that will be retrieved from the supplier
Exit strategy	Ends the cloud service and allows the client to retrieve data from the cloud service
Monitoring	How the cloud customer will monitor the service
License	Software license and its cost
Support	Requirements for the CSP's service to manage customer concerns and questions
Authorization/authentication	Requirements for issuing, validating, and canceling a cloud service user's access/use privileges
External connectivity	Describes the cloud service's ability to connect to systems and services that are not part of the cloud service
Auditing/security	Processes for acquiring independent proof or verification that the cloud service fulfills predetermined requirements
Redundancy level	Considers the percentage of components or services that have failed over mechanism
Timely incident reports percentage	Descriptions of several types of data connected with a cloud service, as well as the authorized uses of each type of data by the client and the CSP
Data usage/classification	Descriptions of many types of data linked with a cloud service, as well as the customer and CSP's rights to use each type of data
Data life cycle	Data handling, storage, and deletion
Response time	Maximum time to get the response for data request
Documentation	List of the documents that the provider makes available, to demonstrate compliance with data protection requirements and obligations
Data portability format	The electronic format(s) in which client data from a cloud service may be transported to and accessible from the cloud service

we mentioned and shows that if the resource is not used, the result of its control will be "not used". If the unused resources will be used later then when the model runs again will check and determine if these resources are controlled or the level of risk while it is not controlled.

Now there are many auditing tools for the cloud. Their purpose is to ensure that all controls are applied to mitigate all risk areas. In this section, we will compare the proposed model with other tools to show the number of controls that our model contains compared with the other tools as given in Table 4. From the comparison,

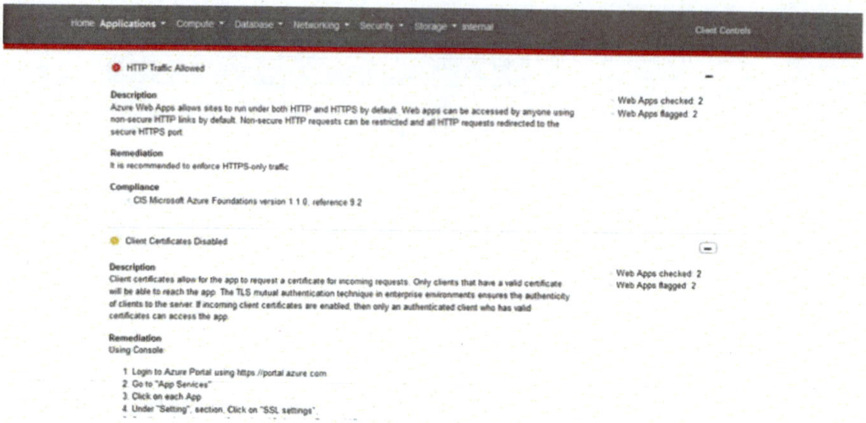

Fig. 4 Report example

we found that even some tools provided more rules but none of them guides SLA parameters. Also, the other tools focus on testing the controls that are applied on the AWS environment without consideration to the controls that must be applied on the side of CC. Other tools, on the other hand, test a greater number of options, therefore our model will need to incorporate more controls.

5 Conclusion and Future Work

Many firms have already begun to utilize cloud computing services as a result of their benefits. Other tools, on this, is accessible in a variety of delivery and deployment options, which will require testing a greater number of options, therefore our model will need to incorporate more controls. Using cloud computing services might put the organization at risk for a variety of issues. Some of the organization's previously implemented controls are lost when moving to the cloud. The cloud service provider will be responsible for implementing these restrictions.

Every service model faces unique dangers. Furthermore, adopting cloud computing services poses several risks. Clients that use the cloud must concentrate on all of them. The priorities of CCs vary depending on the type of business and its requirements. When employing cloud services, risk areas must be determined based on two factors: (1) CC rules for ensuring business process needs, and (2) adherence to legislation.

The CC will be protected from various hazards by implementing governance when adopting cloud computing services. When the necessary controls are in place, governance guarantees that the business plan is implemented and that rules are followed. SLA is crucial for achieving governance. The SLA flaws will result in a violation of the regulations, a violation of the controls, and/or additional fines.

Table 3 Results sample

#	Risk	Risk level	Group	Category
1	EBS volume not encrypted	Critical	EC2	Application
2	Potential secret in instance user data	Critical		
3	Default security groups in use	Medium		
4	Security group allows ICMP traffic to all	Controlled		
5	Security group opens NFS port to all	Controlled		
6	Lack of deletion protection	Medium	ELBV2	
7	Load balancer allowing (HTTP) communication	Critical		
8	Instance storage not encrypted	Medium	RDS	Database
9	RDS instance has a deprecated certificate authority assigned to it	Controlled		
10	Security group allows all IP addresses	Not used		
11	Cluster database encryption disabled	Critical	Redshift	
12	Role passed to stack	Controlled	CloudFormation	Management
13	CloudTrail service not configured	Critical	CloudTrail	
14	Data events logging not configured	Not used		
15	Alarm without action	Controlled	CloudWatch	
16	Domain transfer lock not supported by TLD	Not used	Route53	Network
17	Network ACLs allow all egress traffic	Medium	VPC	
18	Subnet with "Allow All" egress NACLs	Medium		
19	Subnet without a flow log	Medium		
20	ACM certificate expiring in < 7 days	Not used	ACM	Security
21	ACM certificate with transparency logging set to disabled	Not used		
22	Managed policy allows "IAM:PassRole" for all resources	Critical	IAM	
23	Data events logging not configured	Not used	CloudTrail	
24	Alarm without action	Controlled		
25	Log file validation is disabled	Not used	CloudWatch	
26	Bucket access logging disabled	Medium	S3	Storage
27	Bucket allowing clear text (HTTP) communication	Medium		
28	Bucket without MFA delete	Medium		
29	All actions authorized to all principals	Controlled		
30	Bucket's permissions world-readable	Controlled		
31	Put actions authorized to all principals	Controlled		

Table 4 Evaluation by other tools

Group	Prowler	CS-suite Aws	CloudSploit	Proposed model
SLA parameters	✗	✗	✗	30 Parameters
Support rules				
ACM	2	2	2	2
IAM	35	37	Check	37
API gateway	12	✗	20	✗
AutoScaling	8	✗	2	✗
Athena	2	✗	8	✗
CloudFront	6	✗	5	✗
CloudFormation	5	1	5	1
CloudTrail	12	8	11	8
DynamoDB	3	✗	3	✗
EC2	69	14	72	18
EKS	2	✗	5	✗
ELB	7	3	7	3
ELBv2	9	5	9	5
EMR	7	1	3	1
RDS	13	8	12	8
Redshift	13	6	12	5
Route53	5	3	7	3
S3	15	14	14	12
SES	1	4	1	4
SNS	4	4	4	✗
SQS	5	7	5	✗
External rules	✗	✗	✗	30
Supported operating systems				
Windows	✗	True	✗	True
Linux	True	True	True	True
OSX	True	True	True	True

The implemented model tests the AWS environment while checking if there are risk areas not controlled related to all resources used and makes a recommendation to mitigate these risks. It provides the CC with guidance for the controls (policies, and standards) to be applied on the client's side which will lead to more security and achieving governance. Also, it provides the client with some guiding SLA articles that he may need to check to assure the quality of service, availability of the service, service, and security.

The proposed future work is to implement a model working on all CSPs. That will enable the CC to evaluate all environments of all CSPs which enable the CCs to choose the best CSP for their usage. Another thing is to increase the rules that will be tested that will make a better control for cloud services.

References

1. Gouda KC, Radhika TV, Akshatha M (2014) Priority based resource allocation model for cloud computing
2. Ghahramani MH, Zhou M, Hon CT, Member S (2017) Toward cloud computing QoS architecture: analysis of cloud systems and cloud services. IEEE/CAA J Autom Sin 4(1):6–18
3. Saini H, Upadhyaya A, Khandelwal MK (2019) Benefits of cloud computing for business enterprises: a review. SSRN Electron J 1574:1003–1007
4. Mell P, Grance T (2011) The NIST-national institute of standards and technology definition of cloud computing. NIST Spec Publ 800–145:7
5. Jansen W, Grance T, Jansen W, Grance T (2011) Leica microsystems: multi-photon microscopy
6. Ashraf I (2014) An overview of service models of cloud computing. Int J Multidiscip Curr Res 2:779–783
7. Diaby T, Rad BB (2017) Cloud computing: a review of the concepts and deployment models. Int J Inf Technol Comput Sci 9(6):50–58
8. Lainhart JW (2012) "COBIT 5: a business framework for the governance and management of enterprise IT COBIT 5
9. Lourenco J, Santos-Pereira C, Rijo R, Cruz-Correia R (2014) Service level agreement of information and communication technologies in Portuguese hospitals. Proced Technol 16:1397–1402
10. Sultana A, Raghuveer K (2017) Security risks in cloud delivery models
11. Amazon (2022) https://aws.amazon.com/products/. Accessed 15 Jan 2022
12. CIS (2018) CIS Amazon web services foundations benchmark v1.2.0, pp 0–157. https://www.cisecurity.org/cis-securesuite/cis-securesuite-membership-terms-of-use/
13. CIS (2022) https://www.cisecurity.org/cis-benchmarks/cis-benchmarks-faq. Accessed 15 Jan 2022
14. Ramachandra G, Iftikhar M, Khan FA (2017) A comprehensive survey on security in cloud computing. Proced Comput Sci 110:465–472
15. Singh S, Jeong YS, Park JH (2016) A survey on cloud computing security: issues, threats, and solutions. J Netw Comput Appl 75:200–222
16. Giri S (2019) Cloud computing and data security challenges a Nepal case related papers cloud computing and data security challenges a Nepal case. Int J Comput Trends Technol 67:146–150
17. Aljoumah E, Al-Mousawi F, Ahmad I, Al-Shammri M, Al-Jady Z (2015) SLA in cloud computing architectures: a comprehensive study. Int J Grid Distrib Comput 8(5):7–32
18. Tan CB, Hijazi MHA, Lim Y, Gani A (2019) A survey on proof of retrievability for cloud data integrity and availability: cloud storage state-of-the-art, issues, solutions, and future trends. J Netw Comput Appl 110:75–86
19. Yamin MM, Katt B, Sattar K, Ahmad MB (2019) Implementation of insider threat detection system using honeypot based sensors and threat analytics. Future of information and communication conference. Springer, Berlin, pp 801–829
20. Kaur EP (2017) An objective of service level agreement (SLA) in cloud computing environment. In: Proceedings of the 4th international conference on recent advanced engineering science on management, pp 67–77
21. Forrester (2015) API management solutions, Q3 2014. Technical Report. Forrester
22. Alhamad M, Dillon T, Chang E (2010) Conceptual SLA framework for cloud computing, pp 606–610

23. Tunc C et al (2017) Cloud security automation framework. In: Proceedings of the 2017 IEEE 2nd international workshop are founded in the applied self* system FAS*W 2017, pp 307–312
24. Cayirci E, Garaga A, Santana de Oliveira A, Roudier Y (2016) A risk assessment model for selecting cloud service providers. J Cloud Comput 5(1):1–12
25. Tsaregorodtsev AV et al (2018) Information security risk estimation for cloud infrastructure. Int J Inform Technol Sec 10:4
26. Medhioub M, Hamdi M, Kim TH (2017) Adaptive risk management framework for cloud computing. In: Proceedings of the 2017 IEEE 31st international conference on advanced information networking and applications (AINA). IEEE, pp 1154–1161
27. Patel K, Alabisi A (2019) Cloud computing security risks: identification and assessment. ERA J 17(2):11–19
28. Palos-Sanchez PR, Robina-Ramirez R (2019) SS symmetry what role does corporate governance play in the intention to use cloud computing technology, pp 1–19
29. Adelmeyer M, Beike L, Buggenthin M, Osada S, Teuteberg F (2018) RisCC: a risk management tool for cloud computing environments. In: American conference on information system 2018 digital disruption, AMCIS 2018, pp 1–10
30. Faizi SM, Rahman SSM (2019) "Securing cloud computing through IT governance. SSRN Electron J 7(1):1568
31. Al-Ruithe M, Benkhelifa E (2017) Analysis and classification of barriers and critical success factors for implementing a cloud data governance strategy. Proced Comput Sci 113:223–232
32. Anisetti M, Ardagna CA, Damiani E, Gaudenzi F (2017) A security benchmark for OpenStack. IEEE Int Conf Cloud Comput CLOUD 2017:294–301
33. Gnana Singh DAA, Priyadharshini R, Jebamalar Leavline E (2018) Cuckoo optimisation based intrusion detection system for cloud computing. Int J Comput Netw Inf Secur 10(11):42–49
34. Sharma V, Nigam V, Sharma AK (2020) Cognitive analysis of deploying web applications on Microsoft windows azure and amazon web services in global scenario. Mater Today Proc

Digital Healthcare: Reimagining Healthcare With Artificial Intelligence

Assessing and Auditing Organization's Big Data Based on COBIT 5 Controls: COVID-19 Effects

Iman M. A. Helal, Hoda T. Elsayed, and Sherif A. Mazen

Abstract Nowadays, exploiting big data can generate valuable insights and open new directions for business intelligence. Thus, organizations should maintain and assess their big data to improve their capabilities. However, there is still no structured approach with consistent and detailed guidelines that adopt policies and standards to assess and audit big data capabilities in a structured way. In this paper, we aim to provide a generic approach to collecting, managing, and analyzing the organization's big data capabilities. First, we propose an integrated maturity assessment model that aligns the controls, policies, and best practices to comply with the most recent standard (COBIT 5) throughout the entire big data lifecycle. It can help assess the organization's big data. We adjusted COBIT 5 practices, and controls to support big data capabilities. These controls were reviewed by professional auditors. Then, we used our proposed integrated maturity assessment model to establish our auditing approach to support the big data auditing process in an efficient way. Finally, we applied our approach to COVID-19 data to support the process of assessing and auditing patients' data. As proof of concept, we provide an assessment and auditing tool called BMA, and we evaluate it by an auditing expert and apply his recommendations for improvement.

Keywords Big data lifecycle · Maturity models · Auditing · COVID-19 · COBIT 5

I. M. A. Helal (✉) · H. T. Elsayed · S. A. Mazen
Faculty of Computers and Artificial Intelligence, Cairo University, Giza, Egypt
e-mail: i.helal@fci-cu.edu.eg

H. T. Elsayed
e-mail: h.tawakl@fci-cu.edu.eg

S. A. Mazen
e-mail: s.mazen@fci-cu.edu.eg

D. Magdi et al. (eds.), *Green Sustainability: Towards Innovative Digital Transformation*, Lecture Notes in Networks and Systems 753,
https://doi.org/10.1007/978-981-99-4764-5_9

1 Introduction

Nowadays, massive amounts of data can be acquired easily from a large scale of diverse data sources. The data can be generated from various data sources such as commercial (e.g., data collected at shopping malls, and credit card vendors), social networks (e.g., data generated from tweets, posts, pictures, and videos), and the Internet of Things (IoT) (e.g., sensor recordings from computer peripherals, home appliances, and vitals readers such as heart rate and oxygen saturation). These types of data can be treated as big data [1].

The 5 Vs [2] (Volume, Velocity, Variety, Veracity, and Value) of big data require a customized architecture for big data auditing, management, and control. This architecture serves auditing organizations to control and review big data. Auditing is a formal process of examining an organization to check whether it complies with its standards and control [3, 4]. The audit process entails that the auditor gathers evidence and evaluates the strengths and weaknesses of internal controls based on the data gathered through audit tests.

Due to the COrona VIrus Disease-19 (COVID-19) pandemic, a lot of concerns were raised regarding the disease, its causes, how to manage and control its effects, and how to overcome it through medications or control its spread through vaccinations. This crisis has generated a huge amount of data that not only need analytics but needs close auditing. The value of big data has inspired some organizations to advance their capabilities to best leverage big data and gain an advantage. However, not many organizations, especially medical facilities, can handle the complete assessment of their big data and apply auditing procedures in all departments. Thus, we need to assess their big data maturity to decide the required level of assessment and audit controls.

To get more value from big data, organizations should assess and audit their capabilities' readiness or maturity. Maturity assessment is an important factor for organizations to identify their current state of capabilities and draw a new line for improvement processes that affect the efficiency and effectiveness of the organization. Hence, it supports the auditing process which evaluates the organization's behavior. Nowadays, maturity assessments and auditing are becoming challenging due to the heterogeneity of incoming data from different sources. The auditor needs to use aiding CASE tools to assess and audit these data. Then, the auditor will need a structured approach to assess the organization's maturity level and how it can be enhanced to support the auditing process. It is still quite challenging how to fully utilize the use of big data techniques in auditing and how the organization's maturity affects the audit process. Although there are many maturity assessment models and solutions for big data auditing [5–17], these models and solutions are still generic, unstructured, not detailed, and not dependent on a well-known standard.

In this paper, we propose an integrated maturity model to provide the auditor with a set of controls, metrics, and activities that aid in determining the organization's big data maturity level. These controls can provide a set of best practices to enhance and maintain the organization's maturity level. This model provides insights into the existing business to gain a competitive advantage. Also, it supports the big data auditing process as it adapts the most suitable Control Objectives for Information and Related Technologies (COBIT 5) metrics, controls, and best practices for each big data phase in each maturity level. Then, we applied our approach to the COVID-19 pandemic. The rest of this paper is organized as follows: an analytical study to evaluate existing big data maturity assessment models and big data lifecycle auditing approaches by revealing their strengths and weaknesses.

Then we illustrate our proposed integrated maturity model for assessing the current maturity level of the organization and also propose a big data auditing approach to the requirements revealed in that analytical study. Then, an evaluation of the proposed approach is presented. Finally, we conclude our work with an outlook on future work.

2 Literature Review

To evaluate the existing maturity assessment and auditing efforts in big data and reveal their strengths and weaknesses to tackle them in our proposed approach. we conduct an analytical study in which we divided the existing related works into two main directions. The first direction is how to decide the maturity level of the organization's big data using big data maturity models, while the second part is how to decide the required lifecycle to audit big data.

2.1 Big Data Maturity Models

Big data maturity models are assessment models for assessing the organizational progress from the starting point to the desired completion point [2]. One of the first steps in assessing an organization's big data maturity is to determine its big data readiness. There are multiple maturity models [2, 5–7, 9, 10, 12, 13, 18] that considered capability assessment tools to measure the organization's readiness for big data initiatives and then determine its maturity level; see Table 1. However, these maturity models still have many concerns.

We aim to study these models by mapping their levels and highlighting their weaknesses and concerns. Then, we will provide a proposed integrated maturity model to overcome these concerns and propose an improved maturity model to support the maturity assessment. The existing maturity models can be divided into three types: descriptive, comparative, and prescriptive.

Descriptive Models (D)

They assess the organization's current maturity level. Recent research provides models to assess the organization's current capabilities and status to generate value from big data investments [2, 6, 7]. IBM model [6] is designed to assess the current

Table 1 Comparison between recent big data maturity models (D: Descriptive, C: Comparative, P: Prescriptive)

Model	Maturity levels						Type	Governance	Concerns
	1	2	3	4	5	6			
Knowledgent [2]	Infancy	Technical adoption	Departmental adoption	Enterprise adoption	Data as a service	–	D	–	• No clear recommendations for improvement and value creation • Does not have metrics or practices to be used in the assessing process
Infotech [7]	Explorer	Analyzer	Integrator		Innovator	–		–	
IBM [6]	Ad-hoc	Foundational	Competitive	Differentiating	Breakaway	–		✔	• No specific practices or processes to be applied in the organization to get more value from big data and to move to a higher maturity level
Dhanuka [8]	Aware	Exploring	Optimizing	Transforming		–	C	✔ + Budgeting	• Only identifies the big data capability domains to empower the big data platform in the organization
Comuzzi [12]	Initial	Repeatable	Defined	Managed	Optimizing	–		–	• Implicit description of maturity levels • No method to identify the maturity indicators of the assessed capabilities
Halper [14]	Nascent	Preadoption	Early adoption	Corporate adoption	Mature/ visionary	–		✔	• Does not define the improvement practices for the capabilities

(continued)

Table 1 (continued)

Model	Maturity levels						Type	Governance	Concerns
	1	2	3	4	5	6			
Farah [9]	Initial	Defined	Managed	Optimized	Strategic	–	P	✔	• No assessment technique to use independently by organizations to assess big data maturity capabilities • Does not identify the capabilities of big data security and integration
Schmarzo [10]	Business monitoring	Business insights	Business optimization	Data monetization	Business metamorphosis	–		–	• No specific practices or actions to help organizations move from one level to the next
Sulaiman [13]	Ignorance	Coping	Understanding	Managing	Innovating			✔	• Does not cover most big data capabilities • Provides a low level of guidance with an implicit description • Provides some generic action plan for improvement
Radcliffe's [2]	In the dark	Catching up	First pilot	Tactical value	Strategic leverage	Optimize and extend		✔	
IDC [5]	Ad-hoc	Opportunistic	Repeatable	Managed	Optimized			✔ + Management	

(continued)

Table 1 (continued)

Model	Maturity levels						Type	Governance	Concerns
	1	2	3	4	5	6			
El-Darwiche [19]	Performance management	Functional area excellence		Value proposition enactment	Business model transformation			✔	• Adopts some generic steps to move from the current maturity level to the higher one • Provides an implicit description of some factors (people, technologies, etc.) • Does not cover the quantitative capabilities and an improvement value

maturity state and measure the abilities of an organization to pursue big data initiatives. Also, it aims to determine the required technologies to help the organization toward achieving the maturity of big data. The researchers provide a descriptive big data maturity model to assess the organization's current capabilities and status to generate value from big data investments. But the researchers didn't provide the needed practices or processes to be applied in the organization to get more value from big data and to move to a higher maturity level. They also provide a low level of guidance. Additionally, Infotech and Knowledgent models [2, 7]were used only to diagnose an "as-is" state of big data adoption capabilities. But these models don't provide any recommendations for improvement.

Comparative Models (C)

They benchmark the organization's current status to its industrial peers [8, 12, 14]. In Dhanuka [8], researchers assess the organization's big data capabilities and compare the organization to others in the industry along with four key capability domains. But this model only provides the organization with the needed big data capability domains. Besides, the TDWI model proposed by Halper [14]also served for benchmarking purposes. It assesses the organization's maturity and compares it against other analytics initiatives. However, their model did not define the improvement practices for the capabilities.

Also, the Comuzzi model [12] aims at recognizing the current state of building comprehensive big data management to generate value for the organization. It implicitly describes maturity levels and did not provide an assessment mechanism to assess and identify the maturity indicators of the assessed capabilities.

Prescriptive Models (P)

They plot a path toward improving the big data maturity level. Some researchers provide a road map for enhancing and maintaining the organization's maturity level such as Farah, Schmarzo, Sulaiman, Vesset, Braun, and El-Darwiche maturity models [2, 5, 9, 10, 13, 19]. In the research [9], the authors presented a model to assess the current maturity level and plot a path toward improving the big data maturity level. The authors recommend improvements for enhancing and maintaining the organization's maturity level. They did not provide an assessment technique to use independently by organizations to assess maturity capabilities and did not identify the capabilities of big data security and integration.

In the Schmarzo [10] model, the authors aim at helping organizations to measure how effective they were at leveraging data to support their models. However, they do not provide aiding actions to help organizations move from one level to the next. In Sulaiman et al. [13] model, the authors provide an implicit description of selected factors (e.g., people, technologies) of big data qualitative capabilities. However, their model did not cover the quantitative capabilities and did not add clear improvement values. In Vesset et al. [5], the authors provide a big data maturity framework that enables organizations to assess their big data and recommend some general actions for improvement. However, this framework covers some of the capabilities that were implicitly described for the maturity levels. Thus, it acts like a general readiness assessment framework with a low level of guidance. Additionally, the authors argued in Braun [2] model that big data maturity models are the key tools for setting the direction and monitoring the organization's big data program. In El-Darwiche [19] model, the authors provide a framework that assesses big data adoption in the public sector. However, it adopts only some generic steps to move from the current maturity level to a higher one.

As presented in Table 1, each maturity model is related to a defined type with some levels of maturity, whether it supports governance or not, and the main concerns/limitations. Braun has argued that these models have some weaknesses such as a poor theoretical foundation [2]. We can also remark that they lack a well-defined context and structured mechanism with the required practices to help in moving the organization to the next higher level. Also, there are no consistent and detailed guidelines with specific practices, matrices, assessment factors, and processes to support each maturity level through the overall organization's big data phases.

Most of the concerns and problems in related work will be addressed in our proposed approach; see the contribution and evaluation sections. In [11], researchers aim to assess the big data analytics phase by providing some generic practices to help the auditor in the assessment process. Still, their framework is not adopting any specific standard and their main focus was not to assess big data phases. To our knowledge, there is no research that focuses on assessing big data maturity based on a well-defined standard to help the auditor to identify/pinpoint the required practices and metrics in the assessment process.

In conclusion, measuring the current maturity models can be very difficult. These models have different perspectives and objectives. Also, their ways of assessing lack completeness, a structured model with well-defined practices, matrices, or processes that help the auditor in the assessment and auditing processes. Moreover, there are no consistent and detailed controls to support each maturity level assessment process in each big data phase. The current maturity models are based on the Capability Maturity Model Integration (CMMI). The basis of CMMI's key performance indicators is the IT organization which lacks the full support of big data-rich characteristics. Our research addresses these shortcomings by adopting and adjusting COBIT 5 practices, processes, and metrics to provide an approach that is detailed and structured.

2.2 *Auditing Big Data Life Cycle*

Big data has many challenges associated with storing and managing it. In response, there is a growing need for auditors to self-adapt their auditing practices to handle big data challenges that confront the audit community. Unfortunately, auditing big data practices are not as widespread as in other related fields [17]. Also, it is hard to govern big data with its characteristics, insights, and decisions.

Auditing big data starts from the data created until it is archived. Recent research provides a generic framework that aims at auditing a big data lifecycle [17, 20]. Their framework consists of three stages: (1) selection, which selects big data phases, such as planning, management, and collection, (2) organization, which divides big data phases into three types: (operational, management, and support). The operational type has a direct effect on big data. The management type defines how big data operational phases work. The support type helps operational phases to work appropriately and securely.

Finally, (3) evaluation, which consists of yes/no questions. Unfortunately, the researchers only ask general questions that do not follow any standards, or guidelines to show how to audit big data phases. In our proposed approach, we will address these concerns and provide the most suitable controls and practices for each phase in each maturity level to help the auditor in the auditing process based on COBIT 5.

In recent research [21], researchers use machine learning technologies to improve the auditing and decision-making process. But they conclude that machine learning can be applied for only planning and execution phases in big data, as humans are needed in the decision-making formulation and audit report development.

In the big data context, data may come from different sources with different formats such as data resulting from sensors. To audit this IoT data, the organization needs to identify its risks, and then put some controls for handling these risks. These risks may be business risks (e.g., regularity compliance), operational risks (e.g., performance), and/or technical risks (e.g., device vulnerabilities). In the research [16], the authors provide a framework that consists of three risk layers each layer provides some controls. This ensures the alignment between (IoT) implementation and business needs to gain more benefits. Current research is categorized into either an expert-based approach [17, 20] or a semi-automated approach [21]. Thus, researchers prefer to have domain experts while auditing an organization's big data.

In [22], researchers provide empirical studies to analyze the legality production process of big data analytics in financial auditing activities. They highlight the factors affecting the use of big data analytics to improve the audit process. Their research is based on an investigative study that consists of 16 interviews with senior managers of some audit companies. Researchers [23] provide a systematic study of the advantages of big data analytics in internal auditing. The paper further outlines several challenges in obtaining big data analytics in internal auditing such as audit staff may not be trained to realize the exact nature of data to make correct inferences, and there are no regulations or guidelines that deal with all the uses of data analytics in the auditing process. Current research provides a generic auditing framework [16, 17, 20, 21, 24]. Their auditing process does not follow specific standards and guidelines related to big data and its lifecycle. So, our proposed approach can help the auditor to determine the organization's maturity level while conducting the audit process, our structured audit approach follows COBIT 5 controls, practices, and metrics with detailed guidelines to support the big data phases and the organization overall. See the next section.

3 Contribution

With the advance in information technology and the increased amount and complexity of data flowing in and out of organizations daily, there is a need for efficient ways of managing data to make efficient decisions at the right time. Based on the analytical study presented in the previous section; we propose an integrated maturity model.

This model integrates most of the existing maturity models and tackles the concerns of most previous maturity models presented in Table 1, and also it supports the auditing process in a structured way.

Then we will provide an auditing approach to help the organization during the auditing and decision-making processes. Our proposed approach's main target is to avoid the current research weaknesses and provide more improved practices and controls that are based on a well-defined standard. We have divided our contribution into two directions: big data maturity and big data auditing.

3.1 Proposed Integrated Maturity Model

First, we propose an integrated maturity model for assessing the organization's big data maturity. This model integrates the three types of maturity models presented in Table 1 and covers most of their concerns such as these models don't provide detailed guidelines with the practices, metrics, and activities that are based on well-defined standards, and these models don't cover all big data phases and capabilities. Our model has a clear perspective with a set of predefined levels that suits any organization. For each level to provide consistent and detailed guidelines with specific practices, matrices, and processes, we adopt a set of controls based on the well-defined standard for each big data phase. Finally, we utilize this proposed integrated maturity model in an auditing process and provide an auditing approach.

Nowadays, there are an increasing number of standards and compliance guidelines to be adopted in the organization to gain a competitive advantage. These standards such as ISO 27001, COBIT, International Standards on Auditing (ISA), and General Data Protection Regulation (GDPR) have different objectives.

- International Organization for Standardization (ISO 27001): It is the international standard that focuses on establishing and implementing an Information Security Management System.
- COBIT [24] It is developed by the Information Systems Audit and Control Association (ISACA). It focuses on information technology management and governance by applying some metrics, and best practices.
- GDPR [25]: It is developed by the European Union (EU). It provides a set of standards for personal data protection.
- ISA [26]: It is developed by the International Federation of Accountants (IFAC). It focuses on financial information auditing.

In this paper, we choose to adopt COBIT 5 [24] policies and standards. As COBIT 5 focuses on information technology management and governance, it links the organization's goals with its resources and IT infrastructure by providing various maturity metrics that assess the achievement while identifying associated organization goals.

The focus of COBIT 5 was illustrated with a process-based model subdivided into four specific domains for achieving an efficient management process:

- Align, Plan, and Organize (APO) domain; identifies the organization's current activities and the strategy used.
- Build, Acquire, and Implement (BAI) domain; addresses currently implemented solutions concerning the organization processes.
- Deliver, Service, and Support (DSS) domain; identifies secured delivery of IT services.
- Evaluate, Direct, and Monitor (EDM) domain; identifies the needed practices to achieve optimal value to support the decision-making process.

We adjust the first domain [Align, Plan, and Organize (APO)] for ad-hoc, defined, early adoption, and strategic maturity levels. It identifies and controls current activities, processes, and strategies. While the optimized level is aligned with the fourth domain [Evaluate, Direct, and Monitor (EDM)]. It evaluates and optimizes the organization's value. But COBIT 5 also doesn't concentrate on big data capabilities, but it focuses on information technology and information in general, so we need to work on this standard and go in-depth to be able to adjust and modify the most related practices, processes, and metrics to be suitable with big data and hence, support the maturity and auditing process.

The proposed big data maturity model is a perspective model. This model is an integrated model that integrates the current maturity models and proposes to cover their concerns and aims at improvements. The proposed integrated maturity model consists of five maturity levels; see Fig. 1.

These levels are integrated with the big data maturity types: descriptive (D), comparative (C), and perspective (P), as follows:

- Level 1: Ad-hoc (D): The initial state of maturity in which the organization is still using traditional tools without considering big data capabilities.
- Level 2: Defined (D): The next level in which the organization has realized big data benefits. It initiates the effort to integrate its data sources by creating unified information architecture. Also, the organization identifies the required skills to adopt big data. Then, it trains the employees on big data capabilities to enable a business revolution in big data management.
- Level 3: Early adoption (C): As the organization's maturity level shifts from "defined" to "early adoption", the organization adopts the needed technologies for big data, such as Hadoop, Spark, and Apache HBase [27–29] for big data storage and analytics.
- Level 4: Optimized (P): It represents the overall process management and how to utilize both predictive and descriptive-analytical methods to optimize operational performance and support the decision-making process.

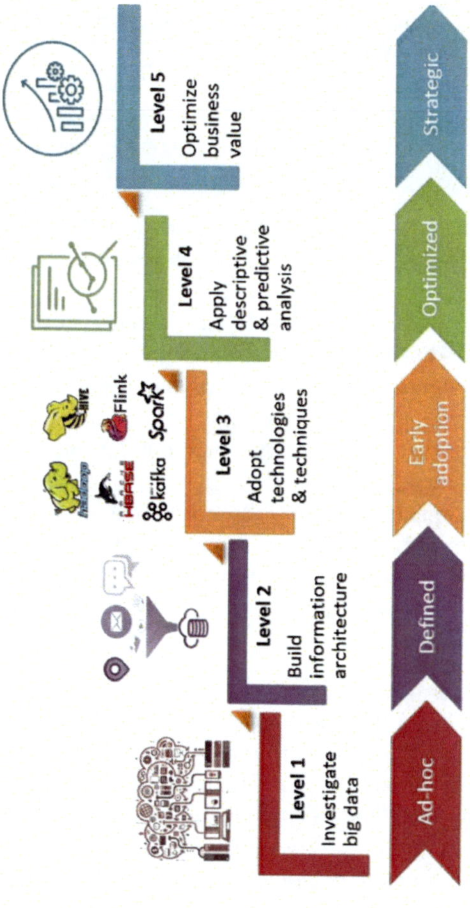

Fig. 1 An integrated model for assessing big data maturity

- Level 5: Strategic (P): This level represents the highest level of maturity in which the organization provides new capabilities to improve the overall process and gain new revenue opportunities.

As presented in Fig. 1, the proposed perspective big data maturity model consists of five maturity levels. Each maturity level assumes meeting a set of capabilities. In level 1, the organization discusses big data and its capabilities without creating information architecture. In level 2, the organization specifies a suitable big data architecture model and repository. It also trains its employees to meet the required level of skills to deal with big data. In level 3, the organization specifies the required infrastructure and tools to adopt big data. In level 4, the organization utilizes the available resources to meet the management requirements. Moreover, it aims for increasing the organization's revenue. Finally, in level 5, the organization draws a road map to achieve its innovation plan. In our model, we will handle issues, concerns, and limitations of the previous maturity models described before.

After we provide a general proposed integrated maturity model that consists of five levels presented in Fig. 1, we will identify the practices, metrics, and activities for each maturity level that will help any organization assess their maturity state and take some comments to move to the next maturity level based on the well-known standard. This standard will help the organization in the assessment process and make the proposed integrated model more structured and detailed. In our model, we adjust COBT 5 [3, 16, 24]domain with its best practices, metrics, and activities for each big data maturity level after adjusting it to comply with big data. as it aims the IT and information in general, not big data, see Table 2. As mentioned before, in our model, we adjust Align, Plan, and Organize (APO) for ad-hoc, defined, early adoption, and strategic maturity levels and Evaluate, Direct, and Monitor (EDM) for the optimized maturity level. It evaluates and optimizes the organization's value. The second and third domains are not directly associated with big data maturity. Table 2 can guide the auditors to assess the maturity level of an organization in a structured way. This can provide an action plan to either enhance or maintain its levels.

As presented in Table 2, we identified the needed practices, metrics, and activities for each maturity level based on COBIT 5 controls and best practices after the adaption and adjusting it as it does not rely on big data itself, so we worked on it to extract the needed information to be adopted. Now we will adopt our proposed maturity model to big data; see Tables 3, 4, and 5. For each big data phase, we identify the activities, processes, and controls needed at each maturity level.

As presented in Tables 3, 4, and 5, we divided the big data phases into three types; then we work on COBIT 5 and adapt it after applying some changes to support big data capabilities. For each big data phase, we provide the needed practices, controls, and metrics to be applied in each maturity level. Organizations can efficiently assess and audit their phases per maturity level.

Table 2 Alignment between COBIT 5 and the proposed big data maturity levels

Level	Domain	Practices	Metrics	Activities
Ad-hoc	Align, plan and organize (APO02.01)	1. Understand organization direction	1. Level of knowledge of the current organizational context 2. Strategy used 3. Organization goals	1. Understand the organization's architecture such as data, information, and technology domains
Defined	Align, plan and organize (APO07.03, APO03.02)	1. Identify IT personnel 2. Define the architecture	1. Available key skills 2. Needed skills 3. Current training 4. Architecture standards	1. Compare the needed and available skills 2. Recommend technical and behavioral training 3. Identify the information architecture model and repository
Early adoption	Align, plan and organize (APO04.03)	1. Scan the technology environment	1. Emerging technologies 2. Infrastructure used 3. Budget	1. Identify and set up needed technology 2. Determine the acceptance level for technology innovation
Optimized	Evaluate, direct and monitor (EDM02.02, EDM02.04)	1. Evaluate and monitor the value optimization	1. The value obtained against business goals 2. Investment efficiency 3. Management satisfaction with value 4. Stakeholder satisfaction	1. Identify key elements required for achieving value 2. Optimize the value created
Strategic	Align, plan, and organize (APO11.05, APO04.01)	1. Continue the improvement process—support environment innovation	1. Innovation feedback 2. Value effectiveness	1. Identify and agree on improvement actions 2. Train the employees regularly to encourage and maintain improvements 3. Provide a plan for innovation 4. Support innovative ideas

Table 3 Auditing controls/practices/metrics applied to big data phases (B.D.P) capabilities in each maturity level (M. Lvl)—management phases

B.D.P	M. Lvl					
	Level 1 ad-hoc	Level 2 defined	Level 3 early adoption	Level 4 optimized	Level 5 strategic	COBIT 5 adapted practices
Management	• Storage structures are still not well defined • No data management strategy • Data does not follow specified standards	• Identify the information architecture model and define the needed storage structure • Identify data management functions • Identify the roles and responsibilities to support the management function • Move to leverage unstructured data with real-time insights	• Identify data management and governance practices • Provide a unified information architecture • Data can be accessed easily across the organization • Manage data and allocates the responsibilities to achieve the effectiveness of the management strategy	• Identify the data auditability process • Define data life cycle management and best practices for data management • Establish data management objectives and scope • Verify the overall big data phases' satisfaction • Use big data capabilities with their infrastructure and data governance strategies	• Aim at improving management practices to be semi-automated	• APO14.01 Define and communicate the organization's data management strategy and roles and responsibilities
Planning	• Start to plan for using big data capabilities • Study the effect and value of using big data	• Provides a plan for the management of all big data phases • Identifies a compact road map of what will happen in the organization's long term • Puts a plan for employing a new team for supporting big data capabilities or providing the old team with the needed training	• Identify business processes, and data migration for infrastructure and IT service data, hardware, software, networks, and operating systems	• Establish a recovery plan for any migration failure • Identify risk management and business goals • Identify the organization's business processes and requirements	• Verify the completeness of transaction data	• BAI07.02 Plan business process, system, and data conversion

Table 4 Auditing controls/practices/metrics applied to big data phases (B.D.P) capabilities in each maturity level (M. Lvl)—operational phases

B.D.P	M. Lvl					
	Level 1 Ad-hoc	Level 2 defined	Level 3 early adoption	Level 4 optimized	Level 5 strategic	COBIT 5 adapted practices
Collection	• Still using traditional collection methods • The collected data has several formats and different data sources	• Identify and classify data sources according to big data architecture • Identify shared data sources and, create relationships among data sources	• Establish a platform to share data between organization sections	• Establish a data collection method • Identify risk taxonomy • Provide validation criteria for big data sources • Gather the data in the form that enables analysis and risk detection	• Improve the data collection method to be a fully automatic method • Convert and verify data by handling any mistakes found through conversion	• APO12.01 Collect data
Analysis	• Still uses traditional analytics tools but try to enhance them to support big data capabilities	• Identify the gap between current and needed business capabilities and the organizational environment • Identify impact changes in organization architecture domains • Increase the speed of big data analytic • Processes	• Capture and prioritize stakeholder functional and technical requirements • Ensure alignment between requirements and organization standards • Identify organization needs, and requirements to accomplish the business goals	• Validate stakeholders' requirements by adopting a peer review method, or prototype • Identify information risk • Identify the needed procedures to exploit the data to get insights and support the decision-making process	• Conduct a post-exercise analysis to consider the achievement • Depending on the results, enhance the continuity analysis plans	• BAI02.01 Define and maintain business functional and technical requirements

(continued)

Table 4 (continued)

B.D.P	M. Lvl					
	Level 1 Ad-hoc	Level 2 defined	Level 3 early adoption	Level 4 optimized	Level 5 strategic	COBIT 5 adapted practices
Filtering	• There are individual efforts on filtering structured current data using available tools	• Distinguish between critical data (useful) and noncritical data (useless), information • Identify big data filtration methods	• Provide the tools and techniques needed for the filtering process • Remove unnecessary data that does not add value to the organization and does not affect the other data or output overall	• Choose a specific big data filtering algorithm • Identify the needed big data filtering techniques	• Improve the filtering techniques to be fully automated • Enhance the execution speed of the filtering	• APO01.07 Define information (data) and system ownership
Visualization	• Visualize data as "snapshots" • Data is presented as a static report by using software like spreadsheets, bar charts, histogram charts, and frequency distributions	• Identify the relationships between different data objects using visualization tools such as text phrase analysis, and correlation matrices	• There is a higher level of visualization to support the analysis and illustrate data over time, hence predicting the future	• Aim for an interactive user interface and real-time visualizations that reflect the change in one factor in the data immediately • Establish concise performance reports to enhance the decision-making process • Identify the root cause of deviations against business goals	• Data experts use analytics tools to create insightful visualizations to provide additional value-adding • Integrate performance and compares it to business goals to analyze performance trends and recommend them	• MEA01.04 Analyze and report performance

(continued)

Table 4 (continued)

B.D.P	M. Lvl					
	Level 1 Ad-hoc	Level 2 defined	Level 3 early adoption	Level 4 optimized	Level 5 strategic	COBIT 5 adapted practices
Enrichment	• The enhancement process has not yet started, the organization uses the existing data as it is	• Identify the potential value of innovation and enhancement • Plan for the enhancement of resources and processes	• Align enhancement initiatives with business strategy and goals • Identify resources that support big data • Identify the adding information that supports the driven big data insights	• Big data insights inform organization decisions • Data is leveraged well to get competitive value	• Identify opportunities for improvement • Add information and conduct structural or hierarchical changes to the data to improve its quality	• APO04.06 Monitor the implementa-tion and use of innovation
Integration	• Still uses uniform schemas	• Plan for the integration process • Collect and integrate data sources • Synchronize big data sources	• Use emerging technologies during integration, and adoption to ensure the guaranteed value • Ensure that data integration applies a specific standard	• Ensure the efficiency and appropriateness of the collected data • Ensure data completeness, accuracy, and integrity • Identify methods that enable gathering large volumes of different data sources	• Apply audit process for the big data integration to ensure integrity and availability	• MEA01.03 Collect and process performance and conformance data

(continued)

Table 4 (continued)

B.D.P	M. Lvl					
	Level 1 Ad-hoc	Level 2 defined	Level 3 early adoption	Level 4 optimized	Level 5 strategic	COBIT 5 adapted practices
Storage	• Look at integrating data sources • Plan for identifying needed plan big data capacity for the storage process	• Identify the on-site and off-site storage methods of such data and its backup	• Identify the storage infrastructures and technologies to enable the storage of big data in such a way that it can easily be accessed and managed	• Ensure the availability of data and establishes a backup of all sensitive data • Ensure the Privacy and security of the stored data	• Improve security and privacy to protect the privacy of data and users • Ensure data availability for operational continuity	• APO14.10 Manage data backup and restore arrangements
Destruction	• The organization is still using the traditional destruction method, but it discusses how to change this method to support the big data capabilities	• Identify incident data to be deleted • Investigate problems to identify their effect and their causes	• Authorize transactions by designated management individuals • Use new technologies and tools to enhance the deletion process	• Ensure that destruction policies support the management of data history • Ensure data accuracy • Ensure the deleted data does not affect the organization's data and processes • Identify and deletes unnecessary data that affects big data negatively	• Protect and secures sensitive data from the destruction process	• DSS03.02 Investigate and diagnose problems

(continued)

Table 4 (continued)

B.D.P	M. Lvl					
	Level 1 Ad-hoc	Level 2 defined	Level 3 early adoption	Level 4 optimized	Level 5 strategic	COBIT 5 adapted practices
Archiving	• The archiving method is still a traditional method that depends on individual efforts • Archiving methods are still semiautomated methods	• Enhance the storage system that supports the long-term storage of the data	• Establish a method that guarantees the accessibility of data anytime • Control access to the archived data to support analytics needs • Use more advanced tools to support the archiving process • Ensure the availability of historical data based on legal requirements for data archiving	• Ensure that archiving and retention process follows data regulatory requirements • Provide a standard with the needed processes to support the archiving process	• Improve the archiving methods to be fully automated	• APO14.09 Support data archiving and retention

Table 5 Auditing controls/practices/metrics applied to big data phases (B.D:P) capabilities in each maturity level (M. Lvl)—support phases

B.D.P	M. Lvl					
	Level 1 ad-hoc	Level 2 defined	Level 3 early adoption	Level 4 optimized	Level 5 strategic	COBIT 5 adapted practices
Quality	• Still use the traditional quality standard	• Establish a plan for data quality tasks • Discuss the changes in quality processes to support big data capabilities	• Define a data quality strategy concerning business objective and technology, the infrastructure used, roles, and stakeholders' needs approved by management	• Identify and monitors quality issues • Ensure that the policies, guidelines, and processes are adopted in the data quality strategy • Establish a data quality to achieve the quality of big data phases to support the business goals	• Improve techniques of measurement for organization big data quality metrics • Establish a plan for data quality improvement	• APO14.04 Define a data quality strategy
Security	• Physical security • Physical security follows the individual efforts • Security has poor measuring and inconsistent metrics • The staff has a low level of awareness of the security roles	• Front-end security • At this level, there are security procedures with some level of confidence • Skills and tools are still inadequate • Allow only authorized users to access organization data • Establish data security procedures by identifying individual authority to achieve confidentiality	• Back-end security • Start to apply a security procedure that adopts an information management architecture • Identify roles and responsibilities for each activity concerning job descriptions • Identify administrative privileges	• Security Awareness • The security practices are identified and completed • Provide a security testing approach to identify security threats • Ensure the validity of access privileges with predefined roles • Identify privileges based on predefined roles	• Proactive security • The organization provides a fully automated security control and aims to improve it	• DSS06.03 Manage roles, responsibilities, access privileges, and levels of authority

3.2 Proposed Auditing Approach

We propose an auditing approach; see Fig. 2. It uses our proposed integrated maturity model. This approach provides a structured and detailed auditing guideline that supports big data capabilities for each big data phase in each maturity level. We conduct the auditing process in a new form that concentrates on big data itself with its phases in each maturity level. For each phase, we identify the needed practices, controls, and metrics in each big data maturity level based on a well-defined adjusted standard COBIT5. See Table 4. Big data has thirteen phases [17, 20]: Collection, Planning, Management, Integration, Filtering, Enrichment, Analysis, Visualization, Storage, Destruction, Archiving, Quality, and Security. We reused these phases in our auditing approach. We adopt the relevant controls, metrics, and best practices [16] of the COBIT 5 framework to match the big data capabilities over all phases. For example, in the storage phase, we adjusted APO14.10. Manage data backup and restore arrangements controls to ensure the availability and accessibility of the stored data. Thus, the organization should identify on-site, and off-site storage and backup methods.

As presented in Fig. 2, we aim at proposing an integrated maturity model to assess the current maturity level of the organization and use it to support the auditing process. After assessing the organization's current maturity level and identifying and understanding the organization's overall behavior while applying the auditing process. Then, we use the yes/no questions presented in Table 6. These questions facilitate using the proposed auditing approach in an organization in each big data phase based on COBIT 5 standards, processes, practices, and metrics for each big data phase.

We provide some questions for each big data phase; see Table 6. These questions are adjusted from COBIT 5 metrics and based on our auditing approach presented in

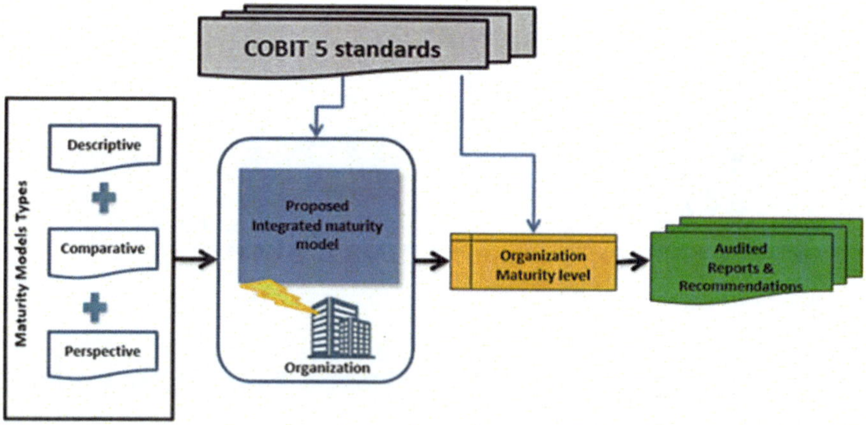

Fig. 2 A proposed approach for assessing and auditing big data

Table 6 Proposed approach evaluation (yes/no questions)

#	Phase	Questions based on COBIT 5 controls	
1	Collection	Q1	Is there a data collection method?
		Q2	Is there a method to classify receiving data inputs based on organization architecture?
		Q3	Is there a method to identify relevant data?
		Q4	Is there a risk taxonomy?
2	Planning	Q5	Is there a Planning strategy for all big data phases?
		Q6	Is there a planning team?
3	Management	Q7	Is there a strategy to manage and verify big data phases?
		Q8	Is there a management team?
4	Integration	Q9	Is there a method for integrating receiving data with the created data?
		Q10	Are there automatic tools to integrate this heterogeneous data?
		Q11	Are there data integrity procedures?
		Q12	Is the organization able to handle this streaming data?
5	Filtering	Q13	Is there a method to filter receiving data based on organization architecture?
		Q14	Are there automatic filtering tools?
		Q15	Is there a method to distinguish between critical data and noncritical data?
		Q16	Is there a filtration team?
6	Enrichment	Q17	Is there a team responsible for identifying improvement opportunities?
		Q18	Is there a strategy to align innovation initiatives with business goals?
		Q19	Is there an enrichment strategy?
7	Analysis	Q20	Is there a method to make the analysis more automatic?
		Q21	Is there a strategy to identify the gap between current and needed business capabilities?
		Q22	Is there a method to ensure alignment between requirements and organization standards?
		Q23	Is there a method to identify risk?
		Q24	Is there a method to Validate the stockholder's requirements?
8	Visualization	Q25	Are there concise performance reports to support the decision-making process?
		Q26	Is there an automatic visualization method?
		Q27	Is there a visualization method?
9	Storage	Q28	Is there a method to ensure the availability of stored data?
		Q29	Is there a backup strategy for all sensitive data?
		Q30	Is there a storage strategy?
10	Destruction	Q31	Is there a strategy to delete this data?

(continued)

Table 6 (continued)

#	Phase	Questions based on COBIT 5 controls	
		Q32	Is this process conflict with the archiving phase?
		Q33	Is there a method to identify the affected processes?
11	Archiving	Q34	Is there an archiving plan?
		Q35	Is there a method that guarantees the accessibility of archived data?
		Q36	Is there an archiving method based on specific parameters to support analytics needs?
12	Quality	Q37	Is there a data quality strategy?
		Q38	Is there a plan for data quality improvement?
		Q39	Is there a method to identify quality issues?
		Q40	Is there a quality team?
		Q41	Is there a method to verify data quality?
13	Security	Q42	Is there a method to encrypt the received data?
		Q43	Is there a strategy for adopting security for all big data phases?
		Q44	Are there privileges for user access control?

Tables 3, 4, and 5. They support the auditing process. For example, in the collection phase, we need to ask, is there a data collection method? Is there a method to classify receiving data inputs based on organization architecture? Is there a method to identify relevant data? and is there a risk taxonomy? All these questions will help the auditor to audit big data phases based on predefined well-known standards.

Nowadays, the COVID-19 pandemic presents a revolution all over the world that has become more intertwined. Auditors must realize how the pandemic is affecting their organizations, how to deal with the changes, and how to help get the full potential of available data. By utilizing the proposed framework, data play a vital role in assisting the world in curbing the spread of the disease. Effective insights cannot be realized without access to huge, timely, and trusted data which is vital for the identification of the most susceptible populations. After COVID-19, many organizations are forced to operate remotely and use digital technologies whether they were ready or not. This causes an uncontrolled expansion of their data, due to time stream data. Thus, organizations are keen to integrate their data sets into big data. This can enrich the data and enhance the decision-making process. However, due to the sensitivity of data that affects human life, this raised more challenges for management. They need to deal with sensitive data in an efficient way to get better insights.

Big data can provide fresh insights and open new directions for business intelligence. Thus, organizations need to maintain and assess their big data to better understand virus status, incubation period, risk factors, the average number of cure days, side effects, origin, and diagnostics and to conclude the whole population [30, 31]. In the next section, we provide a map of how to apply several controls and practices from COBIT 5 [16, 24] to audit any organization.

4 Case Study: Applying the Proposed Auditing Approach to Medical Organization/Institute

In the following, we adjust our proposed big data auditing approach to the data related to the COVID-19 pandemic. In each big data phase, we provide a roadmap with the needed activities/best practices and metrics concerning the COVID-19 environment. This roadmap can help in auditing COVID-19 pandemic data for each phase to improve its big data capabilities and output insights. The following phases are mapped to COBIT 5 management practices and controls to be applied to COVID-19.

1. Collection
    - Gather patient data at interim periods to enable analysis, and risk identification.
    - Establish a data collection method retrospectively or concurrently.
    - Identify risk taxonomy and classify the problems such as the spread of the virus, mortality, and inability to test with ensuring data confidentiality.
2. Planning
    - Verify the completeness of the patient data.
    - Identify the infrastructure, IT service data, and hardware that support the analysis of patient data.
    - Provide the necessary resources and data to support the output insights
    - test the risks such as disease effects.
    - Provide an automatic system that supports the validation of data completeness.
    - Deliver audit evidence for the assessed risks and respond to those risks.
3. Management
    - Provide a strategy to manage and verify the patient data and use quality data management software.
    - Identify best practices for supporting data management, e.g., reducing duplications.
4. Integration
    - Provide a method for integrating patient and cure protocol data.
    - Provide automatic tools to integrate this heterogeneous data for symptoms.
    - Ensure data completeness, accuracy, and integrity, especially with disease spread.
5. Filtering
    - Classify the receiving data inputs based on the organizational architecture.
    - Provide a method to identify the relevant symptoms of the disease.
    - Provide a method to filter received data based on the symptoms of the disease.

- Establish a filtration staff team with the needed skills.

6. Enrichment

- Update incomplete, outdated, and corrupted data with correct and up-to-date data.
- Identify the opportunities for data improvement such as segmenting of the patients.
- Identify the common symptoms of COVID-19 and update them regularly.

7. Analysis

- Provide a method to make the analysis more automatic, especially with this huge and heterogeneous data.
- Identify the gap between the needed and available resources, such as the number of ventilators that exist to help patients breathe.
- Prioritize the patient cases based on those who need urgent vaccination.
- Ensure that the collected data is fit for the analysis purpose.
- Identify the needed standard to guarantee that the staff has a unified working view.
- Provide skilled teams, such as a team responsible for identifying therapy opportunities.
- Predict the effects of those environmental factors on the spread rate of COVID-19.

8. Visualization

- Report and view performance intelligently by using graphs and tables for presenting the patients' status, and the disease spread.
- Use data visualization to identify anomaly cell disease and to detect the data quality issues.

9. Storage

- Identify the period of incubation and infection per each COVID-19 patient and keep a prior history of previous infections.
- Store the needed data only.
- Identify the storage period for patient data.
- Identify the backup strategy for sensitive data.
- Ensure the availability of the stored data whenever you need it.

10. Destruction

- Delete the corrupted, inaccurate, and outdated data.
- Establish destruction standards and guidelines.
- Delete the records that negatively affect the outputs and insights.
- Determine if the deleted records will affect the other processes or not.
- Identify the data retention schedule.

11. Archiving
    - Establish an archiving plan for the old patients to use in the prediction process.
    - Provide a method to ensure the accessibility of the archived patient's data in the long term.
    - Provide archiving methods based on specific parameters such as symptoms of the disease to support analytics needs.
12. Quality
    - Make sure that the collected patient data is suitable for what is needed.
    - Identify data quality assessments earlier to discover any issues regarding symptoms of the disease and its side effects.
    - Provide a list of assessment criteria for the data.
    - Review the outputs and check if the quality of data is acceptable or not.
13. Security
    - Check if the patient data is used only for the purpose it is collected for.
    - Ensure secure remote access to prevent hackers' issues.
    - Identify separable privacy rights to protect public health and safety needs.
    - Ensure compliance with specified regulations of data privacy.

As shown before, we have adopted our auditing approach to guide the organization to achieve governance and management for each big data phase from the collection to the security phases. Then, we adjust our approach to support COVID-19 data to help the hospital or any medical institute to assess and audit the patients' data, thus moving from crisis management to issue management. In each phase, we provide the controls, practices, and activities that will help the auditor to audit big data phases based on predefined well-known standards. These controls are suggested based on the COBIT 5 controls and other collected data from COVID-19 recent research [32–35].

5 Approach Evaluation

In Sect. 3, we proposed an integrated maturity model for assessing the organization and showed how to use it to support our proposed auditing approach. The integrated proposed maturity assessment model consists of five levels where the complexity is increased from one level to the next. In this section, we will provide some predefined factors that have valuable effects and support the evaluation of the current maturity level, see Table 7. We choose these factors based on the COBIT 5 standard, a proposed integrated maturity model, and other earlier studies [6, 7, 9, 10]. Then, we provide controls and metrics for each factor per each maturity level see Table 8. These controls are of great benefit to aid both internal and external auditors. These factors provide changes in the organization, such as the current information architecture, people skills, technology used, and best practices for overall process management [12].

To assess the organization's current maturity level, Table 7 summarizes the essential factors for each level to check. It shows the sequential progress between the levels starting from the awareness level to the strategic level. The organization can only upgrade to a certain maturity level after satisfying all the controls from the previous levels. The affecting factors can be explained as follows: an information architecture that fits big data with multiple sources and formats. People skills/training targets the employees' skills to enable them to use big data capabilities and technologies. Technology and infrastructure address the infrastructure, and the level of technologies, and techniques are specified to fit big data initiatives. Process management is responsible for the best practices performed to support the organization's objectives and overall business value. Finally, process improvement establishes a road map for future enhancement and best practices for new revenue opportunities.

In Table 7, it is notable that all levels except level 1 share the existence of information architecture within the organization as well as target employees' training based on the needed skills. This is due to the uncertainty of how to fully utilize big data capabilities while utilizing existing resources. Moreover, each of these factors can be broken down into controls with the metrics and practices based on our proposed approach see to help the auditor in the maturity evaluation. Also, it provides insights into the organization's big data maturity and initiatives for how to maintain and increase their level based on the following factors.

As presented in Table 8, we have provided an assessment of each capability process factor in each maturity level based on our proposed integrated maturity model and COBIT 5 controls and metrics. This table describes each maturity level for each factor by providing existing running practices, controls, and metrics based on COBIT5 to support the evaluation of the maturity level. Next, we will utilize our proposed approach with these factors to provide a prototype that helps the auditors and enhances the assessment and auditing processes.

Table 7 Factors for assessing the organization's current status (✔: supported, ✗: not supported)

Factors	Levels				
	Ad-hoc	Defined	Early adoption	Optimized	Strategic
Information architecture	✗	✔	✔	✔	✔
People skills/training	✗	✔	✔	✔	✔
Technology and infrastructure	✗	✗	✔	✔	✔
Process management	✗	✗	✗	✔	✔
Process improvement	✗	✗	✗	✗	✔

Table 8 Mapping of factors to the proposed integrated maturity model

Factors	Levels				
	Ad-hoc	Defined	Early adoption	Optimized	Strategic
Information architecture	• Still do not have an information architecture program • No single unified architecture	• Start to define big data architecture with clear roles and responsibilities • Information architecture still cannot handle advanced analytics capabilities	• The unified architecture is well-defined • Management identified architectural standards • Most sections of the organization approve of the acceptance of the information technology architecture	• Architecture documents are modified and reviewed for the new architecture standards • Management team participating in the architecture review process • Performance metrics and standards related to the new architecture are obtained	• Work on improving the architecture • Architecture standards under the governance process improvements • Information architecture fully supports big data capabilities
People skills/ training	• Big data tasks are still out of the employee's self-interest • Do not provide training courses or opportunities to improve employees' skills in the big data field • Employees have a low awareness of big data capabilities in the organization • Plan for hiring new employees and/or improve employees' skills	• Internal employees are encouraged to use big data • Start to make a contract with external trainers to train employees in the big data field • Big data is growing within the organization • The managers provide a plan for new big data positions with the mechanisms of recruitment	• Provide internal training for employees • Training may be done by skilled trainers at the organization, or by an external trainer • Most of the employees are fully aware of big data value	• Provide the documentation, resources, and guidelines for the internal training • Provide big data training to other organizations externally • The employees across the organization work all together with big data capabilities • Adopt big data methodologies	• Provide more materials and advanced courses to enhance the employee's skills • The employees across the organization's departments work with enhancing big data value and returns

(continued)

Table 8 (continued)

Factors	Levels				
	Ad-hoc	Defined	Early adoption	Optimized	Strategic
Technology and infrastructure	• Still using traditional tools • Lack of dedicated resources • The organization does not have a dedicated big data infrastructure for working with big data capabilities • No specific method for big data deployment • Configuration is manually performed	• The organization has realized big data benefits • It starts the effort to integrate its data sources using unified information architecture • The tools used are often specific to the big data project/process • Reused tools often extensive configuration requirements • Big data infrastructure does not scale well • The backups are tested	• Adopt the needed technologies for big data, such as Hadoop, Spark, and Apache HBase for big data storage and analytics • There is a dedicated infrastructure for big data storage and analysis • There is a platform for disaster recovery • There are some specialized big data infrastructures used e.g., Hadoop, HBase, and Spark. That may require manual checks	• Represents the overall process management • Organizations use both predictive and descriptive-analytical methods to enhance performance and improve the decision-making process • Big data Infrastructure deployment can be automated • Analysts access data easily • Security and confidentiality are fully integrated	• Offer new capabilities to improve the overall process and hence optimize revenue opportunities • Focus on continuous technological enhancements • Big data infrastructure is scalable and can work with new complicated projects
Process management	See the (management and planning) phases in Table 3				
Process improvement	• Sill has low awareness of continual process improvement	• Realize the continual improvement value	• Plan for process improvement	• Define a continual improvement approach	• Identify best practices for continually improving performance • Improvement approach defined before • The team applied the improvement approach

6 Big Data Maturity and Auditing Tool: A Prototype

6.1 Build BMA: A Prototype Tool

As proof of concept, we implemented a prototype of our approach, Big data Maturity and Auditing (BMA) tool using Java, full implementation is available on.[1] It applies the proposed approach through maturity and auditing processes. This tool can help the auditor in assessing the organization's current maturity level. It also supports big data initiatives based on COBIT 5 standards. In this tool, we use our proposed integrated maturity model with the proposed auditing approach concerning COBIT 5 domains to make it more structured and helpful to the auditor and management department.

BMA tool is divided into two parts: big data maturity and big data auditing. For big data maturity, it applies the proposed integrated maturity model with its controls, practices, and metrics to decide the big data maturity level of the organization. For big data auditing, we examine the thirteen big data phases based on COBIT 5 controls.

In each phase, we concentrate on the issues, limitations, and challenges that the organization faces while adopting big data initiatives. Finally, we provide a detailed auditing report, which states organization's current issues and how to overcome these issues with achieving efficient and effective use of big data. This report can enable the organization to identify its current maturity level with some recommendations for improvement to the next maturity level. Also, it can give some comments and suggest possible improvements for each audited big data phase.

Our intended users here are Chief Information Officer (CIO), Internal Auditor, and Database Administrator (DBA). After the development of our comprehensive assessment tool, we evaluated it from the domain expert's opinion. The interviewed experts have assessed our tool and recognized our tool as a helpful tool for any company that incorporates big data technologies at any level. Also, it assesses the organization's maturity for big data based on well-defined standards with its metrics and guidelines, controls, and processes. Then, it audits the big data phases by providing guidelines with some tips for enhancement. Finally, the experts gave us some ideas for updating our tool to match any organizational environment. All suggestions and updates were included in the final version of our tool, see Fig. 3. which represents a sample of auditing tools in the planning phase taken from the online assessment tool.

6.2 Evaluating BMA Tool Scores

BMA tool consists of two stages that have 45 questions. The first stage is the maturity stage which consists of 10 questions for assessing the current maturity level based on our proposed integrated perspective maturity model, see Fig. 1, Tables 2, 3, 4, and 5.

[1] https://drive.google.com/open?id=16GDgLnugiuULURJxMNOop5u4aKuVhbn.

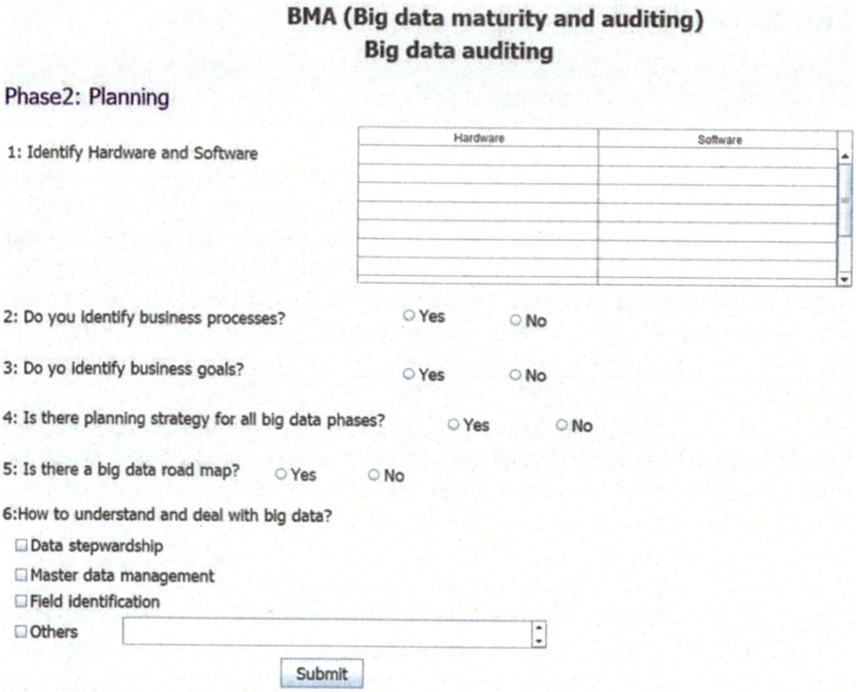

Fig. 3 Sample of the big data maturity auditing (BMA) tool in the planning area

The second stage is the auditing stage which consists of 35 questions for auditing big data phases based on our proposed auditing approach see Fig. 2 and Table 6. These questions are mainly used to assess and audit an organization's big data capabilities.

In the maturity stage, questions are weighted based on their significance. Each question has a high score of 6 points. The scoring percentage will be calculated in the auditing stage, as we score each stage individually. Once the questions have been answered, we produce a customized report which identifies the organization's maturity level and shows complete auditing report containing the result of the auditing process. The analysis of scores for each level/phase is presented in Table 9.

Table 9 Big data maturity level evaluation

Dimension	Score per dimension	Range of the score per dimension
Ad-hoc	12	0–12
Defined	12	13–24
Early adoption	6	25–31
Optimized	24	32–56
Strategic	6	57–63

For example, if the score in a stage is 15, the tool assesses the organization's maturity level as defined. BMA tool does not allow the examination to move forward to the next set of questions unless the organization gets the full score of the current level. Once the maturity level is specified, the auditing process starts. The final report will recommend some best practices to help in increasing the organization's maturity level. The "Auditing" process contains 35 questions. The auditor can customize which big data phases to audit. According to the selected phases, a set of customized questions is tailored. Once the assessment is finished, the tool produces an auditing report to reflect the current status per each phase, state the lacks in any processes to fully support big data capabilities, and provide some recommendations to gain more value and to overcome these lacks. The scoring percentage will be calculated as shown below.

Hint:

NP = number of phases to be audited.

NQ = number of questions per phase.

NA = number of answered questions per phase.

Score = Percentage of the organization's readiness to deal with big data.

$$\text{Score} = \sum_{i=1}^{\text{NP}} \left(\frac{\text{NA} * 100}{\text{NQ} * \text{NP}} \right)$$

7 Conclusion and Future Work

In this paper, we investigated the recent research in big data maturity assessments and their limitations. Our main interest was the auditing process and how big data maturity in the organization can affect the process and its results. Due to the lack of a generic auditing approach that helps any organization to get the full potential of their big data and ensures the organization's maturity, we proposed an integrated model for assessing big data and we use it to support our proposed auditing approach. This approach provides a set of best practices, metrics, and controls based on existing standards to aid the auditor in assessing the organization's big data capabilities and auditing.

As a proof of concept, we provide a big data maturity and auditing tool named BMA tool. This tool uses a scoring function to help in assessing the current level of big data maturity per phase in percentage. Also, it helps in tailoring the auditing to assess a set of specific phases. We mapped our proposed approach with recent research related to the COVID-19 epidemic to get the required standard controls. These controls can help any medical institute to manage and govern their data based on well-defined controls that follow a well-known standard.

In future work, we will investigate how to utilize artificial intelligence techniques to enhance the quality and efficiency of audits. Moreover, the use of cloud-based services can be challenging for auditors due to limited control over service providers. Thus, we may extend our tool to support the auditing process over cloud-based services.

References

1. Dey N, Waghet S, Parikshit N, Pathan M (2019) Applied machine learning for smart data analysis
2. Braun H (2015) Evaluation of big data maturity models: a bench-marking study to support big data maturity as—sessment in organizations. Master, Tampere University of Technology
3. ISACA group (2013) Enabling information superiority, p 90. https://doi.org/10.1001/jama.2015.12763
4. ISACA group (2013) Enabling information superiority, p 90
5. Vesset D, Olofson CW, Brien AO, Woodward A (2015) IDC MaturityScape IDC MaturityScape: big data and analytics 2.0. IDC MaturityScape, pp 1–15
6. IBM (2014) Big data and analytics maturity model. In: IBM big data and analytics hub. Accessed 28 Nov 2019
7. Infotech (2014) Big data maturity assessment tool. https://www.infotech.com/research/ss/leverage-big-data-by-starting-small/it-big-data-maturity-assessment-tool. Accessed 28 Oct 2020
8. Dhanuka V (2016) Hortonworks big data maturity model. Hortonworks, pp 1–2
9. Farah B (2017) A value based big data maturity model. J Manag Policy Pract 18:11–18
10. Schmarzo B (2016) Big data business model maturity index guide. In: EMC Corporation
11. Helmy NY, Mazen M, Elgammal S, Youssef AW (2021) Towards building a comprehensive big data management maturity framework. Int J Inform Decis Sci 1:1576
12. Comuzzi M, Patel A (2016) How organisations leverage big data : a maturity model. Ind Manag Data Syst 21:1–31
13. Sulaiman H, Che Cob Z, Ali N (2015) Big data maturity model for malaysian zakat institutions to embark on big data initiatives. In: Proceedings of the 2015 4th international conference on software engineering and computer systems (ICSECS), Kuantan, Pahang, Malaysia August 19–21, p 384. https://doi.org/10.1109/ICSECS.2015.7333084
14. Halper BF, Stodder D (2016) A guide to achieving big data analytics maturity (TDWI Benchmark Guide), pp 1–16
15. Abreu PW, Aparicio M, Costa CJ (2018) Blockchain technology in the auditing environment. In: Iberian conference on information systems and technologies, CISTI 2018-June, pp 1–6. https://doi.org/10.23919/CISTI.2018.8399460
16. Cooke I, Raghu RV (2018) IS audit basics: auditing the IoT. ISACA J 5:6652
17. Arass Mohamed E, Tikito I, Souissi N (2018) An audit framework for data lifecycles in a big data context. In: Proceedings of the 2018 international conference on selected topics in mobile and wireless networking, MoWNeT 2018, pp 1–5. https://doi.org/10.1109/MoWNet.2018.8428883
18. Aras O (2018) An assessment of big data policies and big data maturity in
19. El-Darwiche B, Koch V, Meer D et al (2014) Big data maturity : an action plan for policymakers and executives
20. Arass M, Tikito I, Souissi N (2017) Data lifecycles analysis: towards intelligent cycle. In: Proceedings of the 2017 intelligent systems and computer vision, ISCV 2017
21. Claudiu B, Muntean M, Didraga O, Brandas C (2018) Intelligent decision support in auditing: big data and machine learning approach. In: Claudiu B, Didraga O, Muntean M (eds) Proceedings of the IE 2018 international conference
22. de Santis F, D'Onza G (2020) Big data and data analytics in auditing: in search of legitimacy. Medit Account Res 29:1088–1112. https://doi.org/10.1108/MEDAR-03-2020-0838

23. Munir A, Shabani N, Mohanty SP (2021) A study of big data analytics in internal auditing. In: Internet of things and hardware security view project epileptic seizure detection in an edge computing paradigm using kriging methods view project a study of big data analytics in internal auditing
24. ISACA group (2019) Governance and management objectives
25. Beslay L, Sanchez I (2018) The right to data portability in the GDPR: towards user-centric interoperability of computer law and security review. Int J Technol Law Pract 34:193–203. https://doi.org/10.1016/j.clsr.2017.10.003
26. IAASB (2017)
27. The Apache Software Foundation (2019) Hadoop. In: The apache software foundation. https://hadoop.apache.org/. Accessed 26 Jan 2021
28. The Apache Software Foundation (2019) Spark. In: Apache Spark. https://spark.apache.org. Accessed 26 Jan 2021
29. The Apache Software Foundation (2019) Apache HBase: Apache HBase Downloads. In: Apache HBase. https://hbase.apache.org. Accessed 26 Jan 2021
30. Jayakumar K (2021) Therapeutics in COVID 19: an update. https://doi.org/10.20959/wjpps20207-16474
31. Limburn J (2015) Master data stewardship and governance master data stewardship and governance MDM Tech Talk
32. Khan S, Siddique R, Bai Q et al (2020) Coronaviruses disease 2019 (COVID-19): causative agent, mental health concerns, and potential management options. J Infect Public Health 13:1840–1844. https://doi.org/10.1016/j.jiph.2020.07.010
33. Koubaa A (2020) Understanding the COVID19 outbreak: a comparative data analytics and study
34. Li L, Yang Z, Dang Z et al (2020) Propagation analysis and prediction of the COVID-19. Infect Dis Model 5:282–292. https://doi.org/10.1016/j.idm.2020.03.002
35. Wolkewitz M, Puljak L (2020) Methodological challenges of analysing COVID-19 data during the pandemic. BMC Med Res Methodol 20:4–7. https://doi.org/10.1186/s12874-020-00972-6

Feature Selection in Medical Data as Coping Review from 2017 to 2022

Sara S. Emam, Mona M. Arafa, Noha E. El-Attar, and Tarek Elshishtawy

Abstract The number of medical applications with large datasets that require great speed and accuracy is continually growing. A large number of features in medical datasets is one of the most critical issues in data classification and prediction models. Furthermore, irrelevant and redundant features have also harmed the complexity and functioning of data classification systems. Feature selection is a reliable dimensionality reduction strategy for identifying a subset of valuable and non-redundant features from massive datasets. This paper reviews the state-of-the-art feature selection techniques on medical data in the last five years.

Keywords High-dimensional dataset · Feature selection · Classification

1 Introduction

Massive data expansion in medical domains has made medical data mining methods challenging. To make medical diagnoses, treatments, predictions, and prognostic schedules on time, doctors and professionals in the field of medicine must examine a vast amount of medical data. As a result, it is critical to provide an intelligent model that can accurately handle an enormous amount of medical data. Therefore, intelligent and machine learning-based techniques have become increasingly important

S. S. Emam (✉) · M. M. Arafa · N. E. El-Attar · T. Elshishtawy
Information Systems Department, Faculty of Computers and Artificial Intelligence, Benha University, Benha, Egypt
e-mail: sara.samir@fci.bu.edu.eg

M. M. Arafa
e-mail: mona.abdelmonem@fci.bu.edu.eg

N. E. El-Attar
e-mail: noha.ezzat@fci.bu.edu.eg

T. Elshishtawy
e-mail: T.shishtawy@fci.bu.edu.eg

D. Magdi et al. (eds.), *Green Sustainability: Towards Innovative Digital Transformation*, Lecture Notes in Networks and Systems 753,
https://doi.org/10.1007/978-981-99-4764-5_10

in medical health care. In various areas of health care, such as diagnosis, screening, prognosis, monitoring, therapy, survival analysis, and hospital management, machine learning classification algorithms are used in the decision-making process. However, machine learning faces a considerable challenge when dealing with medical datasets with a high-dimensional feature space and a limited number of samples [1]. Many features are used to represent data, but only a handful of them are relevant to the desired concept. Thus, the original datasets may have redundancy, which is not required to be included in the modeling process. Dimensionality reduction is one popular strategy for removing irrelevant, redundant, and insignificant features. It is a practical way to increase accuracy, reduce computational complexity, create more generalized models, and reduce storage requirements [1]. Two key strategies for reducing dimensionality have recently been developed: feature extraction and feature selection. Individual features or feature subsets are not searched for during feature extraction. Instead, feature extraction converts the original feature set from a higher to a lower dimensional space. The features are not chosen; instead, they are projected onto a new feature area [2]. Principal component analysis (PCA) is an example of feature extraction. Feature selection is choosing a subset of relevant features to create enhanced prediction models.

1.1 Feature Selection

Feature selection is a preprocessing technique that selects the most critical and relevant features, which may enhance machine learning performance by removing redundant or unnecessary features. As a result of the application of feature selection, modeling accuracy is improved, while the overall computing cost is reduced [2]. Furthermore, feature selection provides various advantages, including [3]:

- Improving the machine learning algorithm's performance.
- Data comprehension, including learning about the process and possibly assisting with visualizations.
- Data compression, limiting storage requirements and lowering processing costs.
- Simplicity and the ability to use simpler models and gain speed.

Feature Selection Approaches

In general, there are three feature selection approaches: filter, wrapper, and embedded, as shown in Fig. 1 [4].

In the *filter approach* algorithms, the classifier is independent. Thus, the feature selection and learning models are also separate. Information Gain (IG), correlation coefficients, Relief method, Relief-F (RF), Fisher score method, Chi-squared (CS), and Gain Ratio are examples of filter approaches. Generally, the filter selection algorithms do not use interrelationships between features to evaluate features [5]. Instead, they employ a scoring method that determines the statistical score of each feature and ranks the most likely highest. The higher a feature's score, the more

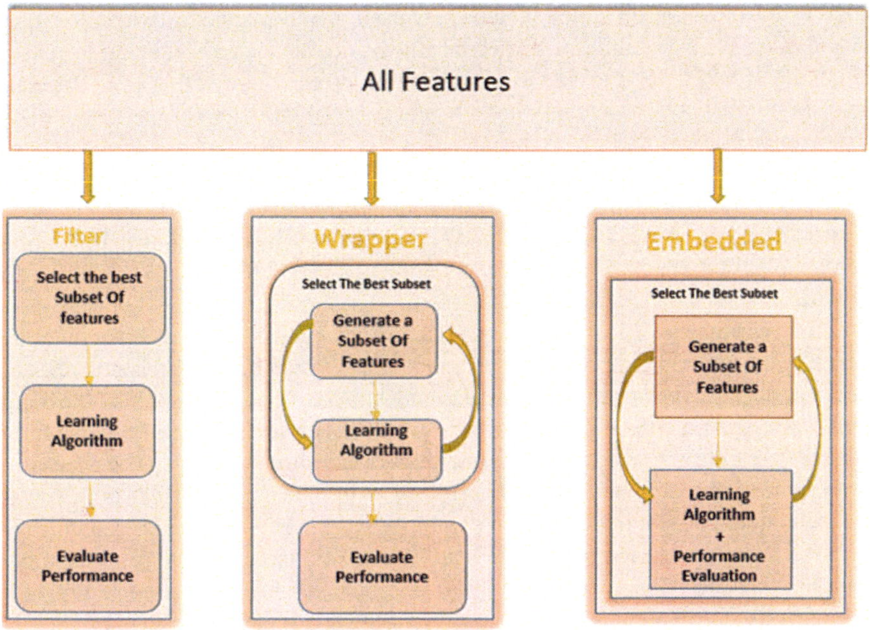

Fig. 1 Feature selection approaches [4]

likely it is to be chosen [2]. The main problem of this approach is that it ignores feature dependencies and relationships across classifiers, resulting in an incorrectly classified model [4].

In contrast, *wrapper-based feature selection* is classifier-dependent. It searches for a combination of features, each of which is referred to as a feature subset. Subsets can be used for prediction, and performance is determined using some metric. The selected feature subset has the best performance metric [2]. Several algorithms are used as wrapper feature selections, such as forward selection, backward elimination, and recursive feature elimination.

Finally, *the embedded model* is primarily concerned with identifying features that rate highly in terms of accuracy. The learning and feature selection processes are inextricably linked, and the feature search process is included in the classification algorithm. Examples of embedded methods are Lasso and Ridge regression algorithms [5]. Wrapper and embedded methods are frequently more accurate at classification than filter methods, although they take longer. As a result, several researchers have proposed hybrid strategies for identifying the essential features [4].

Optimization Algorithms in Feature Selection

Optimization algorithms aim to find the optimal solution for a well-defined problem. It is an iterative process that compares various solutions until an optimum or satisfying one is discovered. Optimization methods are used in multiple fields to identify solutions that maximize or minimize specific study criteria, such as reducing expenses in manufacturing a good or service, maximizing profits, minimizing raw materials in developing a good, or maximizing efficiency. Metaheuristics is an example of an optimization strategy that involves simulating the behaviors of physical phenomena and live creatures to create a general-purpose optimization search framework independent of the task at hand [6]. Genetic algorithms (GA) and evolution strategies such as particle swarm optimization, ant colony optimization, and bee colony optimization are examples of metaheuristics. In recent years, optimization algorithms have been a powerful approach to feature selection. This approach gets the essential feature subset that achieves high classification accuracy according to determined objective function criteria. This approach improves results in most cases than traditional approaches of feature selection. Optimization techniques may be used alone, such as using the brain storm optimization (BSO) in [7], the improved teacher–learner-based optimization (ITLBO) algorithm in [8], and marine predators algorithm (MPA) in [9]. Otherwise, the optimization technique can be hybridized with other traditional feature selection techniques such as in [10] which combined rough set, chaos theory, and binary gray wolf optimization to produce the RS-CBGWO-FS model. In [11], an ensemble of multifilters algorithms such as IG, GR, CS, and RF has been used to utilize a harmonized classification technique based on PSO and SVM.

1.2 Medical Data

Clinical (medical) data include narrative, textual data (HPI, social/family Hx), and numerical measurements (laboratory results, vital signs, and measurement).

- Laboratory tests are a well-known type of medical data. Several types of data can be driven from laboratory tests, like:
  - Hematological or blood tests: This type of test is considered one of the most popular laboratory procedures performed to examine and analyze the hemic system. Microscopes and hematologic analyzers are used in these tests to look at the concentration of HBC in the blood flow (i.e., the oxygen levels in the blood flow), the white blood cell (WBC) count, the red blood cell (RBC) count, the number of platelets (PLTs), the iron concentration, and the number of erythrocytes and leukocytes. Hematological tests are also used to assess and monitor several diseases and disorders, including the prothrombin time and thrombin time, hematocrit (HCT), blood sedimentation, blood coagulation time, fibrin clot lysis time, and bone marrow, among others [12].
  - Urine tests: This test checks the urine using a urinalysis method. Urinalysis examines and analyzes the flow of urine, the gravity of urine, the levels of urine color, and the presence of germs and cellular debris using chemical screening tests and microscopes. Urine tests are commonly used to diagnose kidney and

liver disease, as well as diabetes and prostate cancer, urinary tract infections, and prostatic hypertrophy [12].
  - Histopathological (histological) testing: These tests examine the various types of tissues that indicate the nature of the disease (e.g., muscular, neural, epithelial). The most common way to diagnose cancer is via histopathological tests. A tissue sample is first collected in the least invasive method feasible for a biopsy test, with the amount of the recovered tissue sample varied according to the tissue area under review. A histological report can design a more tailored pharmacological treatment for a malignancy (or metastases) by providing information on the tumor type, hormone responsiveness, and other tumor markers [12].
  - Skin tests: These examinations are used to check the changes in the skin. An allergy, a skin ailment, or even skin cancer could cause these changes. For example, skin tests are commonly used to detect skin redness caused by enlarged blood vessels, non-blanching hemorrhages (such as purpura and palpable purpura), skin carcinoma, and skin lesions that could progress to skin cancer as allergies via a skin prick test [12].

- The vital signs of a live organism are an objective measure of its essential physiological functioning. They are "vital" because assessing and measuring them is the first and most crucial stage in any clinical evaluation. An assessment of the patient's vital signs constitutes the initial set of clinical examinations. Vital signs are the foundation for patient triage in an emergency department or urgent care setting because they can show a doctor how far a patient has deviated from the norm. Temperature, pulse, blood pressure, and respiratory rate are the traditional vital signs. Even though many other indicators besides the standard four vital sign parameters may also be helpful, studies have only demonstrated that pulse oximetry and smoking status are significantly related to patient outcomes [13].

2 Literature Review

This section presents different feature selection approaches on clinical (biomedical) datasets.

In [14], Li et al. have developed a novel prediction framework for medical diagnostics, known as IGWO-KELM, by combining an enhanced gray wolf optimization (IGWO) and kernel extreme learning machine (KELM). It consists of two primary sections. By adaptively looking for the best feature combination in the medical data, IGWO is used to weed out the redundant and irrelevant information in the first stage. In the suggested IGWO, GA is initially used to produce the population's initial positions, and GWO is then used to update the population's current positions in the discrete searching space. Based on the ideal feature subset obtained in the first step, the second stage conducts the effective and efficient KELM classifier. Two typical medical diagnosis issues, including the diagnosis of Parkinson's illness and breast cancer, were looked into in order to assess the proposed IGWO-KELM methodology.

The datasets of them are available in UCI. On the two common disease diagnosis issues, the new solution is contrasted with the original GA and GWO using a set of performance indicators. The results have demonstrated that the proposed method is superior to its two rival competitors.

In [7], Tuba et al. have suggested the brain storm optimization approach for feature selection and optimizing SVM parameters to classify medical datasets. The BSO method was modified for a binary solution to accomplish feature selection. The fitness function combines the number of features selected and classification accuracy. Three well-known medical datasets related to Hepatitis, liver disorders, and diabetes were used to test the suggested methodology. Results were compared to existing state-of-the-art approaches. The proposed method improved classification accuracy for the three datasets under consideration while using fewer features or preserving the same amount of features. The datasets used are Hepatitis (which has two classes, 19 features, and 155 instances), liver (which has two classes, seven features, and 345 instances), and diabetes (which has seven classes, eight features, and 768 cases).

Identifying the ideal feature subset using a feature selection method that is independent of the governing parameters of an algorithm tailored to the particular situation at hand is a difficult task. As a result, in [8], Manonmani et al. have introduced an algorithm based on the original TLBO algorithm's operating principle, which does not call for any algorithm-specific parameters. The improved teacher–learner-based optimization (ITLBO) algorithm, which is the name of the proposed research project, aimed to choose the best feature subset based on the Chebyshev distance formula in the evaluation of the fitness function and standard control parameters (i.e., population size and several generations) to find the ideal feature subset for early diagnosis of chronic diseases. The chronic kidney disease (CKD) dataset was used to test the proposed feature selection technique, resulting in a significant feature reduction of 36% compared to the 25% obtained using the original TLBO algorithm. Furthermore, by assessing the accuracy of classification algorithms (support vector machine (SVM), convolutional neural networks (CNNs), and gradient boosting), the generated optimal feature subset produced from the TLBO algorithm and the feature subset obtained from the ITLBO algorithm are validated. According to experimental results, the suggested feature selection algorithm improves overall classification accuracy for the resulting feature subset compared to the original TLBO approach.

In [10], Azar et al. have introduced a robust hybrid dynamic model for feature selection called RS-CBGWO-FS. Rough set (RS), chaos theory, and binary gray wolf optimization are combined in RS-CBGWO-FS (BGWO) to select the ideal number of features and accomplish an effective classification procedure in the medical area. Ten different chaotic maps are used to estimate and fine-tune GWO parameters. Before moving on to the classification and feature selection process, the process of handling missing values and the max–min normalization process are applied to medical datasets. This proposed strategy is tested on five complex datasets retrieved from the UCI repository. The overall result shows that RS-CBGWO-FS with the Singer and piecewise chaos maps offers greater efficacy, less error, faster convergence, and shorter computation times. The dataset used is cervical cancer (which has 36 attributes and 858 instances), dermatology (which has 33 features and 366 cases), diabetic retinopathy (which has 20 attributes and 1151 instances), arrhythmia (which has 279 attributes and 452 cases), and sonar (which has 60 features and 208 instances).

In [15], Spencer et al. use relevant features selected using various feature selection approaches to evaluate the performance of models produced using machine learning techniques. First, principal component analysis, Chi-squared testing, Relief-F, and symmetrical uncertainty have all been used to examine four widely used heart disease datasets to produce different feature sets. Then, to increase the accuracy of heart condition predictions, various classification algorithms have been employed to develop models that are compared to find the best feature combinations. The advantages of applying feature selection differ based on the machine learning technique utilized to analyze the heart datasets. However, the best model we produced used the BayesNet method and Chi-squared feature selection to reach an accuracy of 85.00% on the datasets under consideration. The dataset combines four heart disease datasets from the UCI ML repository (Cleveland Dataset, Long-Beach-VA Dataset, Hungarian Dataset, and Switzerland Dataset). The new combined dataset has fourteen features and 720 cases.

In [2], Shah et al. have introduced an automatic methodology for diagnosing clinical heart disease. By utilizing feature selection and extraction techniques, the suggested method calculates the essential feature subset. First, mean Fisher-based feature selection algorithm (MFFSA) and accuracy-based feature selection algorithm (AFSA) are introduced to carry out the feature selection. The feature extraction method, principal component analysis, is then used to refine the chosen feature subset further. The suggested approach has been tested with Cleveland, Hungarian, and Switzerland data and a combination of the three. Radial basis function kernel-based support vector machines classify humans as either heart disease patients (HDP) or standard control subjects (NCS). Accuracy, specificity, and sensitivity measures assess the suggested methodology. This paper used three datasets from UCI (Cleveland, Hungarian, and Switzerland). The original dataset comprised 76 features, 14 of which were chosen, including the class label. The following are the details and descriptions of the datasets: Cleveland (which consists of 13 features and 303 instances), Hungarian (which consists of 13 features and 283 cases), Switzerland (which consists of 13 features and 123 instances), and combined (which comprised of 13 features and 709 cases).

In [1], Rostami et al. have combined the multi-objective PSO algorithm and the node centrality methodology to create a feature selection method called MPSONC. This approach is classified as filter-based model feature selection, and its optimization process considers relevance and redundancy ideas. The MPSONC procedure is broken down into three stages: (1) graph presentation, (2) computation of node centrality, and (3) final feature selection utilizing a multi-objective PSO search algorithm. Converting the feature space into an undirected, weighted graph is the goal of the first step. A node in this representation represents each feature, and the weight of each edge reflects the similarity of their associated features. To determine feature popularity, the second phase of the suggested method applies the node centrality criterion to every feature. The initial population in the PSO method will be created using the node centrality criterion in this step. Finally, the most essential and non-redundant features are selected in the last phase using a novel PSO-based search algorithm. Instead of using a single-objective fitness function to evaluate the generated particles as in many earlier PSO-based feature selection approaches, the innovative approach used in this study considers a feature subset utilizing a mix of feature separability index, similarity, and feature subset size. The proposed strategy is creative and performs better than the previous ones for three reasons: Three criteria of relevance, redundancy, and subset size of the chosen feature are considered in the fitness function of the proposed PSO-based technique. To illustrate the performances of the proposed strategy, five medical datasets with various properties are used. The results demonstrated that the introduced method is more efficient and effective than similar prior methods.

In most cases, the data dimensionality and classifier parameters significantly impact the accuracy of a diagnosis system. Because these two procedures are dependent, performing them separately could reduce accuracy. Based on ranking, the filter algorithm is employed to remove unimportant features. On the other hand, independent filters can still not account for feature dependency, resulting in an imbalanced selection of significant features and, as a result, a reduction in classification performance. To address this issue, in [11], Hamid et al. used an ensemble of multi-filters algorithms such as Information Gain (IG), Gain Ratio (GR), Chi-squared (CS), and Relief-F (RF), which takes into account feature intercorrelation. However, kernel parameter values may also influence classification performance. As a result, a harmonized classification technique based on PSO and SVM is used to optimize the simultaneous search for the best relevant features and kernel parameters while maintaining accuracy. As a result, this research proposed an ensemble filter feature selection with PSO and SVM harmonized classification (ensemble-PSO-SVM). The efficiency of the suggested strategy is evaluated using common lymphography and breast cancer datasets compared to other approaches already in use, such as PSO-SVM and standard SVM. Experimental findings show that the suggested method successfully indicates the classifier accuracy performance with the best essential features. As a result, the recommended method can be used as a substitute for selecting the best solution for dealing with high-dimensional data. Two datasets from UCI are used to verify the efficiency of the suggested model. The first dataset is breast cancer, which has 286 cases, nine features, and two predicting classes: class recurrence event and class no-recurrence event. The second dataset is lymphography, which has 148 instances, represented by 18 features and four predictive classes: standard, metastases, malignant, and fibrosis.

In [16], Bania et al. have used the feature-class, feature-feature rough dependence, and feature-significance measures to present a new rough set theory (RST)-based heterogeneous EFS approach (R-HEFS) to select the less repeated and more essential features during the aggregation of varied feature subsets. As a base feature selector, R-HEFS employs five state-of-the-art RST-based filter techniques. The experiments use ten standard medical datasets from the UCI repository. In addition, the *k*-nearest neighbor (KNN) imputation approach and RST-based discretization techniques are used for missing value imputation and continuous feature discretization. They use four classifiers, namely random forest (RF), Naive Bayes (NB), AdaBoost, and support vector machine (SVM). The effectiveness of the proposed R-HEFS technique is assessed and studied. By eliminating irrelevant and redundant features during the aggregation of base feature selectors, the suggested R-HEFS technique proves to be efficient and helps to improve classification accuracy. R-HEFS has obtained improved average classification accuracy on 7 out of 10 diverse medical datasets. As a result, the overall findings strongly show that the suggested R-HEFS method can minimize the dimension of substantial medical datasets, potentially assisting physicians or medical specialists in diagnosing (classifying) various diseases with fewer computational difficulties. The datasets for cancer, heart, skin, liver, thyroid, and cardiac illnesses were gathered from the UCI machine learning repository. And they have a medium to a high level of complexity.

In [5], Omuya et al. presented a hybrid filter approach based on principal component analysis (PCA) and Information Gain for feature selection. By building the main components of the dataset, PCA allows datasets with many linked features to be reduced in size so that the present data can be stated with fewer variables. It is performed by determining the most significant primary components by assessing the association between features. Information Gain Evaluation: This stage uses Information Gain (IG) to examine the feature set selected above to find the most relevant attributes. The final feature set is chosen based on a predetermined threshold, and the IG for features is calculated (t). After that, using machine learning techniques such as the Naive Bayes methodology, the hybrid model is used to support classification (classify breast cancer data). According to experimental results, the hybrid filter model picks relevant feature sets, decreases training time, and minimizes data dimensionality, resulting in higher classification performance as assessed by accuracy, recall, and precision. The dataset used in this paper is the breast cancer dataset. It was created by Zwitter and Soklic of the Institute of Oncology University Medical Center and Ljubljana, Yugoslavia. It has nine features that can be used to detect the existence or onset of cancer.

In [17], Pavithra and Jayalakshmi have proposed a hybrid feature selection technique HRFLC, which combines random forest (RF), AdaBoost (AD), and Pearson coefficient (PC). A subset of features is chosen based on the previous three techniques, and accuracy will be tested for several models. The model's results demonstrate that it effectively predicts diseases and enhances prediction accuracy—the dataset used in this paper was taken from UCI repository. Dataset (heart disease dataset) contains 13 features and includes 280 patient records, 10 of which have missing values that are eliminated during data preprocessing. The dataset is a binary classification problem, with 1 indicating heart illness and 0 indicating no heart disease. The dataset is balanced with 120 heart disease patients and 150 records of those without heart disease patients.

In [9], Elminaam et al. presented a new method for reducing dimension in feature selection. In a seminal attempt, this paper selects the appropriate feature subset to increase classification accuracy using binary variations of the recent marine predators algorithm (MPA). MPA is a brand-new metaheuristic inspired by nature. This study offers the MPA-KNN method, a mix of MPA and k-nearest neighbors (KNN). On medical datasets with feature sizes varying from small to huge, KNN is utilized to evaluate the selected features. The suggested methods are compared to eight well-respected metaheuristic wrapper-based algorithms and tested on 18 well-known UCI medical dataset benchmarks. In MPA, the fundamental exploratory and exploitative processes are modified to choose the best and most significant features for the most accurate classification. The findings show that the suggested MPA-KNN strategy can select the most relevant and optimal features. Furthermore, it outperformed the well-known metaheuristic algorithms that were put to the test. On average, MPA-KNN outperforms all other datasets in terms of accuracy, sensitivity, and specificity.

Based on symptoms and data from patients' electronic medical records, in [4] El-Attar et al. have presented a new multilayer perceptron (MLP) with feature selection (MLPFS) to predict positive COVID-19 instances (EMR). The MLPFS model comprises a layer that determines the most valuable symptoms to reduce the number of symptoms based on their relative value. Using only the most informative symptoms when training the model can hasten to learn and improve accuracy. Three separate COVID-19 datasets and eight different models, including the suggested MLPFS, were used in the experiments. According to the results, MLPFS outperforms all other experimental models in feature reduction across all datasets. It also performs better than the other models regarding classification outcomes and processing speed. In this research, three types of clinical reports served as datasets. The SARS-CoV-2 RT-PCR and further laboratory testing carried out on about 6000 COVID-19 cases during their visits to the emergency room were used to create this dataset. It has one class label and 109 features. Clinical features for symptomatic and asymptomatic individuals are included in the second COVID-19 dataset. There are 34,475 records in this dataset, each with one class label and 41 features. The third dataset uses clinical information to forecast the intensive care unit (ICU) admission for COVID-19 positive cases. There are 1926 cases total, 228 features, and 1 class label.

In [18], Piri and Mohapatra have proposed a multi-objective quadratic binary Harris Hawk optimizer for dealing with the feature selection issue in medical data. The continuous MOHHO is changed to a binary version using four quadratic transfer functions to make the approach practical for the FS problem. As a measure of each Hawk's fitness, two objective functions—the number of features in the candidate feature subset and the KNN classifier's classification accuracy—are considered. The four versions of the proposed MOQBHHO are implemented to extract the best feature subsets. Finally, the crowding distance (CD) value is used as a third criterion for selecting the best non-dominated option. Twelve standard medical datasets are used in this study to measure the performance of the suggested technique. MOBHHO-S (with a sigmoid function), MOGA, MOALO, and NSGA-II are all compared to the proposed MOQBHHO. Compared to deep-based FS approaches, the experimental results reveal that the suggested MOQBHHO effectively discovers a set of non-dominated feature subsets. The used datasets are BreastCancerW, Arrhythmia, Diabetic, Hepatitis, ILPD, Cardiotocography, Lymphography, LungCancer, Primary tumor, Parkinsons, Colon tumor, and SRBCT. The first ten datasets are from the UCI library, and the last two high-dimensional datasets are from Ref [18].

In [19], Gutowski et al. have provided a novel MOFS technique for binary medical classification. It is based on a genetic algorithm and a three-dimensional compass, intended to direct the search to the desired trade-off between the number of features, accuracy, area under the ROC curve (AUC), and accuracy. On several real-world medical datasets, our approach—the genetic algorithm with multi-objective compass (GAwC)—performs better than any other genetic algorithm-based MOFS technique. Furthermore, GAwC guarantees the classification quality of its solution by including AUC as one of the objectives, making it a particularly intriguing method for medical situations where both healthy and ill patients need to be reliably diagnosed. Finally, GAwC is used to solve a real-world medical classification problem, and the results are analyzed and supported from both a medical and a classification quality perspective. The datasets used in this paper are Breast cancer (which consists of 569 instances and 30 feature) from the UCI ML repository, Cardiotocography (which comprises 2126 and 21 features) from the UCI ML repository, Diabetes (which consists of 768 instances and eight features) from UCI ML repository, Kaggle Heart (which comprised of 270 cases and 13 features) from UCI ML repository, Musk1(which consists of 476 instances and 166 features) from UCI ML repository, and ASA-DI (which comprised of 822 cases and 48 features) from University Hospital of Angers (Table 1).

From the above literature review, we see that the feature selection method can be done by using traditional feature selection approaches alone, such as in [4, 15, 16], or by using the optimization technique to optimize the feature selection process, such as in [7–9, 18, 19]. Feature selection method can be done also by using a hybrid model between traditional methods, such as in [2, 5, 17], or making a hybrid model between optimization techniques themselves or between optimization techniques and traditional FS approaches such as in [1, 10, 11, 14].

3 Conclusions

Medical dataset suffers from the curse of dimensionality due to including redundant and irrelevant feature, so feature selection plays a vital role in solving this problem, and it chose most important feature subset. However, the traditional feature selection approaches increase the classification accuracy, but the hybrid model and optimization technique achieve the best classification accuracy.

Table 1 Presents a summary of the above-mentioned related work

References	Year	Method	Classifier	Dataset	Result
[14]	2017	IGWO-KELM	KELM classifier	Parkinson and Wisconsin diagnostic breast cancer	The best accuracy is 97.45 for Parkinson and 95.43 for WDBC
[7]	2019	Brainstorm optimization algorithm (BSO)	(SVM) where the BSO algorithm also tunes parameters	Hepatitis Liver Diabetes	Accuracy is Hepatitis: 97.16% Liver disorder: 84.31% Diabetes: 91.46%
[8]	2020	Improved teacher–learner-based optimization (ITLBO) algorithm	Support vector machine (SVM), convolution neural networks (CNNs), and gradient boosting	Chronic kidney disease (CKD) dataset	Accuracy SVM: 91.75% CNN:95.25% Gradient boosting:94.5%
[10]	2020	RS-CBGWO-FS	KNN	Cervical cancer, dermatology, diabetic retinopathy, arrhythmia, sonar	The highest accuracy achieved for each dataset is, cervical cancer: 97%, dermatology: 96%, diabetic retinopathy: 65%, arrhythmia:71%, sonar: 85%
[15]	2020	Using principal component analysis, Chi-squared testing, Relief-F, and symmetrical uncertainty. Then, several classification methods were used to develop models, which were then compared to find the best feature combinations	BayesNet, logistic, stochastic gradient descent (SGD), KNN (or in WEKA: IBK with K¼21), AdaBoost M1 with decision stump, AdaBoost M1 with logistic, repeated incremental pruning to produce error reduction (RIPPER or in WEKA: JRip), and random forest.29–36	Combination of four heart disease dataset (Cleveland Dataset, Long-Beach-VA Dataset, Hungarian Dataset, and Switzerland Dataset)	The best combination is Chi-squared feature selection with the BayesNet algorithm and achieved an accuracy of 85.00% Keywords

(continued)

Table 1 (continued)

References	Year	Method	Classifier	Dataset	Result
[2]	2020	$T = E(S(F))$, where $E \in$ FET and $S \in$ FST $S = \text{Filter}(F) \cup W(F)$ To accomplish the feature selection, two algorithms are used ((MFFSA) and (AFSA)) To accomplish the feature extraction (PCA)	RBF-based SVM	Heart disease Cleveland, Hungarian, Switzerland, and combined dataset	Cleveland:82.90%, Hungarian: 83.70%, and Switzerland: 91.30% combined dataset: 83.30%
[1]	2020	SIMPSONS	Support vector machine (SVM), Naive Bayes (NB), and AdaBoost (AB)	Colon, SRBCT, Leukemia, Prostate tumor, Lung cancer	SVM, Naive Bayes, AdaBoost Colon: 85.19 (3.21) 80.44 (2.56) 81.87 (1.12) SRBCT: 82.10 (0.25) 78.78 (1.04) 79.52 (1.65) Leukemia: 88.89 (2.08) 83.29 (1.88) 83.89 (2.42), Prostate tumor: 81.67 (0.25) 78.18 (2.83) 77.14 (0.55), Lung cancer: 88.19 (3.21) 88.44 (2.56) 88.87 (1.12)

(continued)

Table 1 (continued)

References	Year	Method	Classifier	Dataset	Result
[11]	2021	Ensemble-PSO-SVM	Harmonize classification of PSO-SVM	UCI breast cancer and lymphography datasets	UCI breast cancer: 96.15 and UCI lymphograph: 96.62
[16]	2021	R-HEFS	Naïve Bayes (NB), random forest (RF), support vector machine (SVM), and AdaBoost	Lung cancer, thyroid, Wisconsin diagnostic breast cancer (WDBC), Indian liver patient (ILP), Dermatology, Arrhythmia, SPECT heart, Hepatitis, SCADI, Hepatocellular carcinoma cancer (HCC)	The SVM accuracy Lung cancer: 90.66, Thyroid: 96.33, Wisconsin Diagnostic Breast Cancer (WDBC): 90.66, Indian Liver Patient (ILP): 58.06, Dermatology: 85.80, Arrhythmia: 74.60, SPECT heart: 83.55, Hepatitis: 85.80, SCADI: 86.18, Hepatocellular carcinoma cancer (HCC): 74.50
[5]	2021	PCA-IG model	Naïve Bayes	Breast cancer dataset	97.81%
[17]	2021	HRFLC, a combination of random forest (RF), AdaBoost (AD), and Pearson coefficient (PC)	Applied to different machine learning technique	Heart disease	79%

(continued)

Table 1 (continued)

References	Year	Method	Classifier	Dataset	Result
[9]	2021	MPA-KNN	KNN	18 dataset from UCI; an example is a lymphography	The best accuracy rate in 77.7% of the dataset
[4]	2022	MLPFS	MLPFS	SARS-CoV-2 RT-PCR dataset, ICU dataset	SARS-CoV-2 RT-PCR: 0.914 ICU dataset: 0.884
[18]	2022	Multi-objective quadratic binary HHO (MOQBHHO)	KNN	Arrhythmia	The heist accuracy using Q4 is 0.95
[19]	2022	GAwC: genetic algorithm with multi-objective compass	Extreme learning machine (ELM)	Breast cancer, cardiotocography, diabetes, heart, Musk1, ASA-DI datasets	The average accuracy for each dataset is: 97.48, 90.7, 79.75, 86.52, 82.2, 77.45

References

1. Rostami M, Forouzandeh S, Berahmand K, Soltani M (2020) Integration of multi-objective PSO based feature selection and node centrality for medical datasets. Genomics 112(6):4370–4384. https://doi.org/10.1016/j.ygeno.2020.07.027
2. Shah SMS, Shah FA, Hussain SA, Batool S (2020) Support vector machines-based heart disease diagnosis using feature subset, wrapping selection and extraction methods. Comput Electr Eng 84:106628. https://doi.org/10.1016/j.compeleceng.2020.106628
3. Sánchez-Maroño N, Alonso-Betanzos A, Tombilla-Sanromán M (2007) Filter methods for feature selection—a comparative study. In: Lecture notes in computer science (including subseries lecture notes in artificial intelligence and lecture notes in bioinformatics), vol 4881 LNCS, pp 178–187. https://doi.org/10.1007/978-3-540-77226-2_19
4. El-Attar NE, Sabbeh SF, Fasihuddin H, Awad WA (2022) An improved DeepNN with feature ranking for Covid-19 detection. Comput Mater Continua 71(2):2249–2269. https://doi.org/10.32604/cmc.2022.022673
5. Odhiambo Omuya E, Onyango Okeyo G, Waema Kimwele M (2021) Feature selection for classification using principal component analysis and information gain. Expert Syst Appl 174(November 2020). https://doi.org/10.1016/j.eswa.2021.114765.
6. Kitagawa S, Takenaka M, Fukuyama Y (2004) Recent optimization techniques. Rev Lit Arts Am 89–93
7. Tuba E, Strumberger I, Bezdan T, Bacanin N, Tuba M (2019) Classification and feature selection method for medical datasets by brain storm optimization algorithm and support vector machine. Procedia Comput Sci 162:307–315. https://doi.org/10.1016/j.procs.2019.11.289
8. Manonmani M, Balakrishnan S (2020) Feature selection using improved teaching learning based algorithm on chronic kidney disease dataset. Procedia Comput Sci 171(2019):1660–1669. https://doi.org/10.1016/j.procs.2020.04.178
9. Elminaam DSA, Nabil A, Ibraheem SA, Houssein EH (2021) An efficient marine predators algorithm for feature selection. IEEE Access 9:60136–60153. https://doi.org/10.1109/ACCESS.2021.3073261
10. Azar AT, Anter AM, Fouad KM (2020) Intelligent system for feature selection based on rough set and chaotic binary grey wolf optimisation. Int J Comput Appl Technol 63(1–2):4–24. https://doi.org/10.1504/IJCAT.2020.107901
11. Hamid TMTA, Sallehuddin R, Yunos ZM, Ali A (2021) Ensemble based filter feature selection with harmonize particle swarm optimization and support vector machine for optimal cancer classification. Mach Learn Appl 5(May):100054. https://doi.org/10.1016/j.mlwa.2021.100054
12. Pezoulas VC, Exarchos TP, Fotiadis DI (2020) Types and sources of medical and other related data
13. Sapra A, Bhandari P (2020) Vital sign assessment, no January, 2020, PMID : 31985994
14. Li Q et al (2017) An enhanced grey wolf optimization based machine for medical diagnosis. Comput Math Methods Med 2017:1–16
15. Spencer R, Thabtah F, Abdelhamid N, Thompson M (2020) Exploring feature selection and classification methods for predicting heart disease. Digit Health 6:1–10. https://doi.org/10.1177/2055207620914777
16. Bania RK, Halder A (2021) R-HEFS: rough set based heterogeneous ensemble feature selection method for medical data classification. Artif Intell Med 114(March).https://doi.org/10.1016/j.artmed.2021.102049
17. Pavithra V, Jayalakshmi V (2021) Hybrid feature selection technique for prediction of cardiovascular diseases. Mater Today Proc 81:336–340. https://doi.org/10.1016/j.matpr.2021.03.225

18. Piri J, Mohapatra P (2021) An analytical study of modified multi-objective Harris Hawk Optimizer towards medical data feature selection. Comput Biol Med 135(June):104558. https://doi.org/10.1016/j.compbiomed.2021.104558
19. Gutowski N, Schang D, Camp O, Abraham P (2022) A novel multi-objective medical feature selection compass method for binary classification. Artif Intell Med 127(March):102277. https://doi.org/10.1016/j.artmed.2022.102277

Machine Learning for Blood Donors Classification Model Using Ensemble Learning

Nora El-rashidy, Amir El-Ghamry, and Nesma E. ElSayed

Abstract Blood transfusion is in constant demand, as it is required for several medical procedures and life-or-death operations. This study is motivated by the fact that the need for blood transfusions is steadily on the rise because of accidents, operations, diseases, etc. such that the ability to accurately predict the number of blood donors allows medical personnel to predict the future blood supply and plan accordingly to recruit enough volunteers to meet demand. We attempt to handle this supply–demand gap by using a predictive model that identifies potential donors. We model the probability of a person donating blood, depending on five specific features, using machine learning techniques. This paper implements a machine learning pipeline that includes data preprocessing and donor classification. Several ensemble models have been developed. The results show that the random forest model led to the best test set accuracy (96.3%), which beat other methods.

Keywords First keyword · Second keyword · Third keyword

N. El-rashidy (✉)
Machine Learning and Information Retrieval Department, Faculty of Artificial Intelligence, Kafrelsheiksh University, Kafrelsheiksh, Egypt
e-mail: noura.alrashidy@ai.kfs.edu.eg

A. El-Ghamry
Faculty of Computers and Information, Mansoura University, Mansoura, Egypt

School of Engineering and Computer Science, Hosted By Global Academic Foundation, University of Hertfordshire, Garden City, Egypt

Faculty of Computer Science and Engineering, New Mansoura University, Mansoura, Egypt

N. E. ElSayed
Delta Higher Institute for Management and Accounting Information Systems, Mansoura, Egypt

D. Magdi et al. (eds.), *Green Sustainability: Towards Innovative Digital Transformation*, Lecture Notes in Networks and Systems 753,
https://doi.org/10.1007/978-981-99-4764-5_11

1 Introduction

Donating blood is crucial since delays in receiving lifesaving transfusions are a leading cause of death for those in need. One pint of blood can save three lives, as we all know. Extreme cases include those involving fatal accidents, people with dengue or malaria, and those awaiting organ transplants. Other causes of death include the inability to obtain blood quickly enough for people with life-threatening illnesses like leukemia or bone marrow malignancy. There is a correlation between patients receiving blood transfusions and a lower chance of death, and better data-driven systems for tracking donations and supply demands can improve the entire supply chain [1].

The new coronavirus is one example of how the expanding global prevalence of viruses and diseases has directly contributed to an already high demand for blood donations. For example, in India, three million units of blood are needed every year, but there is a lack of coordination between blood banks and hospitals. Close to thirty thousand units of blood have been wasted by banks in India over the past five years due to a problem with the blood banking system [2]. The most difficult aspect of managing blood products is the unpredictability of blood demand and supply, necessitating a balance between shortage and waste [3]. Blood demand is rising quickly in affluent nations, such that 10 out of every 100 hospitalized patients require blood products [4]. In developing nations, blood shortages and frequent blood-borne illnesses from unscreened blood account for around 100,000 fatalities every year [4].

There is a lag between the demand for blood by patients with severe blood loss and the blood banks' ability to deliver it. We attempt to reduce this supply–demand gap by using a predictive model that identifies potential donors. In this paper, we examine the application of the binary classification approach to estimate the likelihood that a person will donate blood based on his donation history. We compare four ensemble machine learning algorithms (Bagging, voting, AdaBoost, and Random forest) prediction. We performed data imputation for manipulating data with missing values, then data is normalized. This process consists of changing characteristics so that their mean becomes 0 and variance becomes 1. Standardization aims to place all feature values into the same range, hence accelerating the learning process and preventing bias in feature importance when employing machine learning algorithms. Then feature selection is applied by recursive feature elimination method to generate the optimal set of features that yields the maximum performance in the shortest amount of time. Finally, the prediction phase is accomplished to predict if a person will donate blood based on a recency, frequency, monetary, and time (RFMT) marketing model [5].

The paper is organized as follows. In Sect. 2, we conducted a literature review to see which methods have proven beneficial in comprehending this issue. Next, we make a data description in Sect. 3, analyzing the methodology/design employed and the models examined in Sect. 4. The proposed architecture is discussed in Sect. 5; we present our findings, and discussions in Sect. 6. Section 7 includes our conclusions and future extend to this research.

2 Related Work

In this section, we discuss some recent related work that used classical machine learning algorithms for donor prediction systems. Some research has compared the accuracy of ANNs and other machine learning algorithms with that of traditional time-series models to forecast blood demand [1, 6]. On five years of data from the Taiwan Blood Services Foundation, Shih et al. [7] compared the accuracy of conventional time-series models to that of ANN and multiple regression using machine learning methods. Specifically, they found that traditional approaches to time-series forecasting (such as the Seasonal Exponential Smoothing Method and ARIMA models) performed better than ML methods. As a result, limited data or a short sample size utilized in training these algorithms can result in biased ML predictions in certain situations [8]. According to research by Khaldi et al. [6], ANN performed better than ARIMA models at predicting blood demand. In light of this, they reasoned that these models would be a good way to anticipate monthly blood need. Therefore, traditional time-series approaches are frequently used as reference points for ML algorithms.

Ashqar et al. [9] applied two-step clustering to cluster donors into groups with comparable characteristics based on how they typically arrive. After clusters were created, CART was applied to each one in an effort to boost prediction quality. Their main contribution is the development of a model for a serial queuing network that can be applied to the specific situation of blood center operations, allowing for the estimation of arrival patterns and the subsequent usage of the appropriate number of employees. However, in their research, neither the prediction accuracy of their method nor a comparison of two-step clustering-CART to standard CART algorithm are reported. Given that the blood donation system is also a supply chain, it is possible to optimize it by employing machine learning techniques such as artificial neural networks. Constructing a supply chain framework with hybrid inventory decisions of supply chain numbers in consideration of factors that influence the total supply chain cost. Chen et al. [10] provide a more streamlined approach that influences optimal inventory policies that are derived using multilayer perceptron and backpropagation artificial neural networks.

3 Dataset

The data file blood_donation.csv contains the data used to create the model. For each sample, there are five inputs or features. All of the input variables have a numeric value and correspond to blood donor characteristics. Donation is the target variable, with values of 0 for no blood donation and 1 for the most recent campaign. For each sample, there are five inputs or features. All of the input variables have a numeric value and correspond to blood donor characteristics. Donation is the target variable, with values of 0 for no blood donation and 1 for the most recent campaign.

Table 1 Data statistics of five input donation datasets

	Min	Max	Mean	SD
Recency	0	73	9.52	8.2
Frequency	1	50	5.51	5.84
Time	2	98	34.3	24.4
Quantity	250	1.25e + 4	1.38e + 3	1.46e + 3
Donation	0		0.238	0.426

Maximum (max), minimum (min), standard deviation (SD)
1-recency: Months since the last donation
2-Frequency: Total number of donations
3-quantity: Total blood donated
4-time: Months since the first donation
5-donation: True if the person donated in the last campaign, false otherwise [11]

The information on the variables is compiled in the list below [11]: The following describes the five attributes of the dataset information in Table 1.

4 Methods

Time-series forecasting is performed using ensemble model techniques. Computational intelligence uses preprogrammed algorithms to analyze input data and learn from it through supervised or unsupervised processes to anticipate output values within a reasonable range. Over the past ten years, ML models have excelled as a superior replacement for traditional statistical models for forecasting and other research challenges (such as regression and classification issues). Different classifiers were subsequently trained for the dataset, leading to the use of different data processing pipelines for each of them [12]. Create ML algorithms for regression-type issues, such as Decision Trees, Gradient Boosting, AdaBoost and Random forest. Parallel efforts have been made to compare models, build new ones, and empirically validate those that already exist. The enormous significance of these machine learning breakthroughs for modelers offers a wide range of options and a thorough understanding of the strengths and limitations of the current models for various forecasting issues [12]. Ensemble approach in machine learning is defined as the multimodal system in which multiple classifier and algorithms are carefully blended into a prediction model. Additionally, the ensemble method contributes to more accurate classification and prediction of statistics from complicated situations by minimizing bias in the predictive model and reducing variation in the projected data.

5 Proposed Method

The main objective of this model is to know if the same donor will give blood the next time. This model is a binary classifier, which means that there are only two possible outcomes: 0 indicates that the donor not donate; 1. The donor will donate blood. The number of cases of each specific target value in a dataset is defined as target incidence. Target incidence offers us an indication of how balanced (or imbalanced) is our dataset. Figure 1 gives a general overview of the approach stratifying donors using whole blood donations from the dataset. The application of using ML classification algorithms to predict and understand factors contributing to donor return and find homogeneous groups of blood donors is demonstrated in this section. Testing the chosen classifiers before and after the proposed classification model to allow for the presentation of the effect of return donors' predictions. Classifiers are assessed based on their effectiveness. The proposed model could be summarized in the following steps:

Data imputation: We used a sophisticated multiimputation technique called "miss-Forest," a machine learning-based data imputation tool that uses the random forest algorithm [13].

Data scaling: It is a preprocessing ML step; the standardization and normalization are used in our Donates model to improve the performance of predictive modeling algorithms. Before the modeling, the two most often used methods for scaling numerical data are normalization and standardization. Each input variable is scaled individually to a range of 0–1, which is the range for floating-point values where we have the most precision. Standardization shifts the distribution to have a mean of zero and a standard deviation of one by scaling each input variable separately by subtracting the mean (a process known as centering) and dividing by the standard deviation [13].

Train test split: The total number of data is 794. It divided according to target incidence as 176 will donate (class 1) and 618 will not donate (class 0). Therefore, we decide to make oversampling for class 1 and under sampling for class 0, in order to make the data balanced. The total data after oversampling is 700 with 300 for class

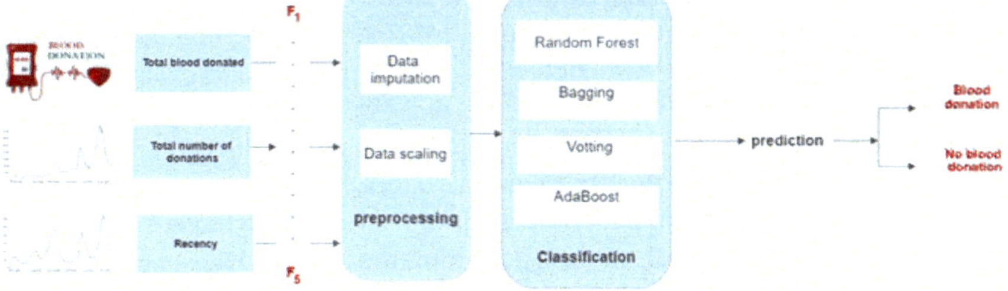

Fig. 1 Blood donation classification model

1 and 400 for class 0, then train test split() method from the scikit learn package with 70% and 30% for training and testing data, respectively.

Classification: We employed three state-of-the-art ensemble machine learning methods: bagging, voting, and Linear Discriminant Analysis. The random forest algorithm proves that data imputation and data scaling preprocessing methods resulted in high-performance classification with Donates blood model. Performance metrics were used to evaluate three classifiers to the proposed model bagging, boosting, voting, and random forest. The evaluation results indicated that the Random forest outperformed efficiency of the proposed Donates blood model classification systems. *K*-fold Cross-Validation: A tenfold cross-validation procedure with the training set was used to train and validate the models.

6 Results and Discussion

As we observe in Table 2, using ensemble machine learning gives adequate performance in terms of different evaluation metrics including accuracy, precision, recall, f measure and area under the roc curve. Bagging gives the least performance among all classifiers (ACC = 0.842, AUC = 0.862) followed by voting with performance of (ACC = 0.882, AUC = 0.886). AdaBoost gives adequate results (ACC = 0.922, AUC = 0.925). The best performance obtained from random forest classifier (ACC = 0.962, AUC = 0.975). The proposed method outperforms all traditional models with difference ranged from (2 to 7%). It achieved 0.962, 0.975 in terms of ACC and AUC. Table 2 shows the classification accuracy in terms of different metrics. To compare between all used algorithms, Fig. 3 compares all algorithms in terms of classification accuracy. To show the importance of each feature and its effect in the output, Fig. 4 shows the feature importance for all used features (Fig. 2).

7 Conclusion

Blood donation is in high demand worldwide, with the greatest unmet need in low- and middle-income countries. According to experts, substantial coordination is essential to lower demand. Although determining true donation rates remains problematic, we tried to predict the donation using machine learning technique. In this paper, ensemble machine learning we carried out ensemble machine learning models for predicting the blood donation based on data aggregated about the donation history of each donator. Several experiments have been learning models have been made using several ensemble ML models. Random forest gave the best performance compared with all ensemble models (ACC = 0.962, AUC = 0.975).

Table 2 Results of the utilized algorithms

Algorithm	Training score	Testing score	ACC	Precision	Recall	F-score	AUC
Bagging	0.842 ± 0.002	0.845 ± 0.002	0.8431 ± 0.01	0.848 ± 0.012	0.8341 ± 0.001	0.831 ± 0.012	0.862 ± 0.011
Voting	0.882 ± 0.011	0.884 ± 0.023	0.881 ± 0.011	0.873 ± 0.001	0.883 ± 0.002	0.883 ± 0.002	0.886 ± 0.001
AdaBoost	0.922 ± 0.001	0.923 ± 0.002	0.933 ± 0.003	0.921 ± 0.001	0.933 ± 0.002	0.944 ± 0.031	0.925 ± 0.033
Random forest	0.962 ± 0.001	0.963 ± 0.002	0.963 ± 0.003	0.971 ± 0.001	0.963 ± 0.002	0.964 ± 0.031	0.975 ± 0.033

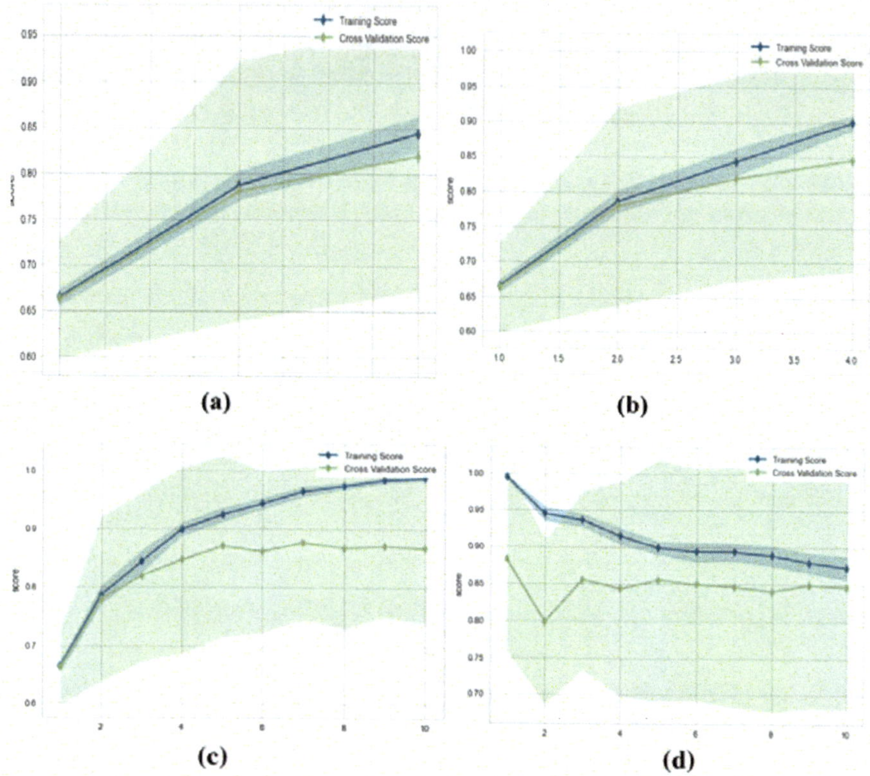

Fig. 2 Validation curves for the of the utilized algorithms, **a** bagging, **b** voting, and **c** AdaBoost and **d** random forest

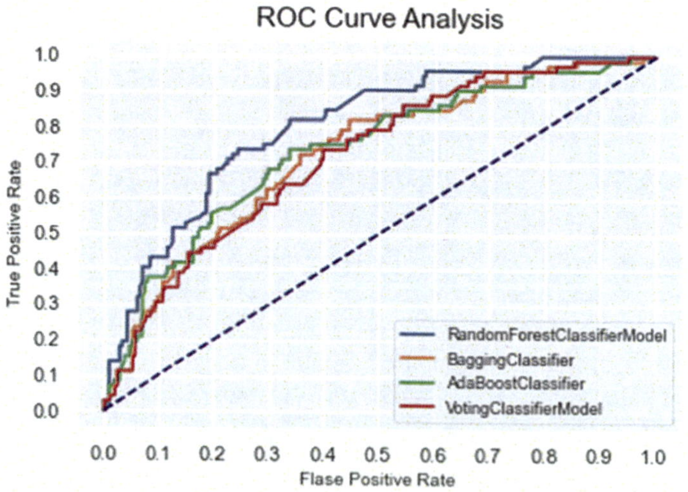

Fig. 3 Comparison between all the used models

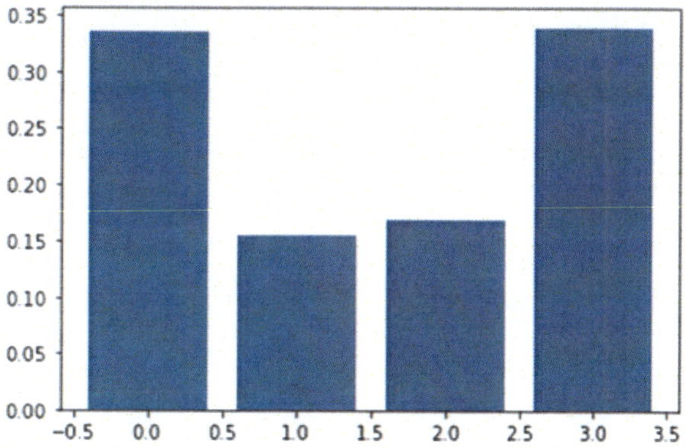

Fig. 4 Feature importance of the features

References

1. Barhoom AM (2019) Blood donation prediction using artificial neural network
2. Selvaraj P, Sarin A, Seraphim BI (2022) Blood donation prediction system using machine learning techniques. In: 2022 International conference on computer communication and informatics (ICCCI). IEEE, pp 1–4
3. Stanger SH, Wilding R, Yates N, Cotton S (2012) What drives perishable inventory management performance? Lessons learnt from the UK blood supply chain. Supply Chain Manag: Int J
4. Lancet T (2005) Blood supply and demand. Lancet 365(9478):2151. https://doi.org/10.1016/S0140-6736(05)66749-9
5. Al-Shayea QK, Al-Shayea TK (2014) Customer behavior on RFMT model using neural networks. In: Proceedings of the world congress on engineering, vol 1, pp 49–52
6. Khaldi R, El Afia A, Chiheb R, Faizi R (2017) Artificial neural network based approach for blood demand forecasting: fez transfusion blood center case study. In: Proceedings of the 2nd international conference on big data, cloud and applications, pp 1–6
7. Shih H, Rajendran S (2019) Comparison of time series methods and machine learning algorithms for forecasting Taiwan Blood Services Foundation's blood supply. J Healthc Eng
8. Vabalas A, Gowen E, Poliakoff E, Casson AJ (2019) Machine learning algorithm validation with a limited sample size. PLoS ONE 14(11):e0224365
9. Ashqar BA, Abu-Naser SS (2018) Image-based tomato leaves diseases detection using deep learning
10. Chen HC, Wee HM, Hsieh YH (2009) Optimal supply chain inventory decision using artificial neural network. In: 2009 WRI global congress on intelligent systems, vol 4. IEEE, pp 130–134
11. Yeh IC, Yang KJ, Ting TM (2009) Knowledge discovery on RFM model using Bernoulli sequence. Expert Syst Appl 36(3):5866–5871
12. Twumasi C, Twumasi J (2022) Machine learning algorithms for forecasting and backcasting blood demand data with missing values and outliers: a study of Tema General Hospital of Ghana. Int J Forecast 38(3):1258–1277
13. Suessner S, Niklas N, Bodenhofer U, Meier J (2022) Machine learning-based prediction of fainting during blood donations using donor properties and weather data as features. BMC Med Inform Decis Mak 22(1):1–7

Machine Learning and Artificial Intelligence

The Effect of Buyer Churn Factors on Buyer's Loyalty Through Telecommunication Service Providers in Egypt

Mohamed Hegazy Mohamed and Dalia Ahmed Magdi

Abstract This study looked into how loyal Egyptian telecoms customers are affected by factors that contribute to customer churn. Descriptive language is used to describe this. The emails used were selected at random from 1500 different telecom service users who had previously used the services. The surveys were disseminated and self-administered for data collection, with a response rate of 25.6%. The responses were subjected to linear regression analyses. The most recent findings suggest that in order to gain a foothold in Egypt's telecom sector, service providers ought to focus on improving the management of customer connections. The discoveries uncovered a genuinely critical connection between client stir counteraction components and client devotion to support degrees of consistency.

Keywords Telecoms customers · Telecom sector · Egyptian telecoms · Customer connections

1 Introduction

In today's digital world, where businesses compete fiercely, customer abandonment is a crucial step for any service provider to take in order to build profitable, long-term relationships with particular clients. Telecom service providers face the challenge of competing for valuable customers; the term for this is "customer abandonment." Managing customer abandonment for the most profitable customers is just one of many recent changes in the telecom industry. According to previous marketing surveys, mobile operators lose 25% of their customer base within a year if

M. H. Mohamed (✉)
Business Information Systems Department, Helwan University, Cairo, Egypt
e-mail: Mhegazy90@hotmail.com

D. A. Magdi
Faculty of Computers and Information Specialties, Sadat Academy for Management Sciences, School of Computer Science, Canadian International College, Cairo, Egypt
e-mail: dalia_a_magdi@cic-cairo.com

D. Magdi et al. (eds.), *Green Sustainability: Towards Innovative Digital Transformation*, Lecture Notes in Networks and Systems 753,
https://doi.org/10.1007/978-981-99-4764-5_12

their average monthly attendance is less than 2%. Given the fierce competition in the market, which includes offering services that are comparable to one another in terms of price and quality as well as discounts on phones and other accessories, cancelation fees are relatively low for the customer. Since many service providers typically match one another's prices, there is no longer any differentiation in the prices at this point. The primary distinction was the addition of services [1]. As a result, the majority of mobile operators face the greatest challenge right now: transitioning from reactive to proactive In other words, before deciding to end the contract, build customer loyalty and identify high-risk customers. As a result, the current body of research investigates the connection between abandonment and customer retention as well as ways to increase the service provider's competitiveness.

Objective of the Research

Any research, study, or report's desired level of performance is the objective. As a result, the specific goals of this study are:

- Determine how churn-related factors affect customer loyalty.
- Determine how customer churn management affects customer retention and churn.
- Search for an answer that will hold purchasers and keep them from changing to contenders [2].

2 Problematic Contention

Customers in markets for telecommunications services may switch providers in search of the best service and rates. In other words, the client can buy into one specialist co-op for a specific timeframe to buy into one more supplier subsequently, in light of the fact that the administrations presented by the broadcast communications specialist organizations are basically the same and simple to be imitated by contenders, particularly with these fast mechanical turns of events. Only price and quality can differentiate service providers in this scenario. As a result, telecom providers should prioritize retaining and acquiring new customers as their primary objectives estimates that online retailers lose 25% of their customers on average each year, and that even a small increase in customer retention can boost revenue by more than 25%. They are prepared to disregard all of their rivals' offers [3].

3 Literature Review

In the information and communication technology (ICT) industry, customers who are about to switch to a new competitor or end their subscription are referred to as "customer abandonment." As shown in Table 1, a review of previous studies in this area revealed a number of factors that influence client transformation [4]: The main factors, as shown in the table, are: cost of conversion, technology and quality rivals, advertising, security size, service cost, and customer satisfaction. Cost of conversion: This is defined as the expense incurred when customers are unable to switch to services offered by competitors. In point of fact, customers who switch to the competition typically waste time, effort, and money, even if they miss out on some unique advantages and opportunities that come with membership in a particular organization. As a result, switching service providers incurs costs and even prevents customers from taking advantage of some benefits. Prices, if the customer is forced to be unusable and experiences high conversion costs [5].

- Goodness: It refers to the quality of a telecommunications provider's text, video, and audio services during a call. Consequently, it is one of the variables affecting the surrender of the telecom supplier's clients.
- Technology and competitors: Every business faces a significant threat from rivals that provide high-speed services. Customers can easily switch providers due to the industry's high level of competition. To put it another way, when customer satisfaction plummets, they are eager to change service providers.
- Marketing: It is defined as a form of fee-based product, service, or idea recommendation. Advertising effectively aids businesses in preventing customer churn and attracting repeat business [6].
- Aspect of safety: It's about data or personal information being lost. Distrust of the service provider is usually the cause of security problems. Additionally, trusting another person is a prerequisite for it. As a result, security anxiety is brought on by customer mistrust [12].
- Cost of service: The amount that the customer is required to pay for the services is referred to as the service price. In point of fact, it is possible to assert that the personal situation improves customer purchase and reduces churn rate. Satisfaction: is the observed average value of a customer divided by their expectations. To put it another way, a satisfied customer is one who perceives that the value they receive meets their expectations. Another way to measure customer satisfaction is how satisfied customers are with the product or service after they use it [13]. You can stay with the company and avoid customer abandonment thanks to this level of customer satisfaction. As of late, progressive things have occurred in media communications, for example, new administrations, advances and the progression of the open market to contend on the lookout. How customers manage quick profits is crucial to the survival and growth of any telecommunications company because

Table 1 Allocation of the test size

Sample terminology	Reaction rate	Example rate (%)	Reply sample	Test
Vodafone	22.4	36.7	129	600
Orange	24.8	30	116	500
Etisalat	21.9	22.3	70	400
We	31.6	10	50	200
Overall	24.7%	100	394	1700

they are the primary source of profit currently reducing abandonment by independently addressing customers. Abandonment can be voluntary or involuntary. Involuntary abandonment occurs when a customer leaves a company and goes to a competitor, whereas voluntary abandonment occurs when a company asks a customer to leave for reasons like not being able to pay [15]. There are three types of abandonment: Customer Abandonment Involuntary: When a customer fails to pay an invoice, the service provider stops providing the service. Abandonment of customers is inevitable: This occurs when customers migrate or die, resulting in their complete exclusion from the market. Abandonment without consent: occurs when customers prefer to switch to a higher-value operator. Customers' actions are influenced by the actions of those around them, not by themselves. It is challenging for operators to acquire new customers in a saturated wireless market, so they must concentrate more on maintaining existing ones. With a very large customer base and a huge, vibrant, and dynamic telecommunications industry, customer acquisition and retention offer a chance to survive and improve profitability. Predicting consumer behavior has been the focus of most communications research. The most significant of these studies are shown in table. The majority of studies have focused on developing a model that can assist decision-makers in predicting customer abandonment among telecom service providers, as shown in the table. The connection between abandonment factors that influence customer loyalty and prevent them from abandoning the service easily. As a result, the purpose of this study is to identify a model for investigating the connection between user retention and influencing factors [16].

Conceptual Structure: Effect of CCF on Client Faithfulness:

From the writing audit, we can track down the quantity of elements that impact both client stir and dependability. We can in this manner propose the applied system: Effect of client stir factors on devotion as in Fig. 1.

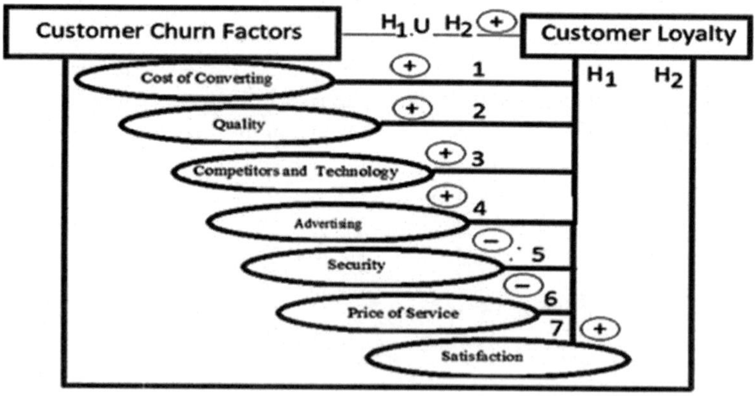

Fig. 1 Effect of client stir factors on devotion

Inquiries and Speculation

Do Client Agitate Variables (CCF) influence client steadfastness of media transmission specialist organizations in Egypt?

(*H*1) There is no critical constructive outcome on the greatness of the variables that impact client stir on client unwaveringness of the telecom suppliers in Egypt. It is separated into the accompanying sub-presumptions:

(1/1) There is no huge effect of the transformation costs on the client reliability of the telecom suppliers in Egypt.
(1/2) There is no huge effect of the Nature of Administration on the client faithfulness of the telecom suppliers in Egypt.
(1/3) There is no critical effect of contenders and trend setting innovation on the client faithfulness of the media transmission specialist co-ops in Egypt.
(1/4) There is no huge impact of publicizing on the client unwaveringness of the media communications specialist organizations in Egypt. (1/5) There is no huge effect of safety on client faithfulness of broadcast communications suppliers in Egypt.
(1/6) There is no massive impact of the cost of the assistance on client reliability of media communications specialist organizations in Egypt.
(1/7) There is no critical fulfillment impact on client unwaveringness of media communications specialist co-ops in Egypt [17].

*Q*3: Does the size of the elements impacting client relinquishment combinedly affect the client faithfulness of telecom specialist organizations in Egypt?
(*H*2) There is no critical beneficial outcome of the extent of the variables influencing the consolidated relinquishment on client reliability of media communications suppliers in Egypt [7].

4 The Exploration Plan

The point of our review is to find, investigate, and comprehend the key variables influencing client stir among media transmission specialist organizations in Egypt. So the specialist originally made a speculation, gathered information and tried it, the scientist gathered information from clients by means of email and self-directed for information assortment and straight relapse examination has been utilized [18]. The scientist embraced the two-step research approach of social exploration: the graphic examination step and the interpretive exploration step. The reason for the initial step is to explain the ideas of the review and reach inferences. To concentrate on past investigations and lead the exploratory review, then recognize the issue and propose speculations. In the subsequent stage, the scientist depended on the causal or logical way to deal with explain the connection between the autonomous and subordinate examination factors and the finish of causal connections between them [8]. Population and Test: In this review, the populace comprises of media communications administration clients working in the Egyptian market. The quantity of clients of the four organizations, as per the most recent insights as of January 1, 2018, is roughly 98.24 million people. Instrument: The specialist involved the study as an information assortment device, which is a completely ready rundown of inquiries addressed to the clients of correspondence specialist organizations in Egypt, requesting that they answer. This rundown has been partitioned into three areas: First area: connects with the variables that influence client beat. Area II: Connecting with client faithfulness [9]. Area III: Individual attributes of the client like nature, age, states of inclusion of the help. Factors affecting client agitate: The specialist embraced the norm on which the review was taken on to gauge the effect of elements impacting client stir. The standard comprises 28 sentences reflecting seven elements of the variables affecting stir at the specialist organization, to be specific, transformation costs (3 sentences), nature of administration (5 sentences), rivalry and innovation (4 sentences), publicizing (4 sentences), security (4 sets), administration cost (5 sets), fulfillment (3 sets).Client dependability: addressing client reliability, comprising of 10 sentences, and the analyst depended on the scale utilized by the review to gauge factors impacting client beat among Egyptian specialist organizations, as they have exhibited elevated degrees of trustworthiness and consistency [19]. Data Assortment: To measure the alternate points of view regarding the matter exhaustively, the scientist depended on two sorts of information to accomplish the targets of the review, to be specific auxiliary information and fundamental information [20]. The following is a show of these two sorts of information: Auxiliary information: the information involved by the analyst in taking shape the issue and inquiries of the review and figuring out speculations, and the arrangement of the hypothetical system of the review and estimation of factors and in deciding society of the exploration and conveyance of jargon, and this data was gathered by reading books, scholarly publications, and magazines that dealt with the subject of client dependability and client stir [21].

Background information: In order to determine the concept of the connection between factors influencing agitation and client steadfastness, starting information was obtained from the example of the review test [22]. Data collection relied upon an outline list that was spread to clients of far off expert associations in Cairo. Data Examination: The hypothetical system, previous investigations, and guidelines adopted and planned by various scientists served as the foundation for the information collection apparatus [10]. Factual projects for the PC, such as factual bundle programs for sociologies like SPSS and AMOS, are used to examine data and test hypotheses with measurable strategies. The ongoing procedures were utilized: (A) Drawing in assessment strategies: the researcher relied upon enchanting assessment procedures, explicitly on math and the standard deviation, during the examination and depiction of the responses of the respondents and the Presentation of the potential gains of the elements inspected [11].

(B) The coefficient of alpha connection: to determine whether this review's multi-component estimates are of consistent quality. (D) Relapse examination strategy: examine the effects of client additions and dependability. The following are the outcomes of the dependability test for the guidelines used in the review and the legitimacy of the norms: Reliability, Legitimacy, and Precision of the Estimates

Table 2 Arithmetical means and standard deviations of the motivating factors in customer stir up

Changing circumstances affecting buyer churn	Math mean	Normal way	Acquisition
Because my friends and family are employed by my current company, I will not join another mobile phone company	4.7812	Agree	2
I sign up for a different mobile service provider due to the advantages I enjoy	5.3047	Totally agree	1
I will switch to a different company if it has a low conversion cost	1.4609	Agree	3
Total	10.8490	Agree	5

Source Statistical analysis results of 0.001 and the value of R2 of 0.113, explains 11.3% of the variation in customer turnover intentions, and the rest is due to another variable not covered by the model and the standard error [26]

[13]. The degree of security and certainty is a proportion of the influencing factors [23].

The results of the dependability test revealed that the alpha coefficient of the factors affecting customer satisfaction had reached 0.93, indicating a significant level of stability or unwavering quality in accordance with the fundamental sciences. "Help out" was the most important factor, and "Quality" was the least stable of the impacting subvariables of client stir. The serious level of reasonableness for each component of the client stir-influencing variables is depicted in the table. Table 4. Stability and Unwavering Quality of Reliability [14] shows that the scale should be approved. After looking at the connection coefficients of the 10 faithfulness sentences that were used for the steadfastness scale, it was decided to exclude the sentence (8), which says, "I won't change to one more specialist organization in spite of the thickness of its promotions and its captivating," because it has a general relationship coefficient with different factors that are the same size of less than 0.3 [22]. The alpha method was chosen after eliminating the assertion to work on the level of dependability for similar estimates [24] (Tables 2 and 3).

5 Hypothesis Testing

The principal hypothesis of the review was tested by the scientist using a variety of relapse examination methods: The components of the variables influencing the withdrawal rate of customers have little effect on broadcast communications specialist organizations' client loyalty in Egypt. The primary hypothesis was isolated into seven sub-hypotheses, each testing the impact of the going with components influencing the expectation to switch compact manager client in Egypt. The second hypothesis of the review examined the relationship between the factors that influence the shifting of customers from one specialist cooperative to the next and their impact on unwaveringness. The elements of the variables affecting combined client stir have

Table 3 Type and strength of the relationship between quality on the customer loyalty (multiple regression analysis)

Excellence of service on loyalty	Worsening coefficient	Amount of correlation	Coefficient of willpower
My ongoing versatile help supplier gives me genuine administrations that meet the administrations you anticipate	– 0.071*	– 0.175	0.038
My current mobile service provider is ready to assist customers and offers prompt service, preventing me from switching providers	0.007	0.173	0.034
If you don't want to leave the service provider, one important factor is the quality of the service	– 0.070	– 0.244	0.040
Coefficient of correlation in model R	– 0.429		
Coefficient of connection in model R	0.208		

* Substantial at 0.05 ** Important at 0.01 *** Considerable at 0.001
Source Statistical analysis results

Table 4 Form and strength of the connection between rivals and knowledge on the buyer's loyalty

Participants and information on loyalty	Decline coefficient	Depth of correlation	Amount of purpose
I can't churn it because I believe my mobile service provider offers the best communication technology at the lowest prices	– 0.019	– 0.198	0.038
I think my portable specialist co-op offer high rates for web association, I can't stir it	– 0.113*	– 0.243	0.058
Measurement of relationship in model R	– 0.288		
Fortitude factor in model R2	0.088		
Computed F value	6.445		
Levels of freedom	(5,265,269)		
Amount of impact	2000 ***		

* Substantial at 0.05 ** Important at 0.01 *** Considerable at 0.001

no significant impact on the customer steadfastness of Egypt's telecom administrators. The test of the review theories to determine their legitimacy is as follows: The Effects of the Review Ha1's Main Speculation Test: the impact of costs associated with changes on dependability and the results of various relapse investigations into the relationship between costs associated with changes and unwaveringness Examine the validity of the primary sub-speculation (1/1) that says [25].

Turning costs have no significant impact on supporter maintenance of Egypt's telecom administrators. Table 2 displays the following.

- The reliability of the model that was used to demonstrate the link between the cost of referring the client to a different supplier as a free factor and the level of client dependability as a dependent variable.
- There is a statistically significant negative correlation between the cost of referring the client to another service provider and client loyalty. The strength of this association is approximately 33.6% (based on the model's R correlation coefficient).
- The most reliable reliability variables are the impact of the switching costs of each service (I join another to take advantage of the advantages I enjoy.) If the switching costs of the other companies are low, I will switch to an other. So (I won't switch to another mobile company because my family and friends are in my current company) [28] service providers in Egypt. Table 3 shows the following.
- The legitimacy of the model used to make sense of the genuine connection between nature of administration as a free factor and loyalty as a reliant variable. The determined worth of f is 6.391, which is critical at a huge degree of 0.001 and the worth of R2 is 10.8%. Hence, dependability is made sense of by 10.8%. In client reliability for provider of portable administrations, and the rest is because of different factors not covered by the model, notwithstanding the standard mistake.
- There is a genuinely critical negative connection between administration quality and client dependability, the strength of which is around 32.9% (as per the relationship coefficients R in the model) [27].
- The best of administration factors made sense of for client faithfulness are (the ongoing cell specialist co-op has top notch lines for making voice, video, and message calls so you can't grunt them). And afterward the my ongoing cell specialist co-op furnishes me with genuine administrations that meet the administrations you anticipate. the effect of Contenders and Innovation on dependability: Table 4 shows the consequences of the different relapse investigation of the connection between Contenders and Innovation and client reliability [29].
- The value of the model is used to figure out the close to home association between the Expense of Organization as a free element and the client Relentlessness as a dependent not entirely set in stone at 357,427, which is basic at 0.001, with a value of R2537. Thus, the assistance cost makes sense of the Second Test Effects of the Review. Table 5 depicts the results of the numerous relapse investigations into the relationship between client dependability variables and relinquishment variables. It states, "In order to test the legitimacy of the second speculation of the review, the size of the variables affecting combined client surrender on client maintenance of media communications specialist organizations in Egypt has no significant positive effect" Table 5 [30].

Table 5 Type and strength of the relationship between advertising on the customer loyalty (multiple regression analysis)

Advertising on loyalty	Worsening coefficient	Measurement of correlation	Factor of determination
The current mobile service provider provides appealing advertisements with novel deals	− 0.004	− 0.232	0.044
I'd like to know what other businesses are offering using this company's latest technology	− 0.158**	− 0.320	0.107
Retention of existing subscribers is affected positively or negatively by the way advertisements are displayed	− 0.078*	− 0.206	0.092
The current mobile service provider offers advertisements solely for the purpose of acquiring new clients	− 0.003	− 0.361	0.067
Coefficient of relationship in model R	− 0.341		
Degrees of independence	(4,265,269)		
Level of impact	0000 ***		

* Substantial at 0.05 ** Important at 0.01 *** Considerable at 0.001

6 Dissection

The outcome showed a positive effect of administration quality, fulfillment, contenders and innovation, change costs, promoting on client beat the board and devotion. Simultaneously, the outcome showed an adverse consequence of safety and administration cost on client stir the board and unwaveringness. Consequently, security variable and cost affect purchaser steadfastness among telecom organizations in Egypt as these two elements are available in all organizations. In any case, different variables have a significant effect. Because of the continuous rivalry in numerous areas, for example, the cell phone industry, they are getting some distance from contenders. That's what his reality is if organizations have any desire to remain in their cutthroat world, they should put resources into client maintenance. Consequently, to all the more likely control the client, we believe organizations should be familiar with client conduct blunders and a client's self-assurance. On the off chance that portable administrators don't offer qualified administrations, the potential for clients to move is higher than in different cases. The last compelling element was the declaration that couple of clients would pass on the portable administrator because of absence of exposure and information.

References

1. Nouza M, Ólafsdóttir R, Sæþórsdóttir AD (2018) Motives and behaviour of second home owners in Iceland reflected by place attachment. Curr Issues Tour 21(2):225–242
2. Topal İ, Ucar MK (2019) Hybrid artificial intelligence based automatic determination of travel preferences of Chinese tourists. IEEE Access 7:162530–162548
3. Tang L, Zhao Y, Duan Z, Chen J (2018) Efficient similarity search for travel behavior. IEEE Access 6:68760–68772
4. Cheng S (2018) Beijing residents heading south to escape winter smog and cold. China Daily. Retrieved from http://www.chinadaily.com.cn/m/guangxi/fangchenggang/2015-04/27content_20554438.htm
5. Vada S, Prentice C, Hsiao A (2019) The influence of tourism experience and well-being on place attachment. J Retail Consum Serv 47:322–330
6. Kim M, Koo DW (2020) Visitors' pro-environmental behavior and the underlying motivations for natural environment: merging dual concern theory and attachment theory. J Retail Consum Serv 56:102147
7. Chi HK, Huang KC, Nguyen HM (2020) Elements of destination brand equity and destination familiarity regarding travel intention. J Retail Consum Serv 52:101728
8. Dong D, Xu X, Wong Y (2019) Estimating the impact of air pollution on inbound tourism in China: an analysis based on regression discontinuity design. Sustainability 11(6):1682
9. Chen H, Zhang L, Chu X, Yan B (2019) Smartphone customer segmentation based on the usage pattern. Adv Eng Inf 42:101000
10. Greenstein-Messica A, Rokach L (2018) Personal price aware multi-seller recommender system: evidence from eBay. Knowl Based Syst 150:14–26
11. Park S, Yu S, Kim M, Park K, Paik J (2018) Dual autoencoder network for retinex-based low-light image enhancement. IEEE Access 6:22084–22093
12. Wu X, Jiang G, Wang X, Xie P, Li X (2019) A multi-level-denoising autoencoder approach for wind turbine fault detection. IEEE Access 7:59376–59387
13. Li X, Katsumata S (2020) The impact of multidimensional country distances on consumption of specialty products: a case study of inbound tourists to japan. J Vacation Mark 26(1):18–32
14. Hair JF, Black WC, Babin BJ, Anderson RE (2018) Multivariate data analysis, 8th edn. Pearson, Harlow
15. Zhang C, Liu Y, Fu H (2019) AE2-nets: autoencoder in autoencoder networks. In: Proceedings of the IEEE/CVF conference on computer vision and pattern recognition (CVPR), pp 2577–2585
16. Kim J-C, Chung K (2020) Multi-modal stacked denoising autoencoder for handling missing data in healthcare big data. IEEE Access 8:104933–104943
17. Prentice C, Chen J, Stantic B (2020) Timed intervention in COVID-19 and panic buying. J Retail Consum Serv 57:102203
18. Tran LTT (2021) Managing the effectiveness of e-commerce platforms in a pandemic. J Retail Consum Serv 58:102287
19. Caber M, González-Rodríguez MR, Albayrak T, Simonetti B (2020) Does perceived risk really matter in travel behaviour? J Vacation Mark 26(3):334–353
20. Miklosik A, Evans N (2020) Impact of big data and machine learning on 900 digital transformation in marketing: a literature review. IEEE Access 8:101284–101292. https://doi.org/10.1109/ACCESS.2020.2998754
21. Çelik Ö (2018) A research on machine learning methods and its applications. J Educ Technol Online Learn 1(3):25–40. https://doi.org/10.31681/jetol.457046

22. Miklosik A, Kuchta M, Evans N, Zak S (2019) Towards the adoption of machine learning-based analytical tools in digital marketing. IEEE Access 7:85705–85718. https://doi.org/10.1109/ACCESS.2019.2924425
23. Wang X, He J, Curry DJ, Ryoo JH (2021) Attribute embedding: learning hierarchical representations of product attributes from consumer reviews. J Mark 1(21):1–21. https://doi.org/10.1177/00222429211047822
24. Huang M-H, Rust RT (2021) A strategic framework for artificial intelligence in marketing. J Acad Mark Sci 49(1):30–50. https://doi.org/10.1007/s11747-020-00749-9
25. Buraga SC, Amariei D, Dospinescu O (2022) An OWL-based specification of database management systems. Comput Mater Continua 70(3):5537–5550. https://doi.org/10.32604/cmc.2022.021714
26. Salamai AA, Ageeli AA, El-kenawy E-SM (2022) Forecasting E-commerce adoption based on bidirectional recurrent neural networks. Comput Mater Continua 70(3):5091–5106. https://doi.org/10.32604/cmc.2022.021268
27. Anastasiei B, Dospinescu N, Dospinescu O (2021) Understanding the adoption of incentivized Word-of-Mouth in the online environment. J Theor Appl Electron Commer Res 16(4):992–1007. https://doi.org/10.3390/jtaer16040056
28. Mariani MM, Wamba SF (2020) Exploring how consumer goods companies innovate in the digital age: the role of big data analytics companies. J Bus Res 121:338–352. https://doi.org/10.1016/j.jbusres.2020.09.012
29. Krafft M, Sajtos L, Haenlein M (2020) Challenges and opportunities for marketing scholars in times of the fourth industrial revolution. J Interact Mark 51:1–8. https://doi.org/10.1016/j.intmar.2020.06.001
30. Kushwaha AK, Kumar P, Kar AK (2021) What impacts customer experience for B2B enterprises on using AI-enabled chatbots? Insights from big data analytics. Ind Mark Manag 98:207–221. https://doi.org/10.1016/j.indmarman.2021.08.011

Plant Disease Detection and Classification Using Machine Learning and Deep Learning Techniques: Current Trends and Challenges

Yasmin M. Alsakar, Nehal A. Sakr, and Mohammed Elmogy

Abstract Every year, all over the world, the major crops are affected by various diseases, which in turn affects agriculture and the economy. The traditional method for plant disease inspection is a time-consuming, complex problem that mainly depends on expert experience. The explosive growth in the field of artificial intelligence (AI) provides effective and smart agriculture solutions for the automatic detection of these diseases with the help of computer vision techniques. This paper presents a survey on recent AI-based techniques proposed for plant disease detection and classification. The studied techniques are categorized into two classes: machine learning and deep learning. For each class, its main strengths and limitations are discussed. Although a significant amount of research has been introduced, several open challenges need to address in this field. This paper provides an in-depth study of the different steps presented in plant disease detection along with performance evaluation metrics, the datasets used, and the existing challenges for plant disease detection. Moreover, future research directions are presented.

Keywords Plant disease · Feature extraction · Handcrafted · Machine learning · Deep learning · Transfer learning · Classification

Y. M. Alsakar (✉) · N. A. Sakr · M. Elmogy
Faculty of Computers and Information, Mansoura University, Mansoura, Egypt
e-mail: yasminmahmoud@mans.edu.eg

N. A. Sakr
e-mail: nehal_sakr@mans.edu.eg

M. Elmogy
e-mail: melmogy@mans.edu.eg

D. Magdi et al. (eds.), *Green Sustainability: Towards Innovative Digital Transformation*, Lecture Notes in Networks and Systems 753,
https://doi.org/10.1007/978-981-99-4764-5_13

1 Introduction

Plants are important for living organisms as they produce oxygen [1]. These plants help in balancing the habitat's biological aspects. Their parts, such as flowers, leaves, fruits, grains, and stems, are consumed and used by animals and humans. Their extracts have also been utilized for mustard oil, medicine, biofuels, food, etc. Plants are categorized with respect to different characteristics such as height/size, color, and shape.

Plants, like humans, also suffer from many diseases [2, 3] that badly affect their normal growth. These diseases infect many parts of plants, such as roots, flowers, fruits, and leaves. The classification and identification of these diseases by plant pathologists are often made on leaves depending on shape, size, color, or texture, as indicated in Fig. 1. Because of the huge number of crops and complexity, there are also many numbers of plant diseases.

Therefore, a timely and precise diagnosis of plant diseases [4, 5] is required to protect the crops from qualitative and quantitative loss. Manual plants diseases identification by the human eyes is time-consuming and requires more monitoring. Therefore, automatic plant disease classification and identification are required, reducing

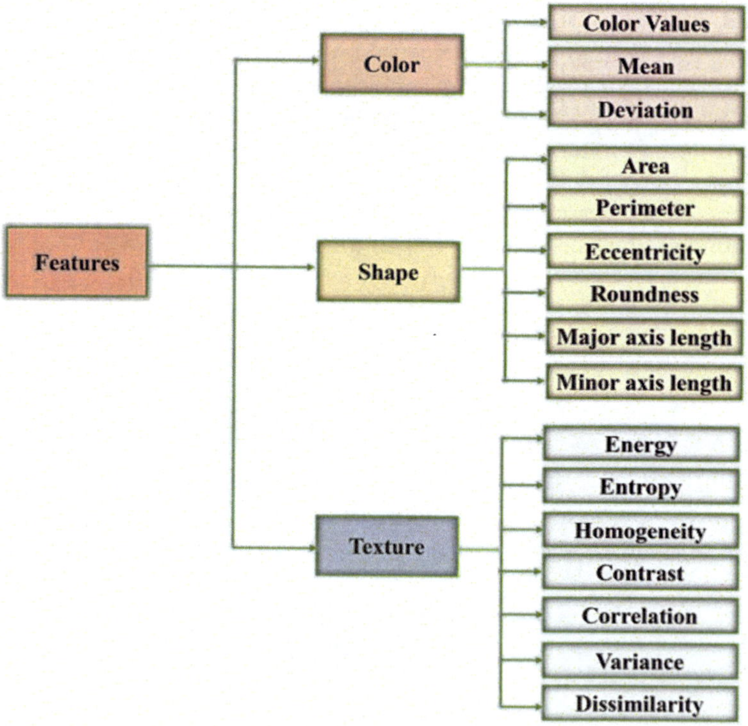

Fig. 1 Different types of features for plant diseases detection

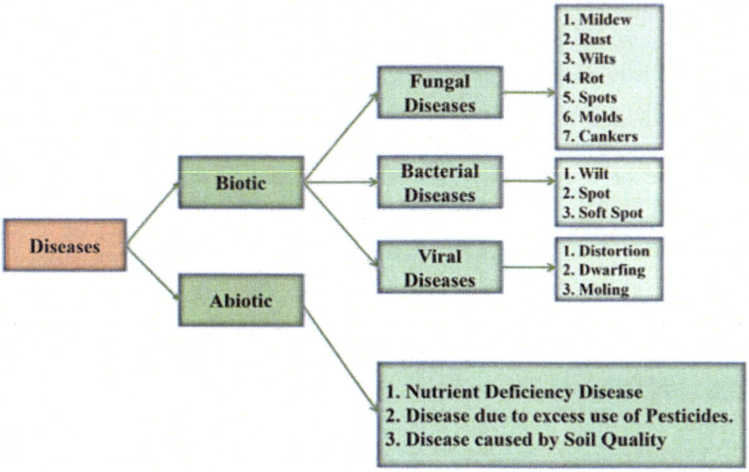

Fig. 2 Classification of plants diseases

human efforts, and providing more accurate results. This plant disease identification is highly important for farmers because they know less about plant diseases.

The damage amount and diseases reasoned from pathogens have widely increased in recent years because of pathogen variation, cultivation changes, and inefficient plant protection. These diseases are severe and greatly impact people's lives and crop protection.

Generally, plant diseases are classified into two main categories: biotic and abiotic, as shown in Fig. 2. Microorganisms, such as viruses, fungi, amoeba, and bacteria in many plants cause biotic diseases. Non-living organisms, such as burning chemicals, hair, weather conditions, cause abiotic diseases. While abiotic diseases are non-infectious, dangerous, and preventable. Spots result from rust, bacteria, fungi, and mildew. Fungal diseases include mildew, rust, molds, rots, spots, etc. While the most common viral diseases are distortion, mottling, and dwarfing.

Identifying plant diseases in their primary stage is very important for managing pesticides in crops, which in turn reduces the effect of economic and agrochemical losses. The pest control decision will depend on two main factors; the infection level and plant growth stage that is obtained by sample checking.

With the advancements, machine learning and deep learning-based artificial intelligence have great attention in computer vision algorithms that are used in plants diseases identification [6, 7]. There are many surveys as [8, 9], that discuss plants diseases detection and identification, but these surveys don't clearly highlight comparative advantages and disadvantages. Therefore, this paper intends to review the most prominent approaches for plants diseases detection and identification.

The main contributions of this survey are as follows:

- The latest artificial intelligence-based techniques proposed for automatic plant disease detection and identification.

- The latest methods used for plants diseases identification and classification are classified into machine learning and deep learning techniques.
- The datasets used in plants diseases identification are discussed.
- The performance evaluation metrics for plants diseases identification are presented.
- Existing limitations and future research directions for plants diseases identification are summarized.

The remainder of this survey is organized as follows. Section 2 introduces the methodology used for research on this topic, including the research keywords and data sources, including, and excluding criteria for selecting and article selection. Section 3 discusses the methods used in plants diseases identification and detection that are categorized into two main classes machine learning and deep learning methods. Section 4 presents the performance metrics used in the evaluation of the plant's disease identification. Section 5 highlights the challenges that researcher face in plants diseases identification. Finally, the conclusion and future directions are presented in Sect. 6.

2 Research Methodology

This section introduces the protocols used to examine the various techniques and methods for plant disease identification and detection during the interval 2002–2022. Search keywords, data sources, inclusion/exclusion criteria, and article selection criteria are presented. Many research attempts have been proposed for plant disease detection within this interval using machine learning and deep learning techniques. The frequency of these attempts is shown in Table 1.

2.1 Search Keywords

The keywords were carefully chosen for the search. Then, various new words found in related articles were used to compile a keyword choice. The basic keywords used in many studies include plant disease identification, plant disease detection, transfer learning, classification, deep learning, and machine learning.

Table 1 Frequency of research attempts for plants diseases detection in the interval 2020–2022

No.	Method type	Method frequency (%)
1	Machine learning-based detection	60
2	Deep learning-based detection	40

Table 2 Academic databases selected for research plant diseases identification

Database name	Link
Science direct	http://www.sciencedirect.com/
Web of science	https://apps.webofknowledge.com/
MDPI	https://www.mdpi.com/
IEEEXplore	https://ieeexplore.ieee.org/
Springerlink	https://link.springer.com/
PeerJ	https://peerj.com/
Scopus	https://www.scopus.com/
PubMed	https://pubmed.ncbi.nlm.nih.gov/

Table 3 Inclusion and exclusion criteria

Inclusion criteria	Exclusion criteria
Our survey only focuses on plants diseases identification articles	Articles not related to this topic are excluded
Only articles related to plants diseases detection	Any articles related to other detection methods are excluded
Only research written in English were taken into consideration	Articles not written in English were excluded

2.2 *Data Sources*

For our research collection, we searched many datasets as indicated in Table 2.

2.3 *Article Inclusion/Exclusion Criteria*

According to our research goal, only one inclusion/exclusion criterion was selected to choose the suitable research for the next review stage. We set a number of research criteria for choosing work related to this survey, denoted as inclusion criteria, and another criterion for excluding research related to our work denoted as exclusion criteria. The set of inclusion/exclusion criteria is presented in Table 3.

2.4 *Article Selection*

We applied the inclusion and exclusion criteria to select suitable articles related to our work. The articles for inclusion criteria were related to the research, and those meeting the inclusion were excluded. The procedure for article selection follows a three-phase process. Only abstracts, titles, and keywords were extracted in the first phase. Then, they were discussed in detail to refine the results from the first phase. Finally, the articles were perused, and thereafter, the article's quality was evaluated according to its research relevance.

3 Plant Disease Detection and Classification

Many methods used for plant disease detection and identification are divided into machine learning and deep learning methods. In this section, we review the recent attempts proposed on this topic. For each attempt, its working methodology, advantages, and limitations are briefly introduced. Initially, the basic steps for plant disease identification and classification are discussed, including data collection, preprocessing, feature extraction, and finally, classification. The general framework for plants diseases identification and detection is shown in Fig. 3.

For effective and accurate plants diseases identification, these steps are discussed below.

1. **Data collection**: Data collection is the first step for plant disease identification and classification. Many standard datasets were tested on this topic, such as the PlantVillage dataset, Hops dataset, Cotton disease dataset, Cassava dataset, and Rice disease dataset. A description of these datasets is shortly presented in Table 4. Some samples of plant images for both healthy and unhealthy plants are shown in Fig. 4.

2. **Preprocessing**: Image preprocessing is considered one of the basic steps in plant disease identification. There are many preprocessing steps, such as image resizing, noise removals, color transformation, morphological operations, and

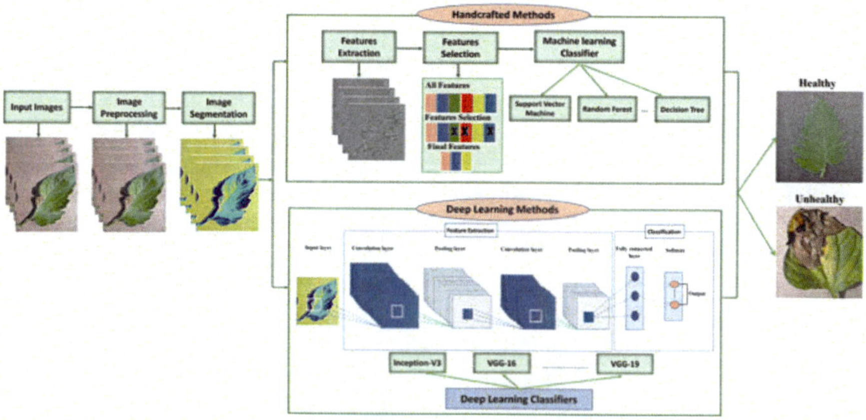

Fig. 3 General framework for plant disease detection and identification

Table 4 Plants diseases datasets

Name	Description	Link
PlantVillage [10]	38 classes of 14 different plant species of fruits and vegetable such as tomato, apple Diseases such as mold, spot	https://github.com/spMohanty/PlantVillage-Dataset/
Hops [8]	Five various classes of diseases such as downy, nutrient, powdery, and pest	https://www.kaggle.com/scruggzilla/hops-classification/
Cotton [8]	Contains diseased and healthy cotton leaves	https://www.kaggle.com/singhakash/cotton-disease-dataset/
Cassava [11]	Contains five various classes of diseases such as bacteria blight, mosaic	https://www.kaggle.com/srg9000/cassava-plant-disease-merged-20192020/
Rice [8]	Contains four various classes of diseases such as tungro	https://data.mendeley.com/datasets/fwcj7stb8r/1/

Fig. 4 Plants diseases images from PlantVillage dataset. **a** Healthy. **b** Late blight. **c** Bacterial spot. **d** Early blight. **e** Leaf mold. **f** Septoria leaf spot

disease region segmentation. There are many techniques for removing noise, such as Gaussian filter [12], median filter [13], and Wiener filter. Various color models have been utilized in image preprocessing, such as RGB, YCbCr, HSV, and CIEL $* a * b *$. There are various segmentation methods such as Sobel edge detector [14], color thresholding [15], K-means clustering [16], and Otsu's segmentation [17].

3. **Image segmentation**: Segmentation of diseases at plant leaves plays a vital role in disease identification and classification. There are many methods used for segmentation, such as K-means clustering, Otsu's segmentation, color thresholding, genetic algorithm-based, and Sobel edge detection.
4. **Feature extraction**: extracting features is considered a basic step in machine learning. It is used to describe important information in mathematical form and

for classification to differentiate the classes. The feature extraction methods are categorized into two categories: handcrafted methods and deep learning methods.

For handcrafted methods it is divided into shape features, color features, and texture features. These methods depend on the manual extraction of features from plant images. Shape features [18] include minor/major axis length, area, perimeter, eccentricity, etc., while color features depend on the different values of color used for identifying the disease region. There are many methods used for texture features, such as [19, 20]: gray-level co-occurrence matrix (GLCM), Gabor texture features, local binary pattern (LBP), and gray-level run-length method (GLRLM).

Regarding deep learning methods, the appropriate features can be found by extracting all contextual and global features. These methods have higher identification accuracy and strong robustness. In the early studies on plant disease identification, some methods depend on deep learning for feature extraction, such as convolution neural networks (CNNs). Firstly, images are input for the CNN model, and then these features are fed into a machine learning classifier such as a support vector machine (SVM).

5. **Feature Selection**: This step is applied to avoid feature redundancy. This is done by discarding and eliminating repeated and irrelevant information and selecting the most discriminant features. There are many methods for feature selection, such as correlation-based feature selection (CFS) and genetic algorithm (GA).
6. **Classification**: Classification is used to organize plant images into some categories and classes. It is categorized into supervised and unsupervised methods. Many classifiers are used in plants diseases identification, such as (SVM) [21, 22] *k*-nearest neighbor (KNN), random forest (RF), logistic regression, naive Bayes (NB), decision tree (DT), probabilistic neural network (PNN) [23], and artificial neural network (ANN) [24]. The plant's disease identification can be made using the pretrained model such as VGG-16, VGG-19, Inception-V3, and EfficientNet-B5.

These phases and their corresponding techniques for plant disease detection and identification are summarized in Fig. 5.

In the following subsection, we study the recent machine learning and deep learning studies proposed for plant disease detection and identify their working methodologies.

3.1 Machine Learning-Based Detection

Many researchers used machine learning in plants diseases detection and identification. These methods are applied with feature vectors and are trained for classifying features related to each disease. The trained algorithm is used to identify features from new plant images. The class step is responsible for matching the given image and one of the learned classes. In the following subsection, machine learning-based

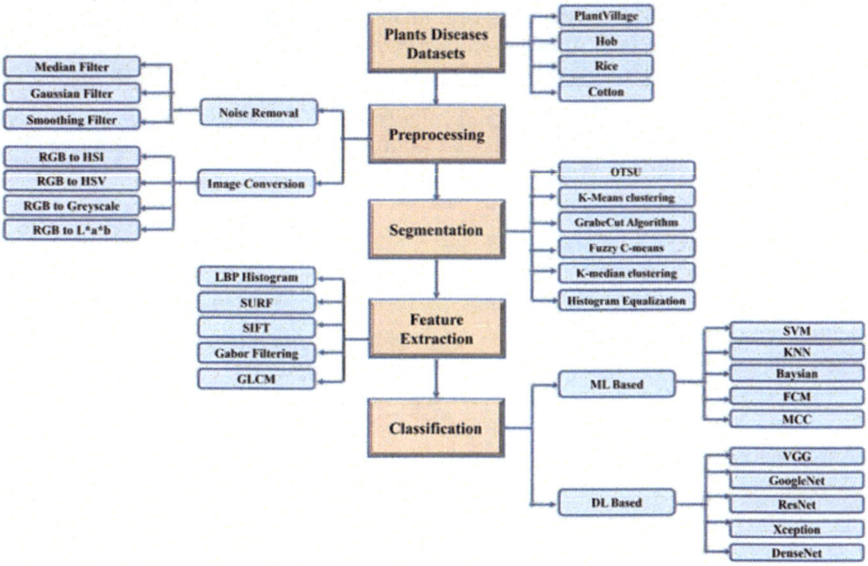

Fig. 5 A summary of main phases of plant disease detection and their corresponding techniques

detection is classified regarding the employed features into color, shape, and texture features.

Color Features-Based Plant Disease Detection

Image color is a basic and distinct feature for plant image representation that has been used in image retrieval [25]. This is due to the fact that the color is invariant with translation, scale, and rotation. The color feature extraction includes color space, similarity measurements, and color quantization. There are many color descriptors, such as color histograms.

Singh et al. [26] proposed the color slicing method for the paddy blast disease. The paddy crop disease effects on crops that are very important in many fields. Firstly, the conversion from RGB into HSI was made, and color slicing was used for diseased area extraction. This method was compared with the canny and Sobel methods and achieved 96.6%. Araujo et al. [27] presented a plant disease identification method. Firstly, a bag of visual words (BoVW) and local binary patterns (LBPs) were used for processing and feature extraction. After that SVM classifier was used for classification. This method achieved 75.8% accuracy.

Shrivastava and Pradhan [28] presented a rice plant disease image classification method that used color features only. This method explored 14 various color features only, and for every color, the channel used 172 features. It used an SVM classifier. The dataset used in this method was collected from original agriculture and belonged to four classes: Rice Blast, Bacterial Leaf Blight, Sheath Blight, and Healthy Leave. This method achieved 94.68% accuracy.

Almadhor et al. [29] presented an artificial intelligence (AI) for detecting and classifying guava plant diseases. Firstly, (RGB, HSV) color histogram, and textural (LBP) features were applied for feature extraction. KNN, Boosted Tree, Complex Tree, Bagged Tree, and SVM classifiers were combined for disease classification. This method identified four guava diseases: Mummification, Canker, Rust, and Dot. The Bagged Tree classifier obtained the best performance metric results, which achieved 99% accuracy.

Pupitasari et al. [30] presented a rice leaf disease detection and identification method using the color histogram. Firstly, the conversion from RGB to HSV for all images was applied. Secondly, the shape features that depended on this method were applied to morphology applied to calculate the image area, perimeter, and diameter of 341 images. This method achieved 85.71% accuracy.

Archana et al. [31] proposed a rice disease identification and detection method. These classifications include brown spot, rice blast, bacterial blight, and healthy. Firstly, *K*-means clustering was applied for the segmentation of plant disease. Secondly, feature extraction methods were applied, such as novel intensity-based color feature extraction (NIBCFE), bit pattern features (BPF), and gray-level co-occurrence matrix (GLCM). Finally, the classification was applied using a support vector machine-based probabilistic neural network (NSVMBPNN). The achieved accuracy for this method is 95.20% for bacterial leaf blight, 99.20% for healthy leaves, 97.60% for brown spots, and 98.40% for rice blasts.

Table 5 presents a summary of color features-based methods used in plants diseases identification.

Shape and Texture-Based Disease Plant Detection

Texture feature is the visual pattern with the homogeneity properties that do not result in only a single-color presence [32]. The texture features include uniformity, coarseness, contrast, and density. One example of texture features is the gray-level co-occurrence matrix (GLCM).

Kurmi et al. [33] discussed a plant disease detection and classification method. They localized the leaf region using the leaf images' color features and then a mixture of model-based country expansion for localization. The features were extracted using a fisher vector according to various orders of Gaussian distribution differentiations. They used (SVM) classifier. The performance of this method was evaluated using PlantVillage databases of potato, common pepper, and tomato leaf. This method achieved 94.35% accuracy in plants diseases classification.

Rao and Kulkarni [34] proposed a method for plant disease classification, and this was divided into three phases: preprocessing, feature extraction, and classification. Firstly, image conversion and enhancement techniques were used. Secondly, features a fusion of features extracted by Gabor, GLCM, and Curvelet feature extraction techniques. Thirdly, the neuro-fuzzy classifier was trained using extracted features and various testing ratios used in testing models. This paper achieved 93.18% accuracy in plants diseases classification.

Kaur and Devendran [35] presented a method for plant disease detection. This paper applied hybrid features of LBP, Law's mask, SIFT, GLCM, and Gabor from

Table 5 A summary of color features-based plant diseases detection

Author	Dataset	Plant	Segmentation	Feature extraction	Classification	Accuracy
Singh et al. [26]	100 captured images	Paddy	Thresholding	Different color values *H*, *S*, *V*, *R*, *G*, *B* values	NA	96.6%
Araujo et al. [27]	Private dataset	Soybean	*K*-means	Local binary patterns (LBP) and bag of visual words (BoVW)	SVM	75.8%
Shrivastava et al. [28]	PlantVillage	Rice	NA	Conversion from RGB to 13 different color spaces	SVM	94.6%
Almadhor et al. [29]	Private dataset	Guava	Delta *E*	(RGB, HSV) color histogram and LBP	Combined classifiers	99%
Pupitasari et al. [30]	Private dataset	Rice	Thresholding	RGB to HSV	NA	85.71%
Archana et.al. [31]	Private dataset	Rice	K-means	(NIBCFE), (BPF), and (GLCM)	NSVMBPNN	95.20% for bacterial blight 99.20% for healthy leaves 97.60% for brown spot 98.40% for rice blast

plant leaf for the feature extraction step. After that ensemble classifier is applied, which contains many classifiers such as ANN, SVM, logistic regression, KNN, and Naïve Bayes. This paper achieved an accuracy of 95.66% in potato (3 classes), 92.13% in bell pepper (2 classes), and 90.23% in tomato (10 classes).

Kurmi et al. [36] proposed a method for classifying the plant's images into two classes diseased and healthy. It applied the fusion way for the extracted information from resources and made optimization for enhancement. The mapping process for low-dimension RGB color images into $L * a * b$ color space provides spectral range expansion. This paper used a random sample consensus (RANSAC) for suitable curve fitting. It extracted a bag of visual words, handcrafted features, and Fisher vectors, and after that, logistic regression, support vector machine, and multilayer perceptron model were applied for classification.

This paper achieved 93.2% accuracy for plant disease identification.

Table 6 presents a summary of shape and texture features-based methods used in plants diseases identification.

Table 6 Summary of shape and texture features-based plant diseases detection

Author	Dataset	Plant	Segmentation	Feature extraction	Classification	Accuracy
Kurmi et al. [33]	Plantvillage	Different types	Thresholding	Fisher vectors (FV)	SVM	94.35%
Rao and Kulkarni [34]	Plantvillage	Different types	NA	Features fusion (Gabor, GLCM, and Curvelet)	Neuro-fuzzy	93.18%
Kaur and Devendran [35]	Plantvillage	Thresholding	Bell pepper, potato, and tomato plant	(LBP, Law's mask, SIFT, GLCM, and Gabor)	Ensemble	95.66% in Potato 92.13% in Bell Pepper 90.23% in Tomato
Kurmi et al. [36]	Plantvillage	Different types	K-means	Bag of visual words and Fisher vectors	SVM, logistic regression, and multilayer perceptron	93.2%

3.2 Deep Learning-Based Detection

Deep learning is done by using a neural network for feature learning. Features have been extracted through many hidden layers. Each of these layers can be seen as a perceptron that is used for low-level feature extraction, and after that, these low-level features can be mixed to obtain high-level features. These methods overcome the traditional disadvantages of only extracting specific feature types. As explained in the following subsections, deep learning-based detection methods are divided into two main classes: training from scratch and transfer learning-based deep learning.

Training from Scratch-Based Detection

This method is applied by adding new layers and then training this model. This type of learning doesn't build on previous knowledge. The model is created by, firstly, data collection and weight initialization; secondly, forward propagation and backpropagation computation; thirdly, weights and bias editing.

Milioto et al. [37] proposed a CNN model for plants diseases identification. This model used multispectral data. After that, it was tested on sugar beet images. Finally, it achieved high results by collecting three convolution layers with two fully connected layers.

Lu et al. [38] presented a method for rice disease identification that depended on deep convolutional neural networks (CNNs) techniques. It applied this model to 500 images from a rice experimental field. This model was trained on 10 rice diseases. It applied tenfold cross-validation. This method achieved 95.48%.

Chen et al. [39] presented a method that depended on CNNs for tea disease identification. In this method, a CNNs model called LeafNet was created with various feature extractor filters that were used for feature extraction from tea plants images. This method achieved an average accuracy equals to 90.16%.

Nkemelu et al. [40] presented a method for plants diseases classification and identification. It applied this model to a dataset that included 4275 plant images that were divided into 12 species. This method improved the efficiency and productivity of plants. It achieved an average accuracy equals to 93%.

Table 7 presents a summary of training from scratch-based deep learning methods used in plants diseases identification.

Table 7 Summary of training from scratch-based DL plant diseases detection

Author	Dataset	Plant	Model	Accuracy (%)
Milioto et al. [37]	Private dataset	Sugar beet	CNN	89.2
Lu et al. [38]	Private dataset	Rice	Deep CNN	95.48
Chen et al. [39]	Private dataset	Tea	LeafNet algorithm	90.16
Nkemelu et al. [40]	Private dataset	Various plants	Deep CNN	93

Transfer Learning-Based Detection

Transfer learning that is called Domain Adaptation, where the model is trained with a dataset. Then, the same model is trained with any other dataset that has many various classes distribution or even with multiple classes in the first using dataset. This model builds on previous parameters and knowledge learned from data.

Ramcharan et al. [41] discussed a method for the identification of cassava plant diseases. It trained a convolution neural network (CNN) model for cassava plant diseases. It tested this model on 720 leaflets in the agriculture field. This method evaluated the mobile CNN performance in realistic plant disease images using multimetrics. This method achieved an average accuracy equals to 80.6%.

Chen et al. [42] proposed the architecture of deep learning called DENS-INCEP for the classification of rice diseases. For transfer learning, the trained models on ImageNet were combined, such as DenseNet and the Inception model. The top layers were truncated using defining the latest fully connected SoftMax layer with the classification's practical number. Moreover, the focal loss function was used instead of the original cross-entropy loss function. The DENS-INCEP enhanced the feature extraction step and decreased the complexity time. This method achieved an average accuracy equals to 98.63%.

Hussain1 et al. [43] presented a cucumber plant leaf disease identification method. It depended on deep learning and made fusion and selecting the best features. It used visual geometry group (VGG) and Inception V3 deep learning models for feature extraction. Feature extracted were fused using the parallel maximum method. After that, the best features were classified through the whale optimization algorithm. The supervised learning algorithm was used for classification. This method was tested on a private dataset and achieved an average accuracy equals to 96.5%.

Lee et al. [44] compared and examined the multiple transfer learning models' performance of depended on various tasks. GoogleNet, VGG16 had 16 layers, GoogLeNetBN with 34 layers, and InceptionV3 with 48 layers. This tested on multiple plants in the Plantvillage dataset and achieved 99.09% (GoogLeNetBN), 99.00% (VGG16), 99.31% (Inception V3), and 99.35% (GoogLeNet).

Atila et al. [45] used a trained model to identify and classify plant diseases. The deep learning model EfficientNet was presented for multiple plant diseases. The PlantVillage was used for training models. 55,448 and 61,486 plant images were tested using this model. This model achieved 99% accuracy and made accurate results.

Nandhini and Ashokkumar [46] presented mutation-based Henry gas solubility optimization (MHGSO) method for hyperparameters optimization of the DenseNet-121 architecture. It was used for computational complexity and CNN error rate reduction. The MHGSO was used to achieve higher accuracy in different plant diseases. It tested its model in the PlantVillage dataset and achieved 98.7% for accuracy, 98.60% for precision, and 98.75% for recall.

Table 8 presents a summary of transfer learning-based detection methods used in plants diseases identification.

Table 8 A summary of transfer learning-based detection of plant diseases detection

Author	Dataset	Plant	Model	Accuracy
Ramcharan et al. [41]	Private dataset	Cassava	MobileNet	80.6%
Chen et al. [42]	Plantvillage	Rice plant leaf	DENS-INCEP	98.63%
Hussain1 et al. [43]	Privately collected dataset	Cucumber	VGG and Inception V3	96.5%
Lee et al. [44]	Plantvillage	Multiple	VGG16, InceptionV3, GoogLeNetBN with Transfer learning and training from scratch	99.09 (GoogLeNetBN) 99.00% (VGG16) 99.31% (Inception V3) 99.35% (GoogLeNet)
Atila et al. [45]	Plantvillage	Multiple	EfficientNet	99%
Nandhini and Ashokkumar [46]	Plantvillage	Various plants	DenseNet-121 architecture	98.7%

4 Performance Evaluation Metrics

Many performance metrics are used for the evaluation of the architecture. This section introduces the mathematical formulations used to compute these evaluation metrics. A healthy plant hasn't had many diseases, and an unhealthy has any plant disease.

TP is the true positive number that represents the correct identification of a healthy plant, FP is the false positive number that represents the false identification of a healthy plant. TN is the true negative number that represents the correct identification of unhealthy plants, and FN is the false negative number that represents the false identification of the unhealthy plant.

Some metrics are utilized for this evaluation, such as accuracy, precision, recall, F1-score, mean square error (MSE), peak-signal-to-noise ratio (PSNR), structure similarity index measure (SSIM), dice score (F1-Score), AuC-RoC, and IoU (Jaccard Index).

- **SSIM** is used for images quality comparison. The larger the SSIM value, the better classification and less error. The SSIM computed by Eq. 1.

$$\text{SSIM}(F, E) = \frac{(2\mu_x\mu_y + C_1) + (2\sigma_{xy} + C_2)}{(\mu_x^2 + \mu_y^2 + C)(\sigma_x^2 + \sigma_y^2 + C_2)} \tag{1}$$

where μ_x, μ_y are the values of means and the ${\sigma_x}^2$, ${\sigma_y}^2$ are the values of standard deviation of x and y patches of pixels. σ_{xy} is the covariance value of x and y

patches of pixels, and $C1 = (k_1L)2$ and $C2 = (k_2L)2$ are the small constant values to prevent the instability. L is the dynamic range value of pixels, $K_1 = 0.01$ and $K_2 = 0.03$.

- **Mean Square Error (MSE)**: computes the squared error between the high- and low-resolution images [47]. The lower the MSE, the higher the quality. The MSE is computed mathematically using Eq. 2.

$$\text{MSE} = \frac{1}{MN}\sum_{i=1}^{M}\sum_{j=1}^{N}[F(i,j) - E(i,j)]^2 \tag{2}$$

where the $M \times N$ is the image size, $F(i,j)$ is the original image, and $E(i,j)$ is the enhanced image.

- **Peak-Signal-to-Noise Ratio (PSNR)**: computes the quality of images [47]. The greater the PSNR, the higher the image quality. It is computed from the MSE using Eq. 3.

$$\text{PSNR} = 20\log_{10}\frac{\text{MAX}_\text{F}}{\sqrt{\text{MSE}}} \tag{3}$$

where MAX_F is the maximum pixel value in an image and is 255 in case of gray-level image.

- **Accuracy** is computed for performance model measure [48, 49]. It is computed using Eq. 4.

$$\text{Accuracy} = \frac{\text{TP} + \text{TN}}{\text{TP} + \text{FN} + \text{FP} + \text{TN}} \tag{4}$$

- **Precision** is computed by the TP ratio to total positives predicted by system [50]. It is calculated by Eq. 5.

$$\text{Precision} = \frac{\text{TP}}{\text{TP} + \text{FP}} \tag{5}$$

- **Sensitivity (Recall)** indicates the ratio of TP to total positives [51]. It is computed by Eq. 6.

$$\text{Sensitivity} = \frac{\text{TP}}{\text{TP} + \text{FN}} \tag{6}$$

- **Specificity** is computed by the TN ratio to total negatives predicted by system [52]. It is calculated by Eq. 7.

$$\text{Specificity} = \frac{\text{TN}}{\text{TN} + \text{FP}} \tag{7}$$

- **Dice Score (F1-Score)** is computed to measure quality of system [53]. It is computed by Eq. 8.

$$\text{F1 - Score} = \frac{2 * \text{TP}}{2 * \text{TP} + \text{FN} + \text{FP}} \tag{8}$$

- **AuC-RoC** is calculated to measure the model performance for a dataset [54]. It is computed by Eq. 9.

$$\text{AuC} = \int_{1}^{\infty} \text{TPR} * d(\text{FPR}) \tag{9}$$

- **IoU (Jaccard Index)** is used to determine the overlap between the predicted output and target output [53]. It is computed by Eq. 10.

$$\text{IoU} = \frac{\text{TP}}{\text{TP} + \text{FN} + \text{FP}} \tag{10}$$

5 Existing Challenges

Plant disease identification in leaves faces many challenges. Challenges resolving is very important for plants diseases detection systems. Below, some existing challenges are discussed in detail.

- **Insufficient variety and size of dataset**: There is a need for a large dataset for better identification of plant diseases. Deep learning needs large datasets with image variety [55]. Collecting the images of plant diseases from the field is very expensive and demands agricultural expertise for precise plants diseases identification.
- **Image segmentation**: Leaves image segmentation from complex backgrounds is challenging, and problem issues for plants diseases identification [56]. The leaf region segmentation can increase performance accuracy. Plants images with multi-illegitimate parts cause difficulties and problems in disease identification.
- **Similar symptoms diseases identification**: Some plant diseases have similar symptoms that even experts fail to identify by eye as one symptom may vary because of crop development, weather condition, and geographic locations [56].
- **Multiple plants diseases**: Most models assume that there is one type of plant disease in the image. In fact, there are many diseases that occur simultaneously [6]. Therefore, we should keep in consideration that various plant diseases and some nutritious may happen simultaneously.

- **Plants leave images problems**: There are many problems in plant images [57] such as illumination, noise, and low contrast. When plant images are taken in real-time conditions with crowded backgrounds, some background features are like to area of interest, so this effect on identification system.

6 Conclusion

Plant diseases have a remarkable impact on agriculture and economics worldwide. Therefore, a comprehensive review of existing research attempts on plant disease detection and classification using AI-based techniques is required. This paper aims to survey the recent research presented for identifying plant diseases using machine learning and deep learning techniques. Although enormous research has been introduced, some open challenges need to be addressed in the future, as summarized at the end of this survey.

In the future, it is recommended to solve the problems faced by this plant's disease identification system. For the insufficient variety dataset, it is recommended to apply a data augmentation technique that makes a variety of plants images and data sharing. Also, the segmentation problems should be solved by applying other techniques for accuracy enhancement. For similar symptoms of disease identification, it is recommended to collect more plant images to increase the accuracy of the system used for plants diseases identification. For multiple plant diseases, the multiclassifier should be used to identify more than one disease in the image. Finally, for problems with plant leaves images, it is recommended to use efficient algorithms for image quality enhancement for higher classification accuracy.

References

1. Chouhan SS, Kaul A, Singh UP, Jain S (2018) Bacterial foraging optimization based radial basis function neural network (BRBFNN) for identification and classification of plant leaf diseases: an automatic approach towards plant pathology. IEEE Access 6:8852–8863
2. Bharate AA, Shirdhonkar M (2017) A review on plant disease detection using image processing. In: 2017 International conference on intelligent sustainable systems (ICISS). IEEE, pp 103–109
3. Ferentinos KP (2018) Deep learning models for plant disease detection and diagnosis. Comput Electron Agric 145:311–318
4. Bock C, Poole G, Parker P, Gottwald T (2010) Plant disease severity estimated visually, by digital photography and image analysis, and by hyperspectral imaging. Crit Rev Plant Sci 29(2):59–107
5. Das R, Pooja V, Kanchana V (2017) Detection of diseases on visible part of plant—a review. In: 2017 IEEE technological innovations in ICT for agriculture and rural development (TIAR), pp 42–45
6. Bhagat M, Kumar D (2022) A comprehensive survey on leaf disease identification and classification. Multimedia Tools Appl 1–29
7. Lee SH, Chan CS, Wilkin P, Remagnino P (2015) Deep-plant: plant identification with convolutional neural networks. In: 2015 IEEE international conference on image processing (ICIP). IEEE, pp 452–456

8. Hassan SM, Amitab K, Jasinski M, Leonowicz Z, Jasinska E, Novak T, Maji AK (2022) A survey on different plant diseases detection using machine learning techniques. Electronics 11(17):2641
9. Wani JA, Sharma S, Muzamil M, Ahmed S, Sharma S, Singh S (2022) Machine learning and deep learning based computational techniques in automatic agricultural diseases detection: methodologies, applications, and challenges. Arch Comput Methods Eng 29(1):641–677
10. Hughes D, Salathé M et al (2015) An open access repository of images on plant health to enable the development of mobile disease diagnostics. arXiv preprint arXiv:1511.08060
11. Ramcharan A, Baranowski K, McCloskey P, Ahmed B, Legg J, Hughes DP (2017) Deep learning for image-based cassava disease detection. Front Plant Sci 8:1852
12. Camargo A, Smith J (2009) Image pattern classification for the identification of disease causing agents in plants. Comput Electron Agric 66(2):121–125
13. Hlaing CS, Zaw SMM (2017) Model-based statistical features for mobile phone image of tomato plant disease classification. In: 2017 18th international conference on parallel and distributed computing, applications and technologies (PDCAT). IEEE, pp 223–229
14. Anthonys G, Wickramarachchi N (2009) An image recognition system for crop disease identification of paddy fields in Sri Lanka. In: 2009 International conference on industrial and information systems (ICIIS). IEEE, pp 403–407
15. Islam M, Dinh A, Wahid K, Bhowmik P (2017) Detection of potato diseases using image segmentation and multiclass support vector machine. In: 2017 IEEE 30th Canadian conference on electrical and computer engineering (CCECE). IEEE, pp 1–4
16. Chuanlei Z, Shanwen Z, Jucheng Y, Yancui S, Jia C (2017) Apple leaf disease identification using genetic algorithm and correlation based feature selection method. Int J Agric Biol Eng 10(2):74–83
17. Al Bashish D, Braik M, Bani-Ahmad S (2010) A framework for detection and classification of plant leaf and stem diseases. In: 2010 International conference on signal and image processing. IEEE, pp 113–118
18. Yao Q, Guan Z, Zhou Y, Tang J, Hu Y, Yang B (2009) Application of support vector machine for detecting rice diseases using shape and color texture features. In: 2009 International conference on engineering computation. IEEE, pp 79–83
19. Elazab N, Soliman H, El-Sappagh S, Islam SR, Elmogy M (2020) Objective diagnosis for histopathological images based on machine learning techniques: classical approaches and new trends. Mathematics 8(11):1863
20. Nader N, El-Gamal FEZ, El-Sappagh S, Kwak KS, Elmogy M (2021) Kinship verification and recognition based on handcrafted and deep learning feature-based techniques. PeerJ Comput Sci 7:e735
21. Helmy M, Eldaydamony E, Mekky N, Elmogy M, Soliman H (2022) Predicting parkinson disease related genes based on pyfeat and gradient boosted decision tree. Sci Rep 12(1):1–26
22. Padol PB, Yadav AA (2016) SVM classifier based grape leaf disease detection. In: 2016 Conference on advances in signal processing (CASP). IEEE, pp 175–179
23. Prasad S, Peddoju SK, Ghosh D (2016) Multi-resolution mobile vision system for plant leaf disease diagnosis. SIViP 10(2):379–388
24. Pujari JD, Yakkundimath R, Byadgi AS (2015) Image processing based detection of fungal diseases in plants. Procedia Comput Sci 46:1802–1808
25. Gevers T, Van De Weijer J, Stokman H (2006) Color feature detection
26. Singh A, Singh ML (2018) Automated blast disease detection from paddy plant leaf—a color slicing approach. In: 2018 7th International conference on industrial technology and management (ICITM). IEEE, pp 339–344
27. Araujo JMM, Peixoto ZMA (2019) A new proposal for automatic identification of multiple soybean diseases. Comput Electron Agric 167:105060
28. Shrivastava VK, Pradhan MK (2021) Rice plant disease classification using color features: a machine learning paradigm. J Plant Pathol 103(1):17–26
29. Almadhor A, Rauf HT, Lali MIU, Damaševičius R, Alouffi B, Alharbi A (2021) AI-driven framework for recognition of guava plant diseases through machine learning from DSLR camera sensor based high resolution imagery. Sensors 21(11):3830

30. Pupitasari TD, Basori A, Riskiawan HY, Setyohadia DPSS, Kurniasari AA, Firgiyanto R, Mansur ABF, Yunianta A (2022) Intelligent detection of rice leaf diseases based on histogram color and closing morphological. Emirates J Food Agric
31. Archana K, Srinivasan S, Bharathi SP, Balamurugan R, Prabakar T, Britto A (2022) A novel method to improve computational and classification performance of rice plant disease identification. J Supercomput 78(6):8925–8945
32. Shahbahrami A, Borodin D, Juurlink B (2008) Comparison between color and texture features for image retrieval. In: Proceedings of the 19th Annual workshop on circuits, systems and signal processing. Citeseer
33. Kurmi Y, Gangwar S, Agrawal D, Kumar S, Srivastava HS (2021) Leaf image analysis-based crop diseases classification. SIViP 15(3):589–597
34. Rao A, Kulkarni S (2020) A hybrid approach for plant leaf disease detection and classification using digital image processing methods. Int J Electr Eng Educ 0020720920953126
35. Kaur N et al (2021) Plant leaf disease detection using ensemble classification and feature extraction. Turk J Comput Math Educ (TUR-COMAT) 12(11):2339–2352
36. Kurmi Y, Gangwar S (2022) A leaf image localization based algorithm for different crops disease classification. Inf Process Agric 9(3):456–474
37. Milioto A, Lottes P, Stachniss C (2017) Real-time blob-wise sugar beets vs weeds classification for monitoring fields using convolutional neural networks. ISPRS Ann Photogramm Remote Sens Spat Inf Sci 4
38. Lu Y, Yi S, Zeng N, Liu Y, Zhang Y (2017) Identification of rice diseases using deep convolutional neural networks. Neurocomputing 267:378–384
39. Chen J, Liu Q, Gao L (2019) Visual tea leaf disease recognition using a convolutional neural network model. Symmetry 11(3):343
40. Nkemelu DK, Omeiza D, Lubalo N (2018) Deep convolutional neural network for plant seedlings classification. arXiv preprint arXiv:1811.08404
41. Ramcharan A, McCloskey P, Baranowski K, Mbilinyi N, Mrisho L, Ndalahwa M, Legg J, Hughes DP (2019) A mobile-based deep learning model for cassava disease diagnosis. Front Plant Sci 272
42. Chen J, Zhang D, Nanehkaran YA, Li D (2020) Detection of rice plant diseases based on deep transfer learning. J Sci Food Agric 100(7):3246–3256
43. Hussain N, Khan MA, Tariq U, Kadry S, Yar MAE, Mostafa AM, Alnuaim AA, Ahmad S (2022) Multiclass cucumber leaf diseases recognition using best feature selection. Comput Mater Continua 70:3281–3294
44. Lee SH, Goëau H, Bonnet P, Joly A (2020) New perspectives on plant disease characterization based on deep learning. Comput Electron Agric 170:105220
45. Atila Ü, Uçar M, Akyol K, Uçar E (2021) Plant leaf disease classification using efficientnet deep learning model. Eco Inform 61:101182
46. Nandhini S, Ashokkumar K (2022) An automatic plant leaf disease identification using densenet-121 architecture with a mutation-based henry gas solubility optimization algorithm. Neural Comput Appl 34(7):5513–5534
47. Hitam MS, Awalludin EA, Yussof WNJHW, Bachok Z (2013) Mixture contrast limited adaptive histogram equalization for underwater image enhancement. In: 2013 International conference on computer applications technology (ICCAT). IEEE, pp 1–5
48. Arjunagi S, Patil N (2019) Texture based leaf disease classification using machine learning techniques. Int J Eng Adv Technol (IJEAT) 9(1):2249–8958
49. Sambasivam G, Opiyo GD (2021) A predictive machine learning application in agriculture: cassava disease detection and classification with imbalanced dataset using convolutional neural networks. Egypt Inform J 22(1):27–34
50. Bonidia RP, Sampaio LDH, Lopes FM, Sanches DS (2019) Feature extraction of long non-coding RNAs: a Fourier and numerical mapping approach. In: Iberoamerican congress on pattern recognition. Springer, pp 469–479
51. Wang B, Zhang C, Du XX, Zhang JF (2021) lncRNA-disease association prediction based on latent factor model and projection. Sci Rep 11(1):1–10

52. Shen W, Le S, Li Y, Hu F (2016) SeqKit: a cross-platform and ultrafast toolkit for FASTA/Q file manipulation. PLoS ONE 11(10):e0163962
53. Chowdhury ME, Rahman T, Khandakar A, Ayari MA, Khan AU, Khan MS, Al-Emadi N, Reaz MBI, Islam MT, Ali SHM (2021) Automatic and reliable leaf disease detection using deep learning techniques. AgriEngineering 3(2):294–312
54. Zhu W, Zeng N, Wang N (2010) Sensitivity, specificity, accuracy, associated confidence interval and Roc analysis with practical SAS implementations. Northeast SAS User Group proceedings, Section of Health Care and Life Sciences, pp 1–9
55. Barbedo JGA (2018) Impact of dataset size and variety on the effectiveness of deep learning and transfer learning for plant disease classification. Comput Electron Agric 153:46–53
56. Li L, Zhang S, Wang B (2021) Plant disease detection and classification by deep learning—a review. IEEE Access 9:56683–56698
57. Barbedo JGA (2016) A review on the main challenges in automatic plant disease identification based on visible range images. Biosyst Eng 144:52–60

A Review for Software Defect Prediction Using Machine Learning Algorithms

Enjy Khaled Ali, M. M. Eissa, and A. Fatma Omara

Abstract One of the most important and expensive stages of the software development lifecycle is software defect prediction. It is considered to be a key component of the software engineering concept and plays a significant role in improving the quality of software systems. In the early stage of the software development process, software defect prediction enhances critical elements including software quality, reliability, and efficiency, and at the same time, decreases development costs. Therefore, finding defects early as possible is becoming more important than having them in the first place. As a result, the management of the software is able to allocate resources for the maintenance and testing stages in an effective manner through the detection and also prediction of software defects. In this paper, most research approaches concerning defect prediction problems are discussed. According to this study, it is observed that the genetic algorithm with the deep neural network, random forest, and artificial neural networks' techniques improve the accuracy of software defect prediction.

Keywords Software defect prediction · Machine learning · Software quality

1 Introduction

Now a day, software has begun more complicated, and purchasing high-quality software is crucial [1]. Therefore, software testing is one of the most critical stages in the software development life cycle. On the other hand, the software bug could result in catastrophes like program crashes or unforeseen outcomes. Most software programs

E. K. Ali (✉) · M. M. Eissa
Software Engineering Department, Ahram Canadian University Giza, Giza Governorate, Egypt
e-mail: Enjy.khaled@acu.edu.eg

M. M. Eissa
e-mail: Mohammed.mamdouh@acu.edu.eg

A. F. Omara
Faculty of Computers and Information, Cairo University Giza, Giza, Egypt
e-mail: f.omara@fci-cu.edu.eg

D. Magdi et al. (eds.), *Green Sustainability: Towards Innovative Digital Transformation*, Lecture Notes in Networks and Systems 753,
https://doi.org/10.1007/978-981-99-4764-5_14

have numerous bugs rated as Critical, High, Medium, and Low bugs which may impact minor or serious issues [2]. Because software testing is the most expensive task for most software development organizations, generating free of bugs software is one of the most time-consuming and difficult activities [3].

As a result, software defect prediction is essential for increasing software reliability. According to the Consortium for Information and Software Quality report [3], it is found that finding and fixing software bugs would be spent ~ $607 billion, highlighting the importance of fixing the bugs before the software releases.

Software defect prediction (SDP) helps software developers to effective resource allocation for testing. SDP aims to locate and validate the most important software modules and can be used for every stage in the software development life cycle including problem identification, planning, designing, building, testing, deploying, and maintaining [4].

A software defect prediction model is created with software metrics and defect data collected from pre-existing systems or related software projects. Additionally, since a software quality prediction model is created using the data from well-known software metrics, selecting a specific collection of metrics becomes an important phase in the model-building process [5]. Currently, the machine learning approach is being utilized extensively in the field of predicting software defects. To determine whether instances introduced in the future will be defective or not, prediction models are trained using previous data that are now available. The research works considering SDP can be divided into various categories, including class imbalance learning, feature selection, and modeling of prediction models. Numerous prediction techniques have been developed over the years based on different learning models, such as Naïve Bayes, decision tree, and support vector machines (SVMs) [6].

Previous studies demonstrated that the classifier and feature representation significantly have emphasized how effectively a prediction model performs. So, any software project will have a strong feature representation that distinguishes it from the others and an effective classifier. Despite the vast variety of used machine learning methods to build the software prediction model, each method has limitations and different levels for predicting defects.

For instance, the support vector machine will perform poorly, for instance, when the number of attributes per data point exceeds the number of training data points. Convolutional neural networks (CNNs) have received a lot of attention in recent years due to their ability to extract semantic features with far stronger discriminative capabilities than standard machine learning techniques, such as semantic Search speech recognition [7].

2 Background

2.1 Software Defect Life Cycle

Defect life cycle, as used in the software development process, refers to the series of states that a defect or bug passes through the course of its whole life. Typically, the term "bug life cycle" refers to a bug's complete state, from when a new defect is identified to when a tester closes the bug. It can also be referred to as a bug life cycle [8].

It is the testing team's responsibility to find as many errors in the software as they can. Either the development team will fix the errors, or the testing team will fix them on their own. Before being published, any software must undergo testing. It is important to keep in mind that not all bugs are affected equally because even the most extensively tested software can have bugs. While some bugs will only affect a small proportion of users, others may affect a large number of users negatively [9].

The bug life cycle is useful for determining when and how to fix bugs. In order to track and evaluate the actual progress of the defect life cycle accurately, the defect status goal is to represent the current condition or progress of a defect or bug. As shown in Fig. 1, all possible states.

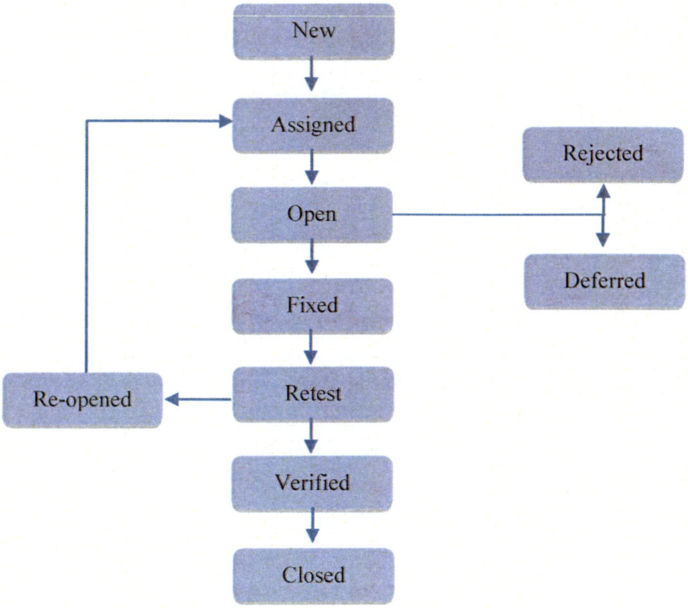

Fig. 1 Software defect/bug life cycle

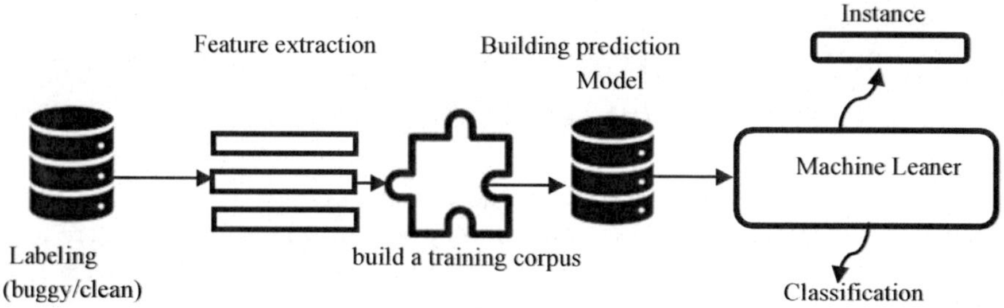

Fig. 2 Software defect prediction process

2.2 *Software Defect Prediction*

Every software testing decision is related to humans, resources, and cost to the developer team. For effective resource management, it is typical to match the software testing effort to the code's perceived criticality and bugs. Figure 2 represents the flow of the process of defect prediction. The first step is to extract modules/files from the software's historical repositories and label them. A file is classified as buggy if it has at least one post-release error. If not, it is classified as clean. By analyzing software code or the development process, the second step is to extract the features that are associated with software defects. Also, features are derived from the software metrics. To train prediction models using machine learning classifiers, the instance features and labels are used. In the final step, new instances are classified as clean or buggy by trained models [10–12].

2.3 *Software Metrics*

Software testing metrics are quantitative measures of the process's development, effectiveness, productivity, and overall health. By giving accurate information about the testing process, software testing metrics aim to improve the process efficiency and effectiveness while also assisting for the decision-making process for the next tests, which play an essential role in identifying software component defects. Design coupling-based metrics for defect prediction in object-oriented (OO) software include coupling between objects (CBO), response for a class (RFC), message passing coupling (MPC), and information flow-based coupling (ICP) [13].

2.4 Machine Learning Algorithms

Machine learning is a programming approach to optimize a performance criterion using example data or past experience. We have a model that has been developed up to a certain point, and learning is the application of a computer program to optimize the model's parameters using training data or prior knowledge. The model may be descriptive to learn from the data or predictive to make future predictions.

Machine learning algorithms can be trained in a variety of ways, each with its own benefits and drawbacks, as with any method. First, we must examine the types of data that each type of machine learning consumes in order to comprehend the benefits and drawbacks of each type. Labeled data and unlabeled data are the two types of data used in machine learning.

Although labeled data have both input and output parameters in a completely machine-readable pattern, labeling the data initially takes a significant amount of human effort. Only one or none of the parameters are present in machine-readable form in unlabeled data. This eliminates the need for human work, but the need for more difficult fixes. Machine learning has two types; supervised and supervised learning (Fig. 3).

Unsupervised learning, which provides the algorithm with no labeled data in order to allow it to discover structure within its input data, and supervised learning are two of the most frequently used machine learning types. Supervised learning involves training algorithms on example input and output data that have been labeled by humans [14–16].

3 Supervised Learning

In supervised learning, the machine is trained using data that are well "labeled". A data scientist would frequently label these training data at the preprocessing stage before it is utilized to train and test the model. Once the model has learned

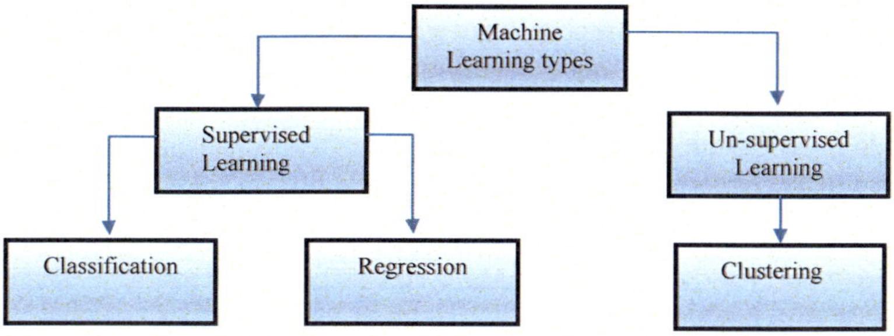

Fig. 3 Machine learning types

how the input and output data are related, it may be used to classify previously unexplored datasets and predict outcomes. The ML algorithm's predicted value is compared to the output, which is already known, and will be modified to make it accurate with respect to the prediction output. To create predictive models, supervised learning employs classification and regression methods. Typical classification algorithms include support vector machine (SVM), decision tree, random forest, k-nearest neighbor, Logistic Regression, and neural networks [15].

3.1 Support Vector Machine (SVM)

(SVM) is a supervised machine learning algorithm that can be used for both classification and regression challenges. However, classification issues are the most frequently utilized. The objective of the SVM algorithm is to find a hyperplane in an N-dimensional space that distinctly classifies the data points. The boundary line must be closest to the border lines of the two classes in order for the choice to be resistant to the new data. Support points are the locations that are most near this boundary line [17].

3.2 Decision Tree (DT)

The decision tree algorithm belongs to the supervised learning subset. They can be applied to both classification and regression issues. Each leaf node of the decision tree corresponds to a class label, and the interior nodes of the tree are used to represent the attributes in order to answer the problem. A decision tree basically enquires and divides the tree into sub-trees according to the response (Yes/No) [12].

3.3 Random Forest

The basis of random forest is the idea of ensemble learning, which is the process of mixing various classifiers to solve a challenging problem and enhance the model's performance. The RF algorithm aims to enhance the classification value by producing several DTs during the classification process. Higher accuracy and over fitting are prevented by the larger number of trees in the forest [18].

4 Unsupervised Learning

Unsupervised learning is a machine learning technique in which models are created without the use of a training dataset. Instead, models discover hidden patterns and insights in the given data. It is comparable to learning in the human brain when learning new things. Discovering features can be helpful for categorization with the use of unsupervised approaches. All of the input data are examined and categorized in front of the students because it is done in real-time. Typical classification algorithms include k-means clustering, principal component analysis, k-nearest neighbors, and neural networks [19].

4.1 K-Means Clustering

In machine learning or data science, clustering issues are resolved using the unsupervised learning algorithm k-means clustering. It combines the unlabeled dataset into various clusters. Here, *K* specifies how many pre-defined clusters must be produced as part of the process. For example, if K = 2, there will be two clusters, if K = 3, will be three clusters, and so on. It gives us the ability to divide the data into various groups and provides a practical method for automatically identifying the groups in the unlabeled dataset without the need for any training [20].

4.2 Principal Component Analysis (PCA)

Machine learning dimensionality reduction uses the unsupervised learning technique of principal component analysis. The use of orthogonal transformation is a statistical process that transforms the observations of correlated features into a set of linearly uncorrelated data. PCA functions by taking into account each attribute's variance since a high attribute demonstrates a strong split between classes, which lowers the dimensionality. Image processing, movie recommendation systems, and power allocation optimization in multiple communication channels are some examples of PCA's practical uses [21].

4.3 Deep Learning

As artificial neural networks will mimic the human brain, deep learning is also a type of mimic of the human brain. Machine learning and deep learning are both types of AI. In short, machine learning is AI that can automatically adapt with minimal human interference. Deep learning is a subset of machine learning that uses artificial neural

Fig. 4 Machine learning versus deep learning

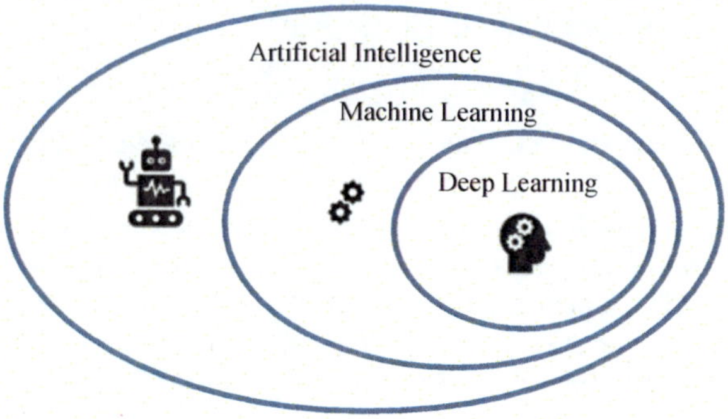

networks to mimic the learning process of the human brain as shown in Fig. 4. Deep learning is a subfield of machine learning that is entirely based on neural networks. It is not necessary to explicitly program everything in deep learning [22].

With the simple perceptron, deep learning begins to work its magic. The perceptron takes in a list of input signals and converts them into output signals, much how a "neuron" in the human brain transmits electrical pulses throughout our nervous system. Typical deep learning algorithms are convolutional neural networks (CNNs), long short-term memory networks (LSTMs), recurrent neural networks (RNNs), and deep belief networks (DBNs).

5 Related Work

In recent years, many papers have discussed the problem of software defect prediction and how to combine machine learning algorithms to achieve the highest accuracy. An extensive survey was done by Pachouly et al. which [18] presents a comprehensive analysis of defect datasets, dataset validation, detection, prediction approaches, and tools for Software Defect Prediction. One hundred and forty-six publications were identified and selected for analysis to address the formulated research questions.

Matloob Faseeha et al. [12] have provided a systematic literature review on using the ensemble learning approach for software defect prediction. This review was conducted after critically analyzing research papers published since 2012 in four well-known online libraries; ACM, IEEE, Springer Link, and Science Direct. In a systematic review, Li et al. [19] also have identified a set of 49 primary studies related to a software defect prediction published between January 2000 and March 2018 which covered a wide range of unsupervised prediction technique families including six different cluster labeling techniques.

Wang et al. [4] proposed a gated hierarchical long short-term memory networks (GH-LSTMs) model, which extracts both semantic features from word embedding of Abstract syntax trees. They constructed a hierarchical LSTMs model composed of a semantic-level LSTM and a traditional-level LSTM, the outputs of semantic-level LSTM and traditional-level LSTM into a gate function separately, where a fully connected layer is used to generate a filter for the information passing through and the gated merge layer and predict whether the file is defective or not.

Manjula et al. [5] proposed an approach that performs optimal feature selection for the deep learner algorithm for software defect prediction. Initially, a random population is created which is divided into various sub-populations resulting in distinct evaluation for each population. The proposed genetic algorithm consists of main stages such as designing the chromosome and fitness function formulation. The rest of the steps remain unchanged and follow the conventional genetic algorithm approach.

Liang et al. [23] proposed an approach that consists of three steps; parsing the source code of project files and extracting token sequences as features, training token vectors and mapping token sequences to real-valued vector sequences, and build the LSTM model with vector sequences in the training sets and predicting defects in the test sets.

Yang et al. [2] have proposed a software defect prediction workflow where experiments are conducted to verify the effectiveness of the proposed method. They applied different machine learning modules and ensemble them to achieve ensemble learning. Before the training, the traditional metric features are normalized.

Pan et al. [24] proposed an improved convolutional neural network model consisting of five steps. First, they parsed the code into abstract syntax trees, and then they mapped the string token vectors into integer input vectors to the convolutional neural network. In the final step, random oversampling is used to handle class imbalance problems.

Zhou et al. [25] proposed a tree network-based software defect prediction using the cascade forest structure. They used a z-score to standardize each feature in our defect datasets and also used cascade forest structure to perform representation learning by processing the raw defect features through the layer-by-layer structure.

Dam et al. [26] proposed a model based on a tree-structured network of long short-term memory units. The model consists of the following steps: first, parsing the source code file into an abstract syntax tree, then mapping the abstract syntax tree nodes to continuous-valued vectors called embedding. The ultimate step was to put the abstract syntax tree into a tree-based network of long short-term memory to obtain a vector representation of the whole source file. This vector is then used by a classifier to predict defect outcomes.

Wang et al. [27] proposed an approach that contains four major steps; parsing source code, mapping tokens to integer identifiers which are the expected inputs to the deep belief network, then leveraging the deep belief network to automatically generate semantic features, and the final step was to build defect prediction models and predicting defects using the learned semantic features of the training and test data. Jayanthi et al. [28] proposed a combined approach to deliver the concept of feature reduction and artificial intelligence where feature reduction is carried out by well-known principal component analysis which was improved to use it. Finally, a neural network-based technique was applied to the data.

Huda et al. [29] proposed an ensemble model which considers the class imbalance problem in software defect prediction. They used a combination of random oversampling, Majority Weighted Minority Oversampling Technique, and Fuzzy-Based Feature–Instance Recovery to build an ensemble classifier.

6 Literature review

In Table 1 and Fig. 5, the related work results are compared according to F1-measure that indicates that the highest results come from Manjula et al. [5] who have created a hybrid model with genetic and deep neural network algorithms and the F-measure was 98%. The second highest result is in Wang et al. [27] who used deep belief networks and the F-measure was 94.2%.

Table 1 Performance comparison of different models

Paper	Learning method	Dataset	Algorithm	$F1$-measure (%)
Wang et al. [4]	DL	Promise repository	Gated hierarchical long short-term memory networks (GH-LSTMs)	The $F1$-measure is 89
Manjula et al. [5]	DL	Promise repository	Genetic algorithm with a DNN	The $F1$-measure is 98
Liang et al. [23]	DL	Promise and Apache repository	LSTM	The $F1$-measure is 59.1
Yang et al. [2]	ML	NASA	ANN, RF, KNN	The $F1$ -measure is 95
Pan et al. [24]	DL	PSC	Improved CNN	The $F1$-measure is 60
Dam et al. [26]	DL	Samsung and Promise repository	LSTM	The $F1$-measure is 80%
Wang et al. [27]	DL	Ant, camel, jedit, log4j, Lucene, Xalan, Xerces, ivy, synapse, poi, avi	Deep belief network	The $F1$-measure is 94.2
Jayanthi et al. [28]	ML	NASA	Feature reduction and artificial-based neural network	The $F1$- measure is 92.7

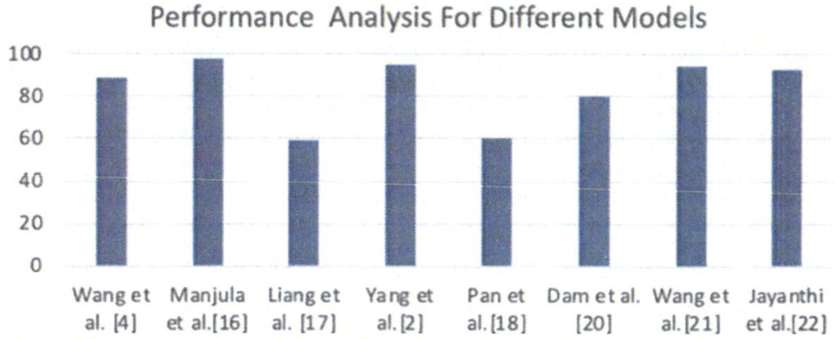

Fig. 5 Overall performance comparison

7 Conclusion

In the field of software development, the detection and correction of software defects are considered critical issues. This paper examines current developments in machine learning for software defect prediction. The performances of the different algorithms were evaluated using F-measure. According to a comparative study, it is observed that the genetic algorithm and deep neural network have the best results with an F-measure of 98%. Also, artificial neural networks gave a high F-measure of 95%.

The performance of other strategies that address additional issues, such as the class imbalance issue that impacts the effectiveness of the current software defect prediction models, will be examined and compared in the future. Also, an observed comparative study can be conducted on several ensemble techniques and feature selection approaches that have been employed for the prediction of software defects.

References

1. Goyal S (2022) Effective software defect prediction using support vector machines (SVMs). Int J Syst Assur Eng Manag 13(2):681–696. https://doi.org/10.1007/s13198-021-01326-1
2. Yang Z, Jin C, Zhang Y, Wang J, Yuan B, Li H (2022) Software defect prediction: an ensemble learning approach. J Phys Conf Series 2171(1). https://doi.org/10.1088/1742-6596/2171/1/012008
3. Krasner H (2020) Member Advisory board consortium for information and software quality TM (CISQ TM) The cost of poor software quality in the us: a 2020 report CISQ consortium for information and software quality I the cost of poor software quality in the US: a 2020 report
4. Wang H, Zhuang W, Zhang X (2021) Software defect prediction based on gated hierarchical LSTMs. IEEE Trans Reliab 70(2):711–721. https://doi.org/10.1109/TR.2021.3047396
5. Manjula C, Florence L (2019) Deep neural network based hybrid approach for software defect prediction using software metrics. Cluster Comput 22:9847–9863. https://doi.org/10.1007/s10586-018-1696-z
6. Jorayeva M, Akbulut A, Catal C, Mishra A (2022) Machine learning-based software defect prediction for mobile applications: a systematic literature review. Sensors 22(7). MDPI. https://doi.org/10.3390/s22072551

7. Li J, He P, Zhu J, Lyu MR (2017) Software defect prediction via convolutional neural network. In: Proceedings - 2017 IEEE International Conference on Software Quality, Reliability and Security, QRS 2017, 318–328. https://doi.org/10.1109/QRS.2017.42
8. IF of E Christ university (Bangalore and institute of electrical and electronics engineers, 2019 international conference on data science and communication (icondsc): faculty of engineering, CHRIST (Deemed to be University), Bangalore, 2019-03-01 to 2019-03-02
9. Farid AB, Fathy EM, Eldin AS, Abd-Elmegid LA (2021) Software defect prediction using hybrid model (CBIL) of convolutional neural network (CNN) and bidirectional long short-term memory (Bi-LSTM). Peer J Comput Sci 7:1–22. https://doi.org/10.7717/peerj-cs.739
10. Pan C, Lu M, Xu B (2021) An empirical study on software defect prediction using codebert model. Appl Sci (Switzerland) 11(11)https://doi.org/10.3390/app11114793
11. Xu Z et al. (2021) A comprehensive comparative study of clustering-based unsupervised defect prediction models. J Syst Software 172. https://doi.org/10.1016/j.jss.2020.110862
12. Matloob F et al. (2021) Software defect prediction using ensemble learning: a systematic literature review IEEE Access Inst Electrical Electron Engineers Inc 9:98754–98771. https://doi.org/10.1109/ACCESS.2021.3095559
13. Gao K, Khoshgoftaar TM, Wang H, Seliya N (2011) Choosing software metrics for defect prediction: an investigation on feature selection techniques. Softw Pract Exp 41(5):579–606. https://doi.org/10.1002/spe.1043
14. Cetiner M, Koray Sahingoz O (2020) A comparative analysis for machine learning based software defect prediction systems
15. IF of E Christ university Bangalore and institute of electrical and electronics engineers, Software bug prediction using supervised machine learning algorithms
16. Professor A Overview of software defect prediction using machine learning algorithms [Online]. Available: http://www.ijpam.eu
17. Mustaqeem M, Saqib M (2021) Principal component based support vector machine (PC-SVM): a hybrid technique for software defect detection. Cluster Comput 24(3):2581–2595. https://doi.org/10.1007/s10586-021-03282-8
18. Pachouly J, Ahirrao S, Kotecha K, Selvachandran G, Abraham A (2022) A systematic literature review on software defect prediction using artificial intelligence: datasets, data validation methods, approaches, and tools. Eng Appl Artif Intell 111:104773. https://doi.org/10.1016/J.ENGAPPAI.2022.104773
19. Li N, Shepperd M, Guo Y (2020) A systematic review of unsupervised learning techniques for software defect prediction. Inf Softw Technol 122, Elsevier B.V. https://doi.org/10.1016/j.infsof.2020.106287
20. Annisa R, Rosiyadi D, Riana D (2020) Improved point center algorithm for k-means clustering to increase software defect prediction. Int J Adv Intell Inform 6(3):328–339. https://doi.org/10.26555/ijain.v6i3.484
21. Survey on software prediction techniques
22. Omri S, Sinz C (2020) Deep learning for software defect prediction: a survey. In: Proceedings - 2020 IEEE/ACM 42nd international conference on software engineering workshops, ICSEW 2020, pp 209–214. https://doi.org/10.1145/3387940.33914630
23. Liang H, Yu Y, Jiang L, Xie Z (2019) Seml: a semantic LSTM model for software defect prediction. IEEE Access 7:83812–83824. https://doi.org/10.1109/ACCESS.2019.2925313
24. Pan C, Lu M, Xu B, Gao H (2019) An improved CNN model for within-project software defect prediction. Appl Sci 9(10), Switzerland. https://doi.org/10.3390/app9102138
25. Zhou T, Sun X, Xia X, Li B, Chen X (2019) Improving defect prediction with deep forest. Inf Softw Technol 114:204–216. https://doi.org/10.1016/j.infsof.2019.07.003
26. Dam HK et al. (2019) Lessons learned from using a deep tree-based model for software defect prediction in practice. In: IEEE international working conference on mining software repositories. vol 2019-May pp 46–57. https://doi.org/10.1109/MSR.2019.00017
27. Wang S, Liu T, Nam J, Tan L (2020) Deep semantic feature learning for software defect prediction. IEEE Trans Software Eng 46(12):1267–1293. https://doi.org/10.1109/TSE.2018.2877612

28. Jayanthi R, Florence L (2019) Software defect prediction techniques using metrics based on neural network classifier. Cluster Comput 22:77–88. https://doi.org/10.1007/s10586-018-1730-1
29. Huda S et al (2018) An ensemble oversampling model for class imbalance problem in software defect prediction. IEEE Access 6:24184–24195. https://doi.org/10.1109/ACCESS.2018.2817572

Using Machine Learning Techniques in Predicting Auditor Opinion: Empirical Study

Ahmed Mahmoud Elbrashy, Amira Mohamed Naguib Abdulaziz, and Mai Ramadan Ibraheem

Abstract For auditors, a key concern is the reliability and quality of the auditing opinion's final decision. The emerging topic of machine learning in auditing is being increasingly explored by developing trusted and efficient algorithms for classifying auditing opinions. The prediction of audit opinions of Egyptian listed companies is crucial to the security market risk mitigation process. Numerous innovations might be put into practice to raise audit efficiency, improve audit quality, and enhance auditor insight by using machine learning techniques. The aim of this paper is to provide a new audit opinion prediction model for financial statements. To this end, a sample of a group of listed Egyptian companies was selected. The model was trained with the aid of auditor opinion labels using two widely used supervised machine learning classifiers (SVM—Support Vector Machine and NV—Naive Bayes). The obtained results were then compared with the trained model that uses the clustering outcomes as a new relative auditor opinion. The results show that the developed method managed to predict the audit opinion with accuracy rates of 83.7 and 83.9 %, respectively. The performance evaluated in terms of overall prediction accuracy, and the Type I and Type II error rates show that the SVM models have higher results than the Naïve Bayes models. This study indicates that traditional methods have a poor performance using two traditional techniques (logistic and probit regressions).

Keywords Audit opinion prediction · Machine learning · Support vector machine · Logistic regression · Naive Bayes

---

A. M. Elbrashy · A. M. N. Abdulaziz (✉)
Accounting Dept, Delta Higher Institute for Management and Accounting Information Systems, Talkha, Egypt
e-mail: amiranaguib1212@hotmail.com

M. R. Ibraheem
Information Technology Dept Faculty of Computers and Information, Kafrelsheiksh University, Kafr Elsheiksh, Egypt

D. Magdi et al. (eds.), *Green Sustainability: Towards Innovative Digital Transformation*, Lecture Notes in Networks and Systems 753,
https://doi.org/10.1007/978-981-99-4764-5_15

1 Introduction

The auditing process is one of the most highly debated subjects in the modern business world. Audit opinions are a crucial component in ensuring the long-term efficacy of economic efficiency. Audit opinions are used to advise stakeholders, such as creditors, investors, stockholders, suppliers, and labor unions, on the financial stability and sustainability of businesses. The information provided in audit reports significantly influences the decisions made by investors [1]. Financial statements and audit reports that are consistently disclosed by listed companies serve as a crucial base for investment decisions to interested parties. The term "audit report" refers to the written document issued by the certified public accountants (CPA) of audit opinion given on the financial statements of the audited company based on carrying out the audit work in compliance with the requirements of the audit standards. The objective of the financial statement audit is to express an opinion on whether the financial statements have been prepared in accordance with applicable accounting standards and whether they are fair in all material respects and accurately reflect the financial position, operational outcomes, and cash flows of the auditee [2].

The final result of the auditing procedures is the audit opinion. Auditors disclose their opinions to the public at the completion of the auditing process According to a standard unqualified audit opinion, the firm's financial statements are fairly presented but do not comply with accounting standards. When the firm's financial statements are not fairly presented and serious misstatements have a material unfavorable effect on the financial statements, an adverse audit opinion is issued [1]. In recent years, demand for auditing has grown significantly due to the significance of the quality and dependability of the audited financial reports for the purpose of achieving the most efficient use of economic resources [3].

Technologies are developing at an unprecedented rate, presenting businesses and other parties, including the accounting profession, with both enormous challenges and opportunities. Companies must respond rapidly to shifting circumstances in the current business environment. Many companies are searching for more effective methods to use new technologies to change the way they do business. In this age of information explosion, new technologies are available that have the power to transform entire industries and business structures. Data is currently one of the most significant assets for many businesses. They collect an incredible amount of data as a result of their regular business operations and work to use analytics to fully realize the potential of the data. Emerging technologies like machine learning, robotic process automation, and data analytics also have a significant impact [4].

Despite these issues, utilizing machine learning techniques offers several benefits, including improved efficiency and effectiveness through faster data processing, high-quality audit, errors reduction, early risk identification, and the creation of a competitive advantage. Specifically, in audit and assurance areas, machine learning will lead to many changes in the foreseeable future [5]. Audit opinion is one such potential change. To address this major issue, this study introduces the idea of applying machine learning techniques for predicting auditor opinion.

In this respect, this research investigates the usefulness of machine learning techniques in auditors' opinion prediction of listed companies in Egyptian Stock Exchange. The classifiers including Support Vector Machine (SVM), Naïve Bayes (NV), and logistic regression were used. In overall, the experimental results of investigating 649 firm-years observations during 2012 to 2016 confirmed the usefulness of machine learning techniques in predicting auditors' opinion. In other words, the application of machine learning method increases the mean of accuracy and reduces the occurrence of Type I and Type II errors.

2 Related Work

Alareeni [6] gave a full study of the performance of and auditors' opinion on going concern. In particular, it demonstrates that neural network models produce superior results and are the most accurate approach to predict a company's future position failure or non-failure. Artificial intelligence technology is superior to auditors' going-concern opinions in this case.

In the study of Awad SS [7], several private commercial banks in Iraq were subjected to a series of financial tests that demonstrated the reality of the situation. Some of them stumbled while others persevered in order to select the most appropriate ratios to support the auditor's opinion in his report on the business's sustainability. The researchers searched for data in 13 banks, 7 of which are failing and 6 of which are not, using a number of technologies such as Decision Tree, ID3, Naive Byes, and Random Forest. The investigation revealed that the ratios Debt Ratio, Liabilities Total Long Term, and Return on Assets and Equity Debt Ratio accurately anticipate the bank's status.

Serrano et al. [8] create a new model for predicting audit opinions for consolidated financial statements. A sample of Spanish firms was chosen, and the multilayer perceptron artificial neural network technique was used. According to the findings, the developed method had an accuracy rate of more than 86% in predicting the audit opinion. The most important variables used to predict the audit opinion for individual accounts differed significantly from those used when using consolidated financial statements, which converted the variables directly related to the industry, group size, auditor, and board members into the main explanatory parameters of the prediction.

Thuy Ha et al. [9] established a link between financial ratios and non-financial data such as firm size, audit firm, going-concern opinion in the previous year, and auditors' opinion on audit report. The research evaluated the financial statements, auditors' opinion, and financial statements notes for listed firms in Vietnam that received a qualified audit report and those that did not. The results of using binary logistic models show that the earnings before taxes (EBT) ratio, the financial leverage ratio, and the preceding year's going-concern opinion are all factors that influence auditors' audit opinions.

Medina et al. [10] provided an approach for objectively identifying and evaluating potential changes in auditor behavior. The process relies on the use of assembled classification trees, specifically bagging and boosting methods. When the results of the two procedures are compared, it is clear that the assembly using bagging generated superior results. These procedures were evaluated using logistic regression. The comparison shows that, despite the former's higher specificity and the latter's higher sensitivity, bagging and logit outcomes are relatively comparable.

In the study of Zarei et al. [11], financial statements from 96 Iranian companies listed on the Tehran Stock Exchange were used over a five-year period (2012–2016). A probit model was used to analyze 480 data over 11 major financial ratios as well as non-financial variables that influence audit report issuance, such as audit firm type, auditor turnover, and corporate performance. Financial ratios and the type of audit company were shown to have a high ability to convey qualifications through audit reports. The estimated model's prediction accuracy was evaluated using regression modeling for the probability of qualified and clean opinions. The model correctly classified the full sample to account for fluctuations in the auditor's performance. With a 72.9% accuracy rate, the model correctly classified the entire sample to account for changes in the auditor's opinion. The model was 72.9% accurate in classifying the entire sample to account for changes in the auditor's opinion.

In the research of Manurung et al. [12], the support vector regression method was used to examine the possibility of bankruptcy. From 2016 to 2018, there are six variables from 17 Indonesian firms. The model developed using support vector regression predicts good performance due to its high coefficient of determination in compared to other studies. The R2 value of 0.5014 shows that the model forecasts the probability of bankruptcy appropriately. According to the study, adding more data and utilizing a more complex machine learning model might possibly enhance performance.

2.1 Research Gap

Prior studies have contributed to the development of models that help predict audit opinions and have applied several methodologies in the search of better predictions models. Nevertheless, and even though the existing literature on the prediction of audit opinions is profuse, the results achieved by the existing modeling are still considerably far from having obtained high levels of prediction. This study addresses several explanatory variables that have not been investigated in the prior Egyptian studies. The existing literature indicates that prior studies attempted to classify the audit opinions into only two groups of unqualified and qualified or predict the going-concern audit reports. The current study attempts to fill the gap by classifying the audit reports into more than two groups, namely unqualified, unqualified with explanatory language, and qualified using a relatively small dataset.

2.2 *Research Motivation*

In the past few years, machine learning techniques have taken center stage in the fields of finance and accounting. The classification power provided via these tools has proven their efficiency when tested in the financial and accounting fields. The initial uses of machine learning in auditing are still in their infancy, and this relationship between auditing and the model is still in its early phases. Single classification models have been actively developed and used in earlier studies [13]. Previous studies have been actively moving toward building and using single classification models. Now, researchers are urging the use of many classifiers because they can reduce the effects of the errors caused by the use of single classifiers.

3 Methodology

3.1 *The Dataset*

The dataset was collected from a sample of 649 Egyptian listed companies for the years 2012 to 2016. Companies operating in the financial and banking sectors, companies with many missing variables, and companies related to service sectors such as tourism and media were excluded. The dataset for classification was constructed by categorizing the companies into three groups: (1) companies with financially correct statements that received an unqualified opinion (coded as 1), (2) companies with financially correct statements but needed clarification, which received an unqualified opinion with explanatory language (coded as 2), and (3) companies with financially risky statements that received a qualified opinion (coded as 3). Financial statements were collected from online platforms that provide financial data, and related indicators were calculated. Table 1 summarizes the frequency of audit opinions of the companies included in our sample.

Table 1 Frequency of audit opinions for the companies' sample

Type of audit opinion	Frequency
Unqualified opinion	535
Unqualified opinion with explanatory language	84
Qualified opinion	30
Total	649

3.2 Variables

To construct the training model for machine learning classifiers, 16 variables were selected. These variables are the most commonly used financial indicators that aid in accurate prediction of auditor opinion about financial statements. The variables can be summarized in Table 2.

Table 2 Most used financial indicators affecting on auditor opinion

Variables	Explanation
Y	Auditor's opinion: dummies coded 1 for unqualified, 2 for unqualified with explanatory language, and 3 for qualified opinions
X1	Working capital = current assets—current liabilities
X2	Total accruals which is the difference between net income and the operating cash flow
X3	Asset intensity, which can be calculated as total assets divided by total revenue
X4	Financial leverage, which can be calculated as total liabilities divided by total assets
X5	Loss, which is dummy variable coded 1 for the loss-making companies and 0 otherwise
X6	Return on total assets: calculated by a company's net profit divided by its total assets
X7	Return on shareholder's equity: calculated by a company's net income divided by its equity
X8	EBIT margin: measured by a company's earnings before interest and tax divided by its sales
X9	Net income: represents the fiscal period income or loss reported by a company
X10	Retained earnings/total assets: measured by a company's retained earnings divided by its total assets
X11	Liquidity ratio (Cash ratio): calculated by a company's total cash and cash equivalents divided by its current liabilities
X12	Quick ratio: calculated by a company's quick assets divided by its current liabilities
X13	Receivables/sales: calculated by a company's receivables divided by its sales
X14	Auditor's size: dummy coded 1 for a Big 4 auditor, 0 otherwise
X15	Log of total assets (Size): natural logarithm of a company's total assets at the end of the fiscal year
X16	Log of net sales (Size): natural logarithm of a company's net sales during the fiscal year

3.3 Classification Algorithms Used in Analysis

Supervised learning techniques are typically used to classify labeled datasets, while unsupervised learning is used to understand relationships within unlabeled datasets. Supervised learning techniques predict a particular output based on a defined set of attributes accompanied by a class label. The dataset that includes the input variables in addition to the class labels is assumed to be a labeled dataset. Machine learning algorithms are used in the classification of the labeled dataset. Unsupervised learning techniques are used for clustering unlabeled data using data mining algorithms.

In this study, the model was trained with the aid of auditor opinion labels using three widely used supervised machine learning classifiers (SVM—Support Vector Machine, logistic regression, and NV—Naive Bayes). The obtained results were then compared with the trained model that uses the clustering outcomes as a new relative auditor opinion. The classification process was performed using three widely used supervised machine learning classifiers (SVM—Support Vector Machine, logistic regression, and NV—Naive Bayes).

A pre-processing step was first applied for data preparation. After that, the dataset was divided into a testing set and a training set, which were used to train the three aforementioned classifiers. The training set, including a set of attributes and the corresponding label in terms of the auditor's opinion, was used for training the three aforementioned classifiers. The test set was also used to assess the performance of the classification algorithms. Then, the fuzzy c-means clustering algorithm was used to predict a new relative auditor opinion based on the labeled data. The learned relationships in terms of the predicted auditor opinion were used for training the three classification algorithms with the same attributes except for the class attribute, and therefore, predictions were generated. Finally, those predictions were compared with the real auditor values to analyze the performance of the proposed classification algorithms. Several evaluation measurement techniques, including average accuracy, AUC, Type II error, and Type I error, were applied to evaluate the performance of each model. Lastly, significance testing was conducted to validate model performance statistically.

3.3.1 Logistic Regression

Logistic regression is a machine learning algorithm used to predict the likelihood of specified classes by computing a sum of the input features in the form of a bias term, which is used to calculate the logistic of the result [14].

For the binary case, the logistic regression output ranges between zero and one.

$$h_\theta = \frac{1}{1 + e^{-\theta x}}$$

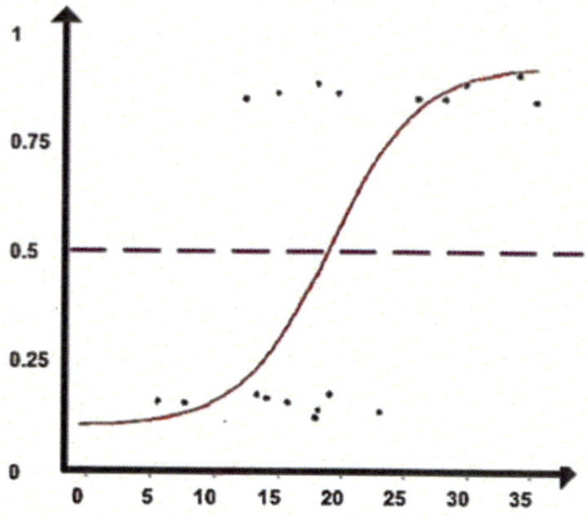

Fig. 1 Sigmoid function for binary predictions based on a cutoff of 0.5

In logistic regression, X represents the input data, and θ represents the parameter to train or optimize. When the result of the prediction is closer to 1, it indicates a positive sample, and when y tends to zero, it refers to a negative sample [15].

The prediction in logistic regression depends on the relationships between variables. In practice, the algorithm analyzes relationships between variables and uses the sigmoid function to assign probabilities in a discrete form [16]. The samples are split into two groups for binary predictions based on a cutoff of 0.5. Group A includes samples above 0.5, and group B includes samples 0.5 (Fig. 1).

The implementation of logistic regression using Python follows these steps.

The data is processed or prepared in this step, and the dependent and independent variables are extracted from the given dataset. Then, the data is split into training and testing datasets, and a fitting model for the training set is applied. The model is well trained using the training set, and the result is predicted using test set data.

Preprocess the data.
Split the data into training and testing datasets.
Create a logistic regression classifier.
Fit the logistic regression classifier on the training set.
Predict the outcome using the test set.
Evaluate the performance of the classifier.

3.3.2 Support Vector Machine

The SVM algorithm generates a decision boundary, also known as a hyperplane, in multidimensional space that divides different classes to enable accurate prediction and classification of new data points [17]. The SVM approach iteratively generates the hyperplane to minimize error and maximize the margin of the hyperplane [18]. In the creation of the hyperplane, SVM selects the extreme points, which are referred to as support vectors [18]. The support vectors describe the extreme cases and are

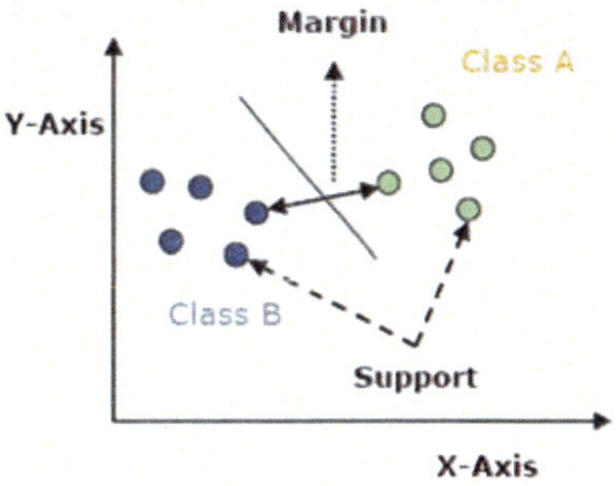

Fig. 2 SVM separating hyperplane in multidimensional space

an essential component of the SVM algorithm. Figure 2 depicts the classification of two different categories using a hyperplane [19].

The SVM algorithm selects the extreme points or vectors that assist in constructing the hyperplane. These extreme values are known as support vectors, and the technique utilized to identify them is called a support vector machine [19].

There are Two Types of SVM

- **Linear SVM**: This type of SVM is suitable for data that can be separated into two groups using only a single straight line, which is called linearly separable data. The classifier used for linear SVM is called the Linear SVM classifier.
- **Nonlinear SVM**: This type of SVM is used for data that cannot be separated using a straight line. Such data is referred to as nonlinear data, and the classification algorithm used is called the nonlinear SVM classifier [18].

The implementation of SVM using Python involves the following steps.

The data is processed or prepared in this step, and the dependent and independent variables are extracted from the given dataset. Then, the data is split into training and testing datasets, and a fitting model for the training set is applied. The model is well trained using the training set, and the result is predicted using test set data [19].

Preprocess the data.
Split the data into training and testing datasets.
Create an SVM classifier.
Fit the SVM classifier on the training set.
Predict the outcome using the test set.
Evaluate the performance of the classifier.

3.3.3 Naïve Bayes

The Naïve Bayes algorithm is so named because it assumes that the presence of one particular feature is independent of the presence of other features [20]. Therefore, each feature contributes to the identification of the class independently. The algorithm is also called Bayes because it is based on Bayes' Theorem. Naïve Bayes is a fast and straightforward machine learning algorithm used for classifying datasets into classes.

It is suitable for both binary and multiclass classification problems. Compared to other algorithms, Naïve Bayes performs better in multiclass predictions [21].

There are Two Types of Naïve Bayes Model

- Gaussian: The Gaussian model assumes that the features are normally distributed. In the case of continuous predictors, the model assumes that the sampled values are from the Gaussian distribution instead of the discrete values.
- Multinomial: The multinomial model is used for multinomial distributed data. It is mainly employed in document classification problems, where the frequency of words is used as predictors [22].

To implement Naïve Bayes using Python, the following steps are typically followed: The data is processed or prepared in this step, and the dependent and independent variables are extracted from the given dataset. Then, the data is split into training and testing datasets, and a fitting model for the training set is applied. The model is well trained using the training set, and the result is predicted using test set data [20].

Preprocess the data.
Split the data into training and testing datasets.
Create a Naïve Bayes classifier.
Fit the classifier to the training set.
Predict the outcome using the test set.
Evaluate the performance of the classifier.

3.3.4 K-Means Clustering Algorithm

K-means clustering is an unsupervised learning algorithm that solves clustering problems by grouping an unlabeled dataset into different clusters. The algorithm is iterative and aims to assign each data point to one cluster with similar properties. Each cluster has a centroid assigned to it, and the algorithm works to minimize the total distance between data points and their corresponding centroids. The algorithm performs two main tasks.

It uses an iterative technique to choose the best k-center or centroids [23]. Each data point is then matched with the nearest k-center, and a cluster is formed by the data points that are close to a specific k-center. This ensures that each cluster is distinct from the others and contains the nearest data points. The K-means clustering algorithm can be better understood from the diagram [24] (Fig. 3).

The K-means algorithm can be implemented using the following steps [23].

Step 1: *Initialize the number of clusters, K.*
Step 2: *Choose K random points to act as the initial centroids.*
Step 3: *Assign each data point to the closest centroid and form k clusters.*

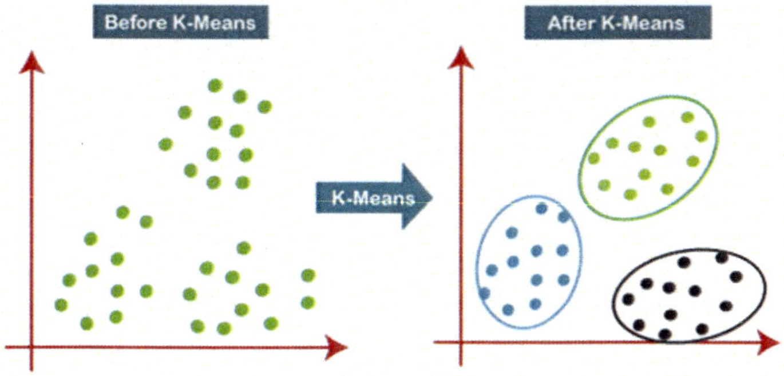

Fig. 3 K-means clustering

Step 4: *Compute the variance of each cluster and replace the centroids with the new ones.*
Step 5: *Repeat steps 3 and 4 until no further reassignments are needed.*
Step 6: *End the algorithm.*

4 Analysis of Data and Results

The results obtained using (SVM—Support Vector Machine and NV—Naive Bayes) machine learning classifiers were compared with the results of traditional techniques (logistic and probit regressions) as follows.

4.1 Machine Learning Techniques Results

Classification results obtained from each algorithm of SVM and Naïve Bayes are presented in Table 3.

Table 3 Classification results for SVM and Naïve Bayes

	Unqualified		
Model	Accuracy (%)	Type I error (%)	Type II error (%)
Naïve Bayes	83.7	16.7	16.3
SVM	83.9	16.2	16.1

Type I error: Refers to the error of predicting "unqualified" when it is an "unqualified with exp. language" or "qualified"

Type II error: Refers to the error of predicting "unqualified with exp. language" or "qualified" when it is "unqualified"

Table 4 Classification results for the both logistic and probit regressions

	Unqualified		
Model	Accuracy (%)	Type I error (%)	Type II error (%)
Logistic regression	83.8	17.7	16.2
Probit regression	25	12.2	75

Type I error refers to the error of predicting "unqualified" when it is an "unqualified with exp. language" or "qualified"

Type II error refers to the error of predicting "unqualified with exp. language" or "qualified" when it is "unqualified"

The classification accuracy rates for unqualified opinion cases under the Naïve Bayes and SVM are 83.7% (Type I error rate is 16.7%; Type II error rate is 16.3%), 83.9% (Type I error rate is 16.2 %; Type II error rate is 16.1 %), respectively. The prediction accuracy and Type I error and Type II error rates show that the Naïve Bayes and SVM performs are better in the classification of unqualified audit reports.

4.2 Traditional Methods Results

The classification results obtained from traditional methods (logistic and probit regressions) are presented in Table 4.

The classification accuracy rates for unqualified opinion cases using logistic and probit regressions are 83.8% (Type I error rate is 17.7%; Type II error rate is 16.2%), and 25% (Type I error rate is 12.2%; Type II error rate is 75%), respectively. The prediction accuracy and Type I error and Type II error rates show that the logistic regression performs best in the classification of unqualified audit reports, but the probit regression has a poor performance for the classification of unqualified audit reports.

5 Comparison Between Machine Learning Techniques and Traditional Methods Results

Based on above results, the machine learning techniques (SVM and Naïve Bayes) is outperformed the traditional methods (logistic and probit regressions). Consequently, we conclude some comparisons between actual opinion and predicted opinion as presented in Table 5.

Table 5 presents pair (1) results of compared means between the actual audit opinion and predicted audit opinion using machine learning indicate that means of predicted audit opinion using machine learning are lower than actual audit opinion which means that actual audit opinion biased toward the "unqualified with exp.

Table 5 Compared means between actual and predicted results of auditor opinion

	Variables	Mean	T	Sig. (2-tailed)
Pair (1)	Actual audit opinion	1.222	5.137	0.000
	Predicted audit opinion using ML	1.177		
Pair (2)	Actual audit opinion	1.222	− 7.404	0.000
	Predicted audit opinion using traditional methods	1.410		
Pair (3)	Predicted audit opinion using ML	1.177	− 10.366	0.000
	Predicted audit opinion using traditional methods	1.410		

language" or "qualified" opinion, where the difference is significant, because of the high accuracy of machine learning techniques. Accordingly, the first hypothesis can be accepted as follows.

H1: ***There is significant difference between actual audit opinions and predicted audit opinion using machine learning (biased actual audit opinion).***

In pair (2), results of compared means between the actual audit opinion and predicted audit opinion using traditional methods indicate that means of predicted audit opinion using traditional methods are greater than actual audit opinion which means that predicted audit opinion using traditional methods biased toward the "unqualified with exp. language" or "qualified" opinion, where the difference is significant, because of the low accuracy of traditional methods techniques. Accordingly, the second hypothesis can be accepted as follows.

H2: ***There is significant difference between actual audit opinions and predicted audit opinion using traditional methods (biased predicted audit opinion using traditional methods).***

In pair (3), results of compared means between predicted audit opinion using machine learning and predicted audit opinion using traditional methods indicate that means of predicted audit opinion using traditional methods are greater than predicted audit opinion using machine learning which means that predicted audit opinion using traditional methods biased toward the "unqualified with exp. language" or "qualified" opinion, where the difference is significant, because of the low accuracy of traditional methods techniques and high accuracy of machine learning techniques. Accordingly, the third hypothesis can be accepted as follows.

H3: ***There is significant difference between predicted audit opinion using machine learning techniques and predicted audit opinion using traditional methods (biased predicted audit opinion using traditional methods).***

6 Summary and Conclusions

Most of previous studies attempted to predict qualified audit opinions in Egypt used classical statistical methods which depend on primary data using questionnaire without depending on secondary data from the Egyptian Stock Exchange. The current study addresses several explanatory variables that have not been investigated in prior Egyptian studies. The existing literature indicates that prior studies attempted to classify the audit opinions into only two groups of unqualified and qualified or predict the going-concern audit reports. The current study attempts to fill the gap by classifying the audit reports into more than two groups, namely unqualified, unqualified with explanatory language, and qualified using a relatively small dataset.

References

1. Özcan A (2016) Determining factors affecting audit opinion: Evidence from Turkey. Int J Acc Financ Rep 6(2):45–62
2. Zeng S, Li Y, Li Y (2022) Research on audit opinion prediction of listed companies based on sparse principal component analysis and kernel fuzzy clustering algorithm. Math Probl Eng 1
3. Saif SM, Sarikhani M, Ebrahimi F (2012) Finding rules for audit opinions prediction through data mining methods. Eur Online J Nat Soc Sci 1(2):28–29
4. Huang F, No WG, Vasarhelyi MA, Yan Z (2022) Audit data analytics, machine learning, and full population testing. J Finance Data Sci
5. Ucoglu D (2020) Current machine learning applications in accounting and auditing. Press Acad Procedia 12(1):1–7
6. Alareeni B (2019) A review of auditors' GCOs, statistical prediction models and artificial intelligence technology. Int J Bus Ethics and Gov 2(1):19–31
7. Awad SS, Wathik IM (2022) Using data mining tools to prediction of going concern on auditor opinion-empirical study in iraqi commercial. Acad Account Financ Stud J 26(S3):1–13
8. Sánchez-Serrano JR, Alaminos D, García-Lagos Callejón-Gil AM (2020) Predicting audit opinion in consolidated financial statements with artificial neural networks. Mathematics 8(8)
9. Ha TT, Nguyen TAT, Nguyen TT (2016) Factors influencing the auditor's going–concern opinion decision. Int Days Stat Econo 10:1857–1870
10. Sánchez-Medina AJ, Blázquez-Santana F, Alonso JB (2019) Do auditors reflect the true image of the company contrary to the clients' interests? An artificial intelligence approach. J Bus Ethics 155(2):529–545
11. Zarei H, Yazdifar H, Dahmarde Ghaleno M (2020) Predicting auditors opinions using financial ratios and non-financial metrics: evidence from Iran. J Account Emerg Econ 10(3):425–446
12. Adler HM, Suhartono D, Hutahayan B, Halimawan N (2023) Probability bankruptcy using support vector regression machines. J Appl Finance Bank 13(1):13–25
13. Cao M, Chychyla R, Stewart T (2015) Big data analytics in financial statement audits. Account Horiz 29(2):423–429
14. Bakumenko A, Elragal A (2022) Detecting anomalies in financial data using machine learning algorithms. Systems 10:130. https://doi.org/10.3390/systems10050130
15. Manglani R, Bokhare A (2021) Logistic regression model for loan prediction: a machine learning approach, Emerging trends in industry 4.0 (ETI 4.0) pp 1–6. https://doi.org/10.1109/ETI4.051663.2021.9619201
16. Guo J, Gao J (2022) Comparison of different machine learning algorithms on cell classification with scRNA-seq after principal component analysis. In: 2022 7th international conference on

intelligent computing and signal processing (ICSP) pp 1476–1479. https://doi.org/10.1109/ICSP54964.2022.9778439
17. Dai T, Dong Y (2020) Introduction of SVM related theory and its application research. In: 2020 3rd international conference on advanced electronic materials, computers and software engineering (AEMCSE) pp 230–233. https://doi.org/10.1109/AEMCSE50948.2020.00056
18. Wang Q (2022) Support vector machine algorithm in machine learning. In: 2022 IEEE international conference on artificial intelligence and computer applications (ICAICA) pp 750–756. https://doi.org/10.1109/ICAICA54878.2022.9844516
19. Ibraheem MR, El-Sappagh S, Abuhmed T, Elmogy M (2020) Staging melanocytic skin neoplasms using high-level pixel-based features. Electronics 9:1443. https://doi.org/10.3390/electronics9091443
20. Sugahara S, Ueno M (2021) Exact learning augmented naive bayes classifier. Entropy 23:1703. https://doi.org/10.3390/e23121703
21. Changpetch P, Pitpeng A, Hiriote S, Yuangyai C (2021) Integrating data mining techniques for naïve bayes classification: applications to medical datasets. Computation 9:99. https://doi.org/10.3390/computation9090099
22. Barandas M, Folgado D, Santos R, Simão R, Gamboa H (2022) Uncertainty-based rejection in machine learning: implications for model development and interpretability. Electronics 11:396. https://doi.org/10.3390/electronics11030396
23. Park J, Choi M (2022) A K-means clustering algorithm to determine representative operational profiles of a ship using AIS data. J Mar Sci Eng 10:1245. https://doi.org/10.3390/jmse10091245
24. Ahmed M, Seraj R, Islam SMS (2020) The k-means algorithm: a comprehensive survey and performance evaluation. Electronics 9:1295. https://doi.org/10.3390/electronics9081295
25. Zhu A, Hua Z, Shi Y, Tang Y, Miao L (2021) An improved K-means algorithm based on evidence distance. Entropy 23:1550. https://doi.org/10.3390/e23111550

A Comparative Study of Features Selection in the Context of Forecasting PM2.5 Concentration

Ayman Aboualnour, Mohamed Shalaby, and Emad Elsamahy

Abstract Air pollution is a critical issue for our world today, the emissions of air pollutants cause serious environmental and health issues. In this the main objective is to forecast one of the most dangerous pollutant on human health named particulate matter that has a diameter of 2.5 μm or less (PM2.5). Improving accuracy of prediction as early warnings of PM2.5 concentration can save individuals from many threats' exposure to pollutant. Beijing Multi-Site Air-Quality (12 air quality stations) dataset was utilized to improve PM2.5 concentration forecasting. One of the most important factors for improving forecasting models is features selection. In this study, distinctive features selection techniques were examined for selecting best features selection method such as by correlation coefficient, Select-K-Best and XGBoost and used the selected features to feed artificial neural networks (ANN) models. Many ANN models were constructed using widely used neural networks architectures that deal with multivariate time series regressions problems, namely bi-direction long short-term memory (BiLSTM), long short-term memory (LSTM), gated recurrent unit (GRU) and convolutional neural network (CNN). To evaluate models' forecasting results, mean absolute error (MAE) and root mean square error (RMSE) were used and results showed that each features selection method produces distinctive features and has direct impact on model performance according to evaluation metrics. Based on experiments, we conclude that Select-K-Best can outperform other features selection methods applied in this work in forecasting PM2.5 concentration utilizing Beijing Multi-Site Air-Quality dataset.

Keywords Forecasting PM2.5 · Time series · Artificial neural networks · Features selection · LSTM · GRU · BiLSTM · CNN

A. Aboualnour (✉) · E. Elsamahy
Arab Academy for Science, Technology, and Maritime Transport, El Moshir Ismail st, P.O. Box 2033, Cairo, Egypt
e-mail: ayman.aghani@gmail.com

E. Elsamahy
e-mail: e.elsamahy@aast.edu

M. Shalaby
Egyptian Armed Forces, Cairo, Egypt

D. Magdi et al. (eds.), *Green Sustainability: Towards Innovative Digital Transformation*, Lecture Notes in Networks and Systems 753,
https://doi.org/10.1007/978-981-99-4764-5_16

1 Introduction

The most hazardous pollutant is PM2.5, or fine particulate matter has a diameter of 2.5 μm or less, since it may pass through the lung barrier and enter the bloodstream, resulting in cardiovascular, cancer and respiratory diseases. It has a greater impact on human over other pollutants and can have negative effects on health even at very low concentrations [1]. Forecasting PM2.5 concentration became a hot area of research as the early warnings from accurate PM2.5 concentration forecasting can protect individuals from exposure to high emissions and save lives of those who are susceptible to high pollution by wearing masks or stay in a safe place.

Many studies performed for PM2.5 concentration forecasting, some of these works use statistical-based model for instance autoregressive moving average (ARMA) [2] and some use deep learning-based models such as artificial neural network (ANN) such as recurrent neural networks (RNN) that is heavily used for time series forecasting. Air pollution is dynamically affected by several factors such as weather, traffic and other factors; it is difficult to accurately predict by the statistical methods and traditional machine learning models [3]. In this paper, different ANN architectures were applied for PM2.5 prediction to examine features selection methods.

RNN is one of ANN models that is the most effective structure for dealing with multivariate time series data because it can capture temporal dependencies over a range of timescales [4]. The most popular deep learning method built on RNN are long short-term memory (LSTM) and gated recurrent units (GRU) as they are the most popular variants of RNN, and improved architecture of LSTM named bi-directional LSTM (BiLSTM). Also, we used convolutional neural network (CNN) that is widely used in image processing; however, one-dimensional CNN can also be used for time series prediction [3]. The deep learning ANN models that were used in this paper deal with multivariate time series regression problems to get the best advantage of including metrological and chemical features affecting concentrations of PM2.5 forecasting.

Guoyan Huanga et al. [2] made a PM2.5 concentration forecasting using GRU based on empirical mode decomposition, the forecast results in their study showed that the EMD-GRU model reduces the RMSE compared with the GRU model. Da Silva et al. [5] made evaluation for the performance of LSTM neural networks for predicting consumption, they compared LSTM with the algorithms of random forest (RF) and extreme boost gradient (XGBoost), the results indicate that the LSTM model showed a tendency of better RMSE performance than other models. Siami-Namini et al. [6] made a study for evaluating time series forecasting using LSTM and BiLSTM, their findings showed that BiLSTM-based modelling, which is based on additional training of data, provides better predictions than regular LSTM-based modelling. RuiYan et al. [7] made a comparative study of air quality index forecasting on hourly and multi-site dataset of Beijing using LSTM, CNN, CNN-LSTM, and spatiotemporal clustering and they found that better performance was provided by LSTM and CNN-LSTM models than back-propagation neural network (BPNN) and the CNN.

Many studies perform features selection by correlations [2], some others by decision tree [4] and others explicitly use specific features, in this paper different features selection methods to observe models' performance and prediction results.

2 Methodologies

The dataset used in this paper is "Beijing Multi-Site Air-Quality" [8], obtained from the "UCI Machine Learning Repository". It includes hourly air pollution data collected from 12 widely dispersed air quality monitoring locations (stations) between 1 March 2013, and 28 February 2017. There are 18 attributes in the dataset, which are listed in Table 1.

The target variable PM2.5 concentration level and dataset's attributes. Each station has 35,064 samples for each one of the 12 stations with total 420,768 samples.

2.1 Data Preprocessing

Missing Data

To treat missing values, the basic action some studies oversaw by dropping samples have missing values. Dropping missing values is an inappropriate action due to the time order of the data and the correlation of observations in time series data. There are many other techniques to deal with missing data by replacement such as forward-filling and backward-filling, but these techniques can affect data statistical properties and consequently affect prediction results. Therefore, choose to estimate missed values by linear interpolation method instead of using mean as illustrated

Table 1 Dataset attributes

Attribute	Description	Attribute	Description
No	Row number	CO	CO concentration (ug/m^3)
Year	Year of data in this row	O_3	O_3 concentration (ug/m^3)
Month	Month of data in this row	TEMP	Temperature (degree Celsius)
Day	Day of data in this row	PRES	Pressure (hPa)
Hour	Hour of data in this row	DEWP	Dew point temperature (degree Celsius)
PM2.5	PM2.5 concentration (ug/m^3)	RAIN	Precipitation (mm)
PM10	PM10 concentration (ug/m^3)	wd	Wind direction

by Noor et al. for estimating PM10 [9]. The authors made comparison between estimating values by linear interpolation method and by using mean and the best results were achieved by linear interpolation in the course of comparing MAE for regression model, the interpolation is computed using Eq. 1.

$$f(x) = f(x_0) + \frac{f(x_1) - f(x_0)}{x_1 - x_0}(x - x_0) \tag{1}$$

where x is the independent variable, x_0 and x_1 are known values of independent variable and $f(x)$ is the value of the dependent variable for a value x_1 of the dependent variable.

For handling missed data in categorical features such as wind direction (WD) feature, forward-fill was used.

New Features

New features were added based on date such as day of week, day of year and "Isholiday" flag. For this purpose, a library in Python called "chinese_holiday" which have a function called "is_holiday" returns boolean value true for official holidays and weekends and false for working days. Moreover, a new meteorological feature was adopted for relative humidity (RHUM) computed from dew point temperature (DEWP) and temperature features.

Data Encoding

For improving neural network model's performance, we encoded all numerical variables using Python sklearn preprocessing object named Min–Max-Scaler that rescales variables into the range 0–1 as in Eq. 2,

$$x = \frac{x - \min}{\max - \min} \tag{2}$$

Categoric data such as WD was encoded using label encoding that encodes categorical data in a numerical sequence.

Data Splitting

Before fitting neural network models, the 12 stations data was divided into 80% for training and 20% for testing and used 20% of testing data for validation purpose. The validation data subset was used for early training termination to avoid overtraining.

2.2 *Features Selection*

Features selection is a key process in machine learning and deep learning models as it has significant effects on model results' accuracy in improving or degrading performance of the model. It is almost rare to use all variables in the dataset to build a model as it can cause overfitting and complicate the achieved model. In this paper, we examine different feature selection techniques that were examined to get the best features for each station of the 12 air quality stations with four different models of ANN. Three common feature selection methods were implemented in Python by correlation, Select-K-Best and XGBoost. A comparison was carried out between these methods and the baseline which has all features.

Correlation Coefficient

Pearson correlation coefficient was used as it is the most widely used method [10] which is computed by Eq. 3.

$$r = \frac{\sum(x_i - \overline{x})(y_i - \overline{y})}{\sqrt{\sum(x_i - \overline{x})^2 \sum(y_i - \overline{y})^2}} \tag{3}$$

where r is correlation coefficient, x_i is the value of x-variable in a sample, $\overline{x}$ is the mean of the values of x-variable, y_i is the values of the y-variable in a sample and $\overline{y}$ is mean of the values of the y-variable.

Select-K-Best

This method calculates metrics between the target variable and all features, sorts them, and then selects the K best features. This method is implemented in Python sklearn package, and the score function "f_regression" was used to compute F-value between label/feature for regression tasks.

XGBoost

Tianqi Chen and Carlos Guestrin [11] introduce a tree boosting model called XGBoost. This algorithm is widely used for classification and regression cases. One advantage of using gradient boosting is that, once the weighted trees are built, it is relatively easy to retrieve importance scores for each attribute. In general, importance provides a score that indicates how valuable each feature was in building the boosted decision trees in the model. An attribute's relative relevance increases more frequently it is employed in decision trees to make important judgements. This importance computed plainly for each attribute in the dataset allows ranking and comparison of attributes against one another. Importance calculated for a single decision tree by the amount by which each attribute split point improves the performance measure, weighted by the number of observations for which the node is responsible. The performance measure can be the purity (Gini index) used to select the split points or another more specific error function.

2.3 Modelling

To examine distinctive features selection methods, four different neural networks architectures efficient for multivariate time series regression problems were built, namely LSTM, BiLSTM, GRU and CNN to evaluate features selection methods provided in this work for forecasting PM2.5 concentration levels. Figure 1 shows the flow of the proposed model. The proposed model starts with air quality data and weather data acquisition, then data preparation is performed by handling missed data and creating new features. The next step is features selection. Here, we use different methods for feature selection and compare them with the baseline where all features are used. After that, the ANN model is fit with the preprocessed dataset. Here, we used different ANN models. Finally, the results are evaluated in the term of accuracy.

LSTM is a deep recurrent neural network-based (RNN) model that is capable of learning long-term dependencies, especially in sequence prediction problems. LSTM has feedback connections; it can process the entire sequence of data such as time series data. LSTM has an input gate, an output gate and a forget gate in addition to a memory cell. The cells are establishing a recurring link. Every gate continuously performs operations on the cells such as write, read and reset. The cell is used to "conveying" values over time intervals [12]. GRU is like LSTM, since it has gating units that regulates the flow of information inside the unit. Without having a separate memory cell, it has two gates, which are called reset gate and update gate. BiLSTM is a sequence processing model that has two of LSTMs, one accepting the input in a forward direction, and the other in a backward direction. Having the input data in both the directions increases the available information to the network. CNN is one of the most successful methods of deep learning. Its network structure that includes 1D CNN and 2D CNN. 1D CNN used for sequencing data, while 2D CNN is often used for text and image recognition. The 1D CNN was utilized as the data is sequential.

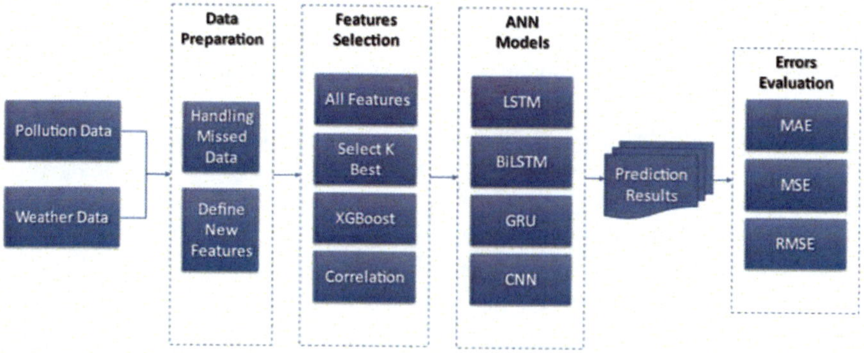

Fig. 1 Proposed model of predicting PM2.5

2.4 Models Training and Testing

Environment Set-up

To perform model training and testing, we used a computer with Intel CoreTM i7-11,370 processor with 16 GB RAM and RTX 3060 Nvidia GPU was used. The ANN implemented in Python using Keras 2.9.0 using TensorFlow 2.9.1 as backend.

Models Hyper-Parameters

The ANN models configured by time steps 30; epochs = 100 and batch size was set to 64 for LSTM, GRU and CNN and 128 for BiLSTM. Each model was trained and tested separately on each station's dataset. LSTM is configured by batch size 32, epochs = 100, early stop after 10 epochs, loss = mse, optimizer "adam", input layer: LSTM (64), dropout (0.3), hidden layer: LSTM (64), dropout (0.3) and output layer: dense (1). GRU is configured by batch size 32, epochs = 100, early stop after 10 epochs without loss = mse, optimizer "adam", input layer: GRU (64), dropout (0.3), hidden layer: GRU (64), dropout (0.3) and output layer: dense (1). BiLSTM was configured by input layer BiLSTM (64), dropout = 0.3, hidden layer BiLSTM (64), dropout = 0.3 and output layer: dense (1) and loss = mse, optimizer = "adam". CNN was configured by, Conv1D with kernel_size = 2, maxpooling1D(), dropout(0.3), flatten(), dense(1, activation = "linear")).

2.5 Evaluation Metrics

Two performance metrics were applied for performance evaluation, namely mean absolute error (MAE) and root mean squared error (RMSE).

MAE finds the difference between the actual value and predicted value that is an absolute error, by summing all the errors and dividing them by a total number of observations, our aim is to get a minimum MAE. Mean square error (MSE) represents the squared distance between actual and predicted values. We perform squared to avoid the cancellation of negative terms and it is the benefit of MSE. RMSE is the square root of MSE in Eqs. 4, 5 and 6.

$$\text{MAE} = \frac{\sum_{i=1}^{n} |y_i - x_i|}{n} \tag{4}$$

where y_i is the prediction and x_i is the true value and n is the total number of data points.

$$\text{MSE} = \frac{1}{n}\sum_{i=1}^{n}\left(y_i - \widehat{y_i}\right)^2 \tag{5}$$

$$\mathrm{RMSE} = \sqrt{\frac{1}{n}\sum_{i=1}^{n}(y_i - \widehat{y_i})^2} \tag{6}$$

where n is number of data points, y_i is the true value and $\hat{y}_i$ is the predicted value.

3 Results and Discussions

Figures 2 and 3 illustrate the average overall stations for MAE and RMSE. The previously mentioned two metrics (MAE and RMSE) were calculated using all proposed features selection methods as well as the proposed ANN models.

The previous results show that the Select-K-Best features selection method is the best method according to both MAE and RMSE (see column 3 in both Figs. 2 and 3). Moreover, the BiLSTM proofed itself as the best model for forecasting the PM2.5 with average MAE and RMSE of 17.6 and 30.11, respectively.

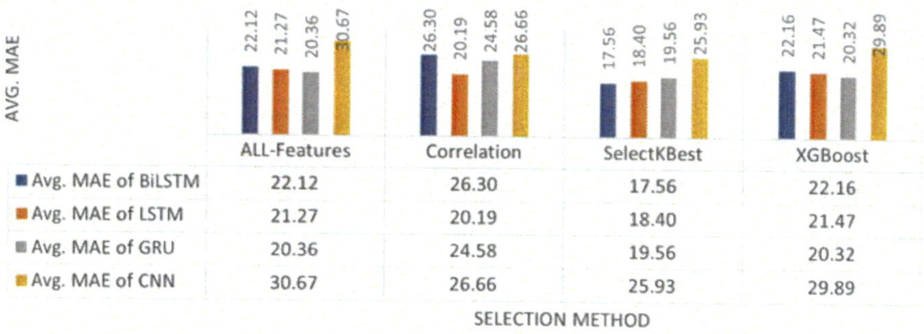

	ALL-Features	Correlation	SelectKBest	XGBoost
Avg. MAE of BiLSTM	22.12	26.30	17.56	22.16
Avg. MAE of LSTM	21.27	20.19	18.40	21.47
Avg. MAE of GRU	20.36	24.58	19.56	20.32
Avg. MAE of CNN	30.67	26.66	25.93	29.89

Fig. 2 Average MAE for all models using all features selection methods

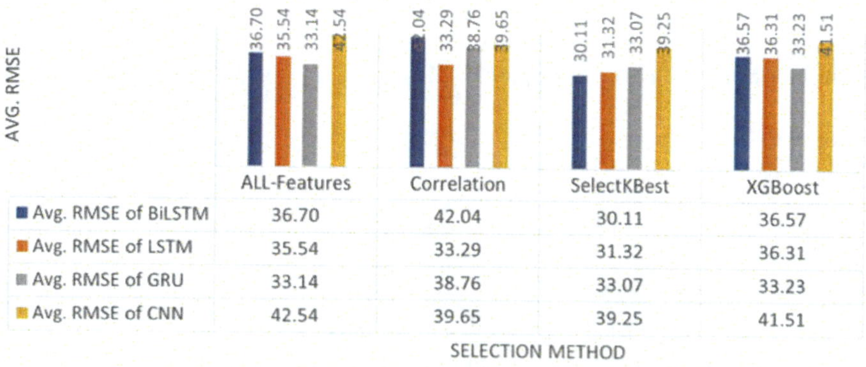

	ALL-Features	Correlation	SelectKBest	XGBoost
Avg. RMSE of BiLSTM	36.70	42.04	30.11	36.57
Avg. RMSE of LSTM	35.54	33.29	31.32	36.31
Avg. RMSE of GRU	33.14	38.76	33.07	33.23
Avg. RMSE of CNN	42.54	39.65	39.25	41.51

Fig. 3 Average RMSE for all models using all features selection methods

Furthermore, to validate the achieved results, Select-K-Best as the best features selection method and BiLSTM as the best ANN model, Figs. 4, 5, 6 and 7 illustrate details for the MAE and RMSE over both ANN models and the features selection methods for each station.

Figures 4 and 5 show MAE and RMSE, respectively, for all ANN models by using features selected by Select-K-Best method for all stations.

Figures 6 and 7 show the performance of BiLSTM using all features selections methods for all stations.

As illustrated in Figs. 4, 5, 6 and 7 that Select-K-Best features selection method has the most stable performance on all stations with the lowest error values in average with BiLSTM model.

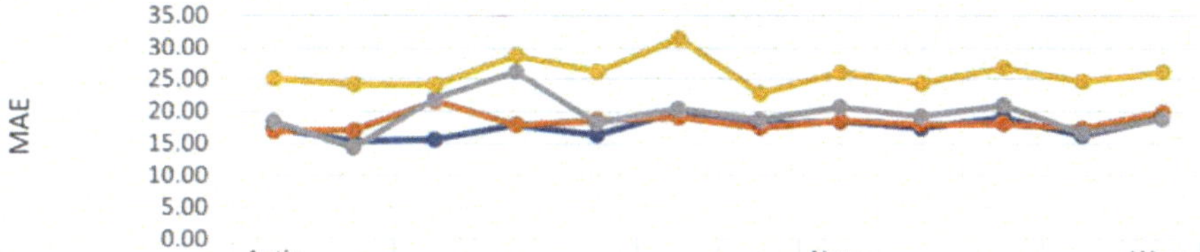

	Aotizhongxin	Changping	Dingling	Dongsi	Guanyuan	Gucheng	Huairou	Nongzhanguan	Shunyi	Tiantan	Wanliu	Wanshouxigong
BiLSTM MAE	17.34	15.33	15.60	17.94	16.40	19.90	18.00	18.49	17.47	19.03	16.28	18.95
LSTM MAE	16.94	17.07	21.58	17.90	18.64	19.08	17.50	18.61	18.02	18.12	17.36	19.93
GRU MAE	18.47	14.32	21.96	26.15	18.24	20.45	18.68	20.70	19.26	20.99	16.69	18.85
CNN MAE	25.09	24.20	24.18	28.78	26.23	31.52	22.82	26.16	24.44	26.85	24.71	26.24

Stations

Fig. 4 MAE for all ANN models using Select-K-Best feature selection method

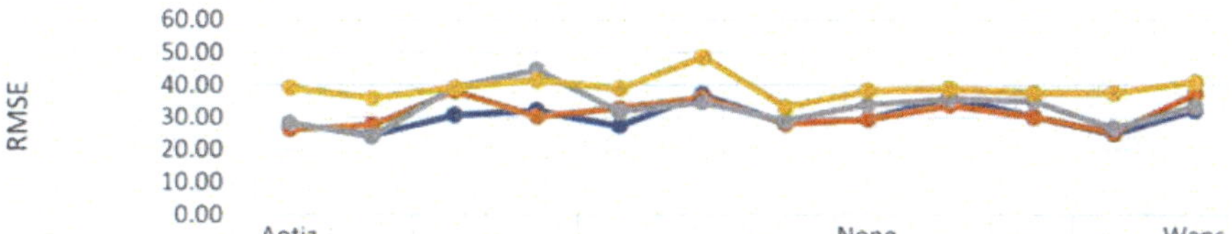

	Aotizhongxin	Changping	Dingling	Dongsi	Guanyuan	Gucheng	Huairou	Nongzhanguan	Shunyi	Tiantan	Wanliu	Wanshouxigong
BiLSTM RMSE	28.10	24.26	30.74	32.18	27.42	37.10	28.45	29.44	36.18	30.27	24.94	32.26
LSTM RMSE	26.17	27.65	38.20	30.35	32.92	36.22	28.11	29.44	33.86	30.18	25.27	37.48
GRU RMSE	28.35	23.88	39.34	44.84	31.84	34.63	29.00	33.92	35.81	35.04	26.79	33.37
CNN RMSE	39.22	35.99	39.29	41.64	38.98	48.62	33.36	38.40	38.98	37.67	37.71	41.14

Stations

Fig. 5 RMSE for all ANN models on all stations using Select-K-Best feature selection method

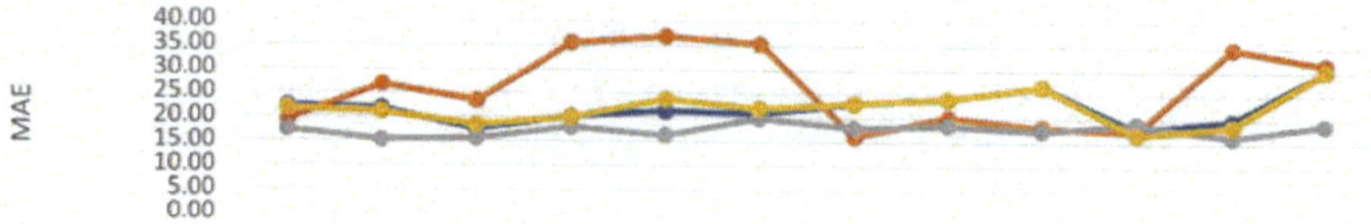

	Aotizhongxin	Changping	Dingling	Dongsi	Guanyuan	Gucheng	Huairou	Nongzhanguan	Shunyi	Tiantan	Wanliu	Wanshouxigong
ALL-Features	22.28	21.82	17.61	20.19	21.30	20.92	22.99	23.95	26.59	17.89	19.80	30.07
Correlation	19.38	26.83	23.41	35.51	36.86	35.44	16.18	20.22	18.17	17.45	34.69	31.44
SelectKBest	17.34	15.33	15.60	17.94	16.40	19.90	18.00	18.49	17.47	19.03	16.28	18.95
XGBoost	21.79	21.10	18.41	19.98	23.81	21.97	22.99	23.94	26.59	16.77	18.51	30.07

Stations

Fig. 6 MAE of BiLSTM by features selections methods for all stations

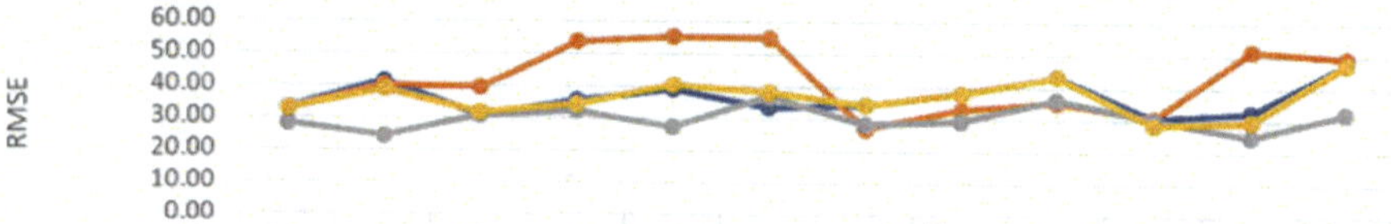

	Aotizhongxin	Changping	Dingling	Dongsi	Guanyuan	Gucheng	Huairou	Nongzhanguan	Shunyi	Tiantan	Wanliu	Wanshouxigong
ALL-Features	32.94	41.55	30.87	35.46	39.04	33.55	34.44	38.09	43.60	30.65	32.47	47.74
Correlation	32.71	40.15	39.61	53.78	55.51	55.19	27.08	33.02	35.37	30.73	51.71	49.66
SelectKBest	28.10	24.26	30.74	32.18	27.42	37.10	28.45	29.44	36.18	30.27	24.94	32.26
XGBoost	32.86	38.87	31.69	34.31	40.56	38.37	34.44	38.16	43.60	28.57	29.68	47.74

Stations

Fig. 7 RMSE of BiLSTM by features selections methods for all stations

4 Conclusion

In this work, distinctive features selection methods were examined against different neural networks models to decide on the best features selection method for each site's station in Beijing Multi-Site Air-Quality data in China. Specifically, there are 12 air pollution stations recorded in Beijing, the data were collected between 1st March 2013, and 28th February 2017. The particulate matter PM2.5 predictions performed using artificial neural networks models and performance evaluated by metrics such as MAE and RMSE. The results show that features selections differ from location to another which reflected on prediction accuracy. Models' prediction results were compared using all features and features selected by different selection methods of Select-K-Best, XGBoost and correlations. The achieved results showed that the best features selection method is Select-K-Best and the best ANN model that can provide the best PM2.5 concentration prediction is the BiLSTM averaged over all stations.

References

1. US EPA (2021) Particulate Matter (PM) Basics _ US EPA. In: Particulate Matter Pollution. https://www.epa.gov/pm-pollution/particulate-matter-pm-basics. Accessed 27 Nov 2022
2. Huang G, Li X, Zhang B, Ren J (2021) PM2.5 concentration forecasting at surface monitoring sites using GRU neural network based on empirical mode decomposition. Sci Total Environ 768:144516. https://doi.org/10.1016/j.scitotenv.2020.144516
3. Du S, Li T, Yang Y, Horng SJ (2021) Deep air quality forecasting using hybrid deep learning framework. IEEE Trans Knowl Data Eng 33:2412–2424. https://doi.org/10.1109/TKDE.2019.2954510
4. Freeman BS, Taylor G, Gharabaghi B, Thé J (2018) Forecasting air quality time series using deep learning. J Air Waste Manage Assoc 68:866–886. https://doi.org/10.1080/10962247.2018.1459956
5. da Silva DG, Geller MTB, Santos Moura dos MS, de Mauro Meneses AA (2022) Performance evaluation of LSTM neural networks for consumption prediction. e-Prime Adv Electr Eng, Electr Energy 2:100030. https://doi.org/10.1016/J.PRIME.2022.100030
6. Siami-Namini S, Tavakoli N, Namin AS (2019) The Performance of LSTM and BiLSTM in forecasting time series. In: Proceedings—2019 IEEE international conference on big data, big data 2019. Institute of Electrical and Electronics Engineers Inc., pp 3285–3292
7. Yan R, Liao J, Yang J et al (2021) Multi-hour and multi-site air quality index forecasting in Beijing using CNN, LSTM, CNN-LSTM, and spatiotemporal clustering. Expert Syst Appl 169:114513. https://doi.org/10.1016/j.eswa.2020.114513
8. Zhang S, Guo B, Dong A et al (2017) Cautionary tales on air-quality improvement in Beijing. Proc Royal Soc Mathe Phys Eng Sci 473. https://doi.org/10.1098/rspa.2017.0457
9. Noor NM, al Bakri Abdullah MM, Yahaya AS, Ramli NA (2015) Comparison of linear interpolation method and mean method to replace the missing values in environmental data set. In: Materials science forum, pp 278–281
10. Hauke J, Kossowski T (2011) Comparison of values of pearson's and spearman's correlation coefficients on the same sets of data. Quaestiones Geograph 30:87–93. https://doi.org/10.2478/v10117-011-0021-1
11. Chen T, Guestrin C (2016) XGBoost. In: Proceedings of the 22nd ACM SIGKDD international conference on knowledge discovery and data mining. ACM, New York, NY, USA, pp 785–794
12. Thakur N, Karmakar S, Soni S (2022) Time series forecasting for uni- variant data using hybrid GA-OLSTM model and performance evaluations. Int J Inf Technol 14:1961–1966. https://doi.org/10.1007/s41870-022-00914-z

Effective E-commerce Based on Predicting the Level of Consumer Satisfaction

Maha Fouad, Sherif Barakat, and Amira Rezk

Abstract COVID-19 has dramatically accelerated the growth of e-commerce, and this has led to increased competition between online sellers. To provide a competitive service, businesses need to achieve a high level of customer satisfaction. This paper tries to enhance the e-commerce's services through the analysis of earlier customer feedback and reviews to predict the level of customer satisfaction. Five classification algorithms—Decision Tree (DT), Random Forest (RF), XGBoost, support vector machine (SVM), and K-Nearest Neighbors (KNN)—were applied to predict the rating before the user gave the rating or review score, which represents customer satisfaction. Data preprocessing was conducted, and new features were developed. The different techniques were evaluated. (RF) model can predict the review score by $F1$-score 0.67 better than the other models. The analysis highlights the important features that affect customer satisfaction, which include location, delivery time, product value, and freight ratio.

Keywords E-commerce · Customer satisfaction · Data analysis · Classification

1 Introduction

One of the strategies that companies may use to efficiently manage and monitor their operations is to gauge how satisfied their customers are with their products and services. As long as there is a high degree of satisfaction among consumers, every product has a chance of lasting a significant amount of time on the market. A high

M. Fouad (✉) · S. Barakat · A. Rezk
Department of Information System, Faculty of Computers and Information, Mansoura University, P.O.35516, Mansoura, Egypt
e-mail: mahafouad@std.mans.edu.eg

S. Barakat
e-mail: sheiib@mans.edu.eg

A. Rezk
e-mail: amira_rezk@mans.edu.eg

D. Magdi et al. (eds.), *Green Sustainability: Towards Innovative Digital Transformation*, Lecture Notes in Networks and Systems 753,
https://doi.org/10.1007/978-981-99-4764-5_17

degree of consumer satisfaction with a product does not depend just on the product's good quality alone. There are other factors at play as well. Customer satisfaction may be significantly influenced by a wide variety of elements, including but not limited to the length of time it takes to deliver the goods, the need for further payments, the purchase price, and so on. Keeping existing customers happy is essential to any business's success. When a client is pleased with a product or service, they are more inclined to consider making further purchases of that product or using that service, which ultimately results in more revenue for the business. A consumer who is unhappy with the service they received from a certain firm may decide not to buy the same product again in the future because of the poor quality of the service they received in the past [1]. The evaluation of the level of satisfaction that a company's customers have with the firm's products or services is essential in order to help companies select the next marketing plan that will result in a profit for the company. Because of the unpredictability of each individual customer's behavior, estimating the level of pleasure that a customer will feel can be challenging. On the other hand, in order to circumvent this challenge, there are a multitude of statistical approaches that, without a doubt, will be of assistance to businesses in the analysis and forecasting of customer satisfaction based on certain characteristics [2–4]. Usually, a customer receives a feedback email from the e-commerce platform once the goods have been delivered. The customer can rate products on a scale of 1–5, as well as leave reviews or comments on the goods he/she has bought. An e-commerce platform ranks the items based on these reviews and ratings, which enables other users to learn more about the product's quality. However, from the seller's standpoint, these evaluations will be extremely important in helping to grow the company. However, buyers frequently choose not to submit any ratings or reviews. How can one anticipate the rating a consumer could provide in a review? This is the issue with the e-commerce industry. The issue may also be expanded to include 'Is it feasible to foresee the review rating that a client may provide before he actually provides the rating?' If this issue is resolved, it will also be feasible to forecast customer ratings, for which none have been provided. The primary focus of this paper will be on forecasting consumer satisfaction based on e-commerce datasets. The purpose of this research is to discover the most important factors affecting e-commerce consumer satisfaction and develop a model for predicting whether those customers will be satisfied.

The rest of this paper will be organized as follows: In Sect. 2, the relevant literature will be discussed. Section 3 presents the proposed model, and Sect. 4 discusses its ability to predict customers' levels of satisfaction. The major conclusions and future prospects are presented in Sects. 5 and 6.

2 Related Work

Since the data of consumers is now being captured by businesses, one of the most significant issues that has been debated by a large number of researchers is the subject of customer satisfaction prediction. The level of satisfaction experienced by customers may be predicted using a variety of methods. In the next few paragraphs, we'll talk about some of the most recent research done in this field.

Khalid et al. [5] examine consumer satisfaction with Saudi Arabia's e-commerce system using the ACSI methodology. The model was evaluated with 149 online survey sample data sets. Client satisfaction with e-commerce is determined by customer expectations, service quality, and perceived value. The data reveal that e-commerce service quality is the major factor determining consumer satisfaction with the system, which is in line with Saudi Arabian online customers, who are more concerned about security and payment methods. However, the sample included 149 Saudi Arabians ages 20–29. The sample may not represent Saudi Arabia overall. Data obtained through social networking sites and e-mail may only reach a selected sample and not the entire Saudi population. Bouzakraoui et al. [6] used machine learning to recognize facial expressions to measure customer satisfaction. They derived geometric characteristics from facial expressions. Then, neutral responses were compared to negative and positive ones. Distances were determined using SVM, KNN, Random Forests, ad hoc clustering, and decision trees. The proposed approach increases SVM accuracy to 98.66%. This study aims to identify the right distances to distinguish between emotional and satisfied customer facial expressions. In the last step, these distances are used to identify a customer's product sentiments as positive, negative, or neutral. This data will help the company understand how customers see the product. Kotsokechagia [7] constructs three distinct predictive models to solve the challenge of predicting review ratings. Particularly, it is suggested that big datasets may be mined using multi-class classification, and other methods to anticipate consumers' numerical review ratings. The results showed that the performance of multi-classification method $F1$ score 60%. The performance is poor, so we recommend adjusting the models to make them more effective. Moon and colleagues [8] investigated the impact of internet shopper satisfaction. Online purchasing depends on product quality, pricing, a flexible return policy, and fast delivery. By evaluating these elements, they've determined customer behavior and online buyer satisfaction. They used 40,000 data points to assess their performance and client satisfaction. This study uses Naive Bayes, Apriori, Decision Tree, and Random Forest. Apriori (88%) and Nave Bayes (87%) gave the best results. They've also studied internet buying and customer behavior. Hussain [9] used Amazon customer reviews to predict customer satisfaction using four supervised machine learning approaches. Naive Bayes, SVM, LR, and Decision Tree (DT) training and testing datasets use TF-IDF vectorization. After lowercasing, lemmatization, stop word removal, smiley removal, and digit removal, the models are applied. Models' accuracy, precision, recall, and $F1$ scores are compared. Support vector machines (SVM) have the highest accuracy rate (83%), followed by Naive Bayes (82%), Logistic Regression (80%), and Decision Tree (76%). Classifiers are evaluated using a confusion matrix. All algorithms classify satisfied customers better than dissatisfied customers. Uneven case assignment results in an unbalanced class classification. The imbalance can also be caused by a poorly performing dictionary or stemmer when applied to a large corpus due to a lack of memory. Zhao and colleagues [10] used 127,629 online reviews to predict customer satisfaction based on textual attributes and reviewers' identities. Subjectivity, readability, and length negatively affect customer ratings, and diversity and sentiment polarity positively affect them. Customer reviews affect ratings. This study shows the relationships between online customer reviews' linguistic style, customer identity, and customer perception and satisfaction. However, it has limitations. First, the sample only includes one city and one website. Second, customers' languages and cultures affect online textual reviews. Another extension is comparing online reviews in different languages and cultures. Third, hotel reviews and rankings can change. This research should be dynamic.

Most studies have been conducted in the area of sentiment analysis, where the reactions of consumers after using a product or service have been analyzed. But we are especially interested in how well machine learning can predict the level of satisfaction of clients.

3 Proposed Model

This paper will introduce a proposed model to predict the customer review score, in order to enhance the user experience with the e-commerce. The proposed model defines the problem as a multi-class classification problem. Predicting the precise value of the review score is the goal. There is a distinct category for each of the ratings that can be given in a review. As a result, there are five possible categories for an instance to be placed in: a rating of 1, 2, 3, 4, or 5. Classes can't share an instance, so an instance can only belong to one class. Figure 1 illustrates the proposed model's construction

3.1 Data Set

The research was done using 112,000 orders from the 'Brazilian E-Commerce Public Dataset by Olist' during a three-year period (2016–2018) [11]. Data has 9 csv files, namely olist_customers_dataset, olist_geolocation_dataset, olist_order_items_dataset, olist_order_payments_dataset, olist_order_reviews_dataset, olist_orders_dataset, olist_products_dataset, olist_sellers_dataset, product_category_name_translation.Data and 45 features. Data tables were merged together in one table as seen in Table 1.

3.2 Methodology

Experiments were run in Jupyter notebooks on a local PC using Python (version 3.7.3). These are the most important Python packages for creating, running, and

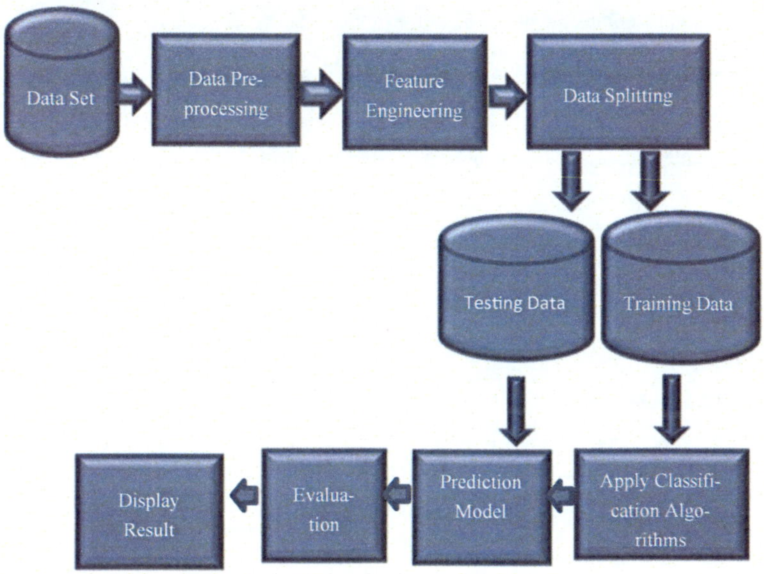

Fig. 1 Proposed model for predicting customer's review score

validating the experiments: tensorflow.keras, Numpy, Matplotlib.pyplot, Sklearn, Pandas, Seaborn, Imblearn, and plotly.express.

3.3 Data Preprocessing

Dataset was created by merging all source files. Due to the small number of null and duplicated values, null and duplicated instances were removed, and just 2.6% of the data was lost, but the remaining 97.4% were all usable. After this preprocessing, there were 113,195 instances left.

Using the train–test split() method, which is imported from the sklearn Python library's class model selection, the dataset is divided into training (0.67) and test (0.33).

StandardScaler(), an object imported from the class preprocessing module of the sklearn Python library, is used to standardize the X and y variables used in training and testing. The standardization of X and y data has resulted in the creation of new objects. The data are fitted, and the values of X and y are transformed into standard form using the fit transform () method.

Table 1 Brazilian e-commerce public dataset by Olist [11]

feature	Data type	Description
customer_id	object	Data key for orders. Each order's customer id is unique
customer_unique_id	object	An individual code for each consumer
customer_zip_code_prefix	int64	The client's zip code's initial five digits
customer_city	object	Name of city of clients
customer_state	object	Customer state. Short state name
order_id	object	Obtain a Numbered ID number via order
order_status	object	Identifying the current order status (delivered, shipped, etc.)
order_purchase_timestamp	object	Indicator of the date and time the item was purchased
order_approved_at	object	Time and date of payment authorization
order_delivered_carrier_date	object	Time and date of order's delivery from seller to logistics partners
order_delivered_customer_date	object	Date and time of delivery
order_estimated_delivery_date	object	This is the anticipated date of arrival
payment_sequential	int64	A consumer is allowed to pay for an order using multiple payment methods. If he follows through, a sequence will be produced
payment_type	object	Mode of payment that the buyer choose. (ballot, credit card)
payment_installments	int64	A customer-specified number of instalments
payment_value	float64	Price paid for a good or service
order_item_id	int64	The number of things contained in the same order is identified by a sequential number
product_id	object	Identification number for the product
seller_id	object	A number that is used to identify the vendor
shipping_limit_date	object	The date by which the seller must send off the order to the logistics partner is displayed
price	float64	The cost of the item
freight_value	float64	Price includes shipping costs for the order
product_category_name	object	A primary classification of products in Portuguese
product_name_lenght	float64	The total number of characters used to create the product name
product_description_lenght	float64	Retrieved character count from product description
product_photos_qty	float64	Total number of published product photo
product_weight_g	float64	The gram weight of the product
product_length_cm	float64	The length of the item is given in centimeters

(continued)

Table 1 (continued)

feature	Data type	Description
product_height_cm	float64	The product's height is indicated in centimeters
product_width_cm	float64	A width in centimeters for a product
review_id	object	ID number for this review
review_score	int64	The customer's rating on a scale from 1 to 5 indicating how satisfied they are
review_comment_title	object	Title taken from the Portuguese customer review
review_comment_message	object	Customer's comment in the review, written in Portuguese
review_creation_date	object	The date the customer's satisfaction survey was sent
review_answer_timestamp	object	A time stamp of the respondent's satisfaction survey answer
seller_zip_code_prefix	int64	Seller's first five-digit zip code
seller_city	object	Name of the seller's city
seller_state	object	Name of the seller's state
product_category_name_english	object	The name of category in English
geo_location_zip_code_prefix	int64	Column for zip code prefixes
geo_location_lat	float64	Location's latitude
geo_location_lng	float64	Location's longitude
geo_location_city	object	The city's name
geo_location_state	object	The state's name

3.4 *Exploratory Data Analysis (EDA)*

EDA is crucial for knowledge discovery [12]. EDA assists in comprehending the dataset and each feature, to proceed with feature engineering and model application. EDA should be dependent on the review score goal variable. No feature is strongly connected with the review score, as shown by the correlation matrix of numerical variables in Fig. 2. A high positive connection between (payment value and price), (product weight g and freight value, as well as product width cm), (product length cm and product width cm), and (product height cm and product weight g) were observed based on the correlation matrix. However, the majority of features do not appear to be useful for classification. So, the problem needed additional informative features to be created for modeling purposes. The data is also imbalanced. The number of 5-star reviews is exceptionally high, followed by 4-star reviews, and finally 1- and 2-star reviews, which are the lowest.

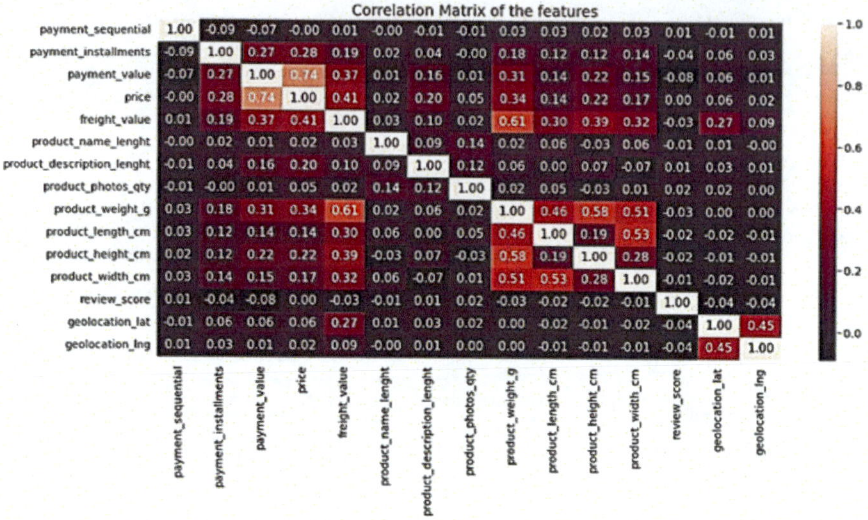

Fig. 2 Correlation heatmap of features before feature engineering

3.5 Feature Engineering

Figure 3 depicts the correlation between the features; it shows that not many columns are associated with the target (review score), indicating that additional useful features should be developed in order to model this issue. Columns will be analyzed for the potential to gather additional data and build novel features. Customers of online retailers may be happier if their orders arrive earlier than expected or less happy if they arrive later than promised. As a result, a new feature called 'Working Days Delivery Time Delta' will be developed to figure out how much time there is between the estimated and actual delivery dates. The following formula will be used if the negative arrives first and the positive arrives later: The difference between the expected and actual delivery times is the 'wd delivery time delta'. Other features were created explained in Table 2.

The new advanced features now have a greater correlation with the target (review score) according to Fig. 3.

3.6 Apply Classification Algorithms

After the null values were dropped, there were still 24 features and 75,640 training data count used to train the model using five different machine learning classification algorithms because of their effectiveness and popularity, all of which have been utilized in previous research published in the academic literature. These classification algorithms are Decision Trees [13], Random Forests [14], *K*-Nearest Neighbors [15],

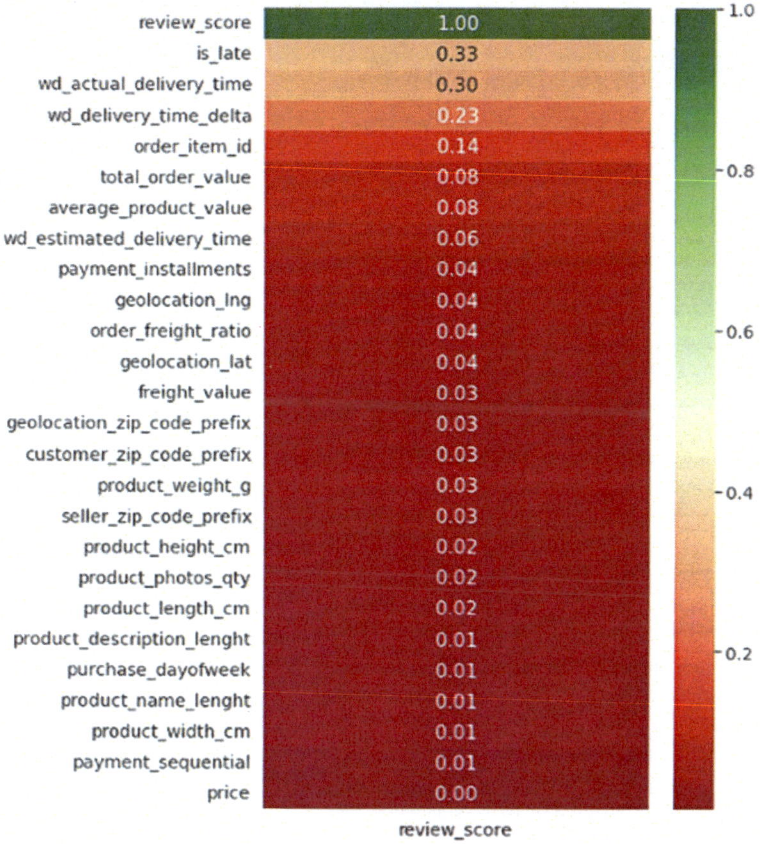

Fig. 3 Relations between new created features and review score

Support Vector Machines (SVMs) [16], and Extreme Gradient Boosting (XGBoost) [17].

4 Results and Discussion

In the case of multi-class classification, for the purposes of calculating precision, recall, and $f1$-scores, an averaging approach should be chosen. Micro, macro, and weighted averaging are the three main kinds of averaging [18]. In micro averaging, the total number of true positive, false negative, and false positive is used to calculate the metrics. Macro averaging involves first calculating metrics for each label independently, and then calculating the unweighted average score across all labels. The weighted average is calculated by separately calculating metrics for each label and then dividing by the number of components in each class. The label imbalance

Table 2 Description of new created feature

New feature	Description
Working days delivery time delta	Compute the amount of delay between the actual and anticipated delivery dates. The positive was delivered late, whereas the negative was given early. Calculated by the next formula: 'wd_delivery_time_delta' = 'wd_actual_delivery_time' -'wd_estimated_delivery_time'
Average product value	Get the mean price of the product. Poor quality at a lesser price might lead to dissatisfied customers. Calculated by the next formula: 'average_product_value' = 'order_products_value/order_items_qty'
Total order value	Compute the total cost of the order. A client may anticipate better order fulfillment if he spends more. Calculated by the next formula: 'total_order_value'] = 'order_products_value' + ' order_freight_value'
Working days actual delivery time	Determine the difference between the predicted and actual number of working days required for delivery. As a result, fewer variables like weekends and holidays could be considered when estimating delivery times. If no date was specified for when the order was delivered to the customer, the value 0 was returned
Order freight ratio	Compute the freight percentage of the order. Calculated by the next formula: 'order_freight_ratio' = 'order_freight_value'/'order_products_value'
Working days estimated delivery time	Compute the predicted and actual delivery times in terms of working days
Is late	This is a binary variable. Compare the estimated delivery date to the actual one and work out the difference. If the answer was negative, it arrived on time; if it was positive, it arrived late. Calculated by the next formula: If 'order_delivered_customer_date' > 'order_estimated_delivery_date'
Purchase day of week	Return the purchase day of the week, does it have an impact on consumer satisfaction?
Working days delivery time delta	Compute the amount of delay between the actual and anticipated delivery dates. The positive was delivered late, whereas the negative was given early. Calculated by the next formula: 'wd_delivery_time_delta' = 'wd_actual_delivery_time'-'wd_estimated_delivery_time'
Average product value	Get the mean price of the product. Poor quality at a lesser price might lead to dissatisfied customers. Calculated by the next formula: 'average_product_value' = 'order_products_value / order_items_qty'

is also considered in this average, which the other two do not. Table 3 summarizes the outcomes for different algorithms across several measures and average methods before feature selection and resampling techniques.

As observed from Table 3, in the case of micro averaging, all metrics have the same value. This is because there are always an equal amount of false positives and false negatives (If a prediction results in a false positive for one class, it also results in a false negative for another class). The most optimistic results come from micro averaging when comparing $f1$-scores (as it just globally counts the metrics). However, macro averaging yields the worst performance. In this scenario, the $f1$-score is an average across classes, and since the model's ability to predict most classes is poor, a low $f1$-score value is appropriate. Last but not least, the weighted average $f1$-score is somewhere in the middle because it takes label imbalance into account.

The training of the SVM algorithms with 24 features was unable to be executed. This confirms that: (1) the SVM does not support multi-class classification in its most basic form with too many features, but it can be used after breaking the multi-class classification problem down into binary classification problems. (2) With proper feature selection, it is possible to maintain model accuracy while reducing computational costs.

Analyzing the data shows that the performance is poor, and the imbalanced data is the reason for this performance as seen in Fig. 4, the distribution of classes in the training set. Class 5 was demonstrated to be associated with 43,504 in the training set, which explains the model's excellent performance when predicting this class. Since there are so few instances of the remainder of the classes in the training set, the model cannot effectively acquire the knowledge it needs to correctly predict them. In very unbalanced datasets, such as this one, this type of scenario frequently arises. We intend to make the classifier even more effective.

Table 3 Comparison of the results between different algorithms

Algorithm	Averaging method	Precision	Recall	$F1$-score	Training time in seconds
DT	Micro	0.48	0.48	0.48	< 1
	Macro avg	0.32	0.32	0.32	
	Weighted avg	0.49	0.48	0.48	
RF	Micro	0.65	0.65	0.65	28
	Macro avg	0.76	0.35	0.39	
	Weighted avg	0.68	0.65	0.57	
KNN	Micro	0.54	0.54	0.54	< 1
	Macro avg	0.35	0.30	0.31	
	Weighted avg	0.49	0.54	0.51	
XGB	Micro	0.62	0.62	0.62	20
	Macro avg	0.71	0.30	0.31	
	Weighted avg	0.65	0.62	0.52	
SVM	–	–	–	–	–

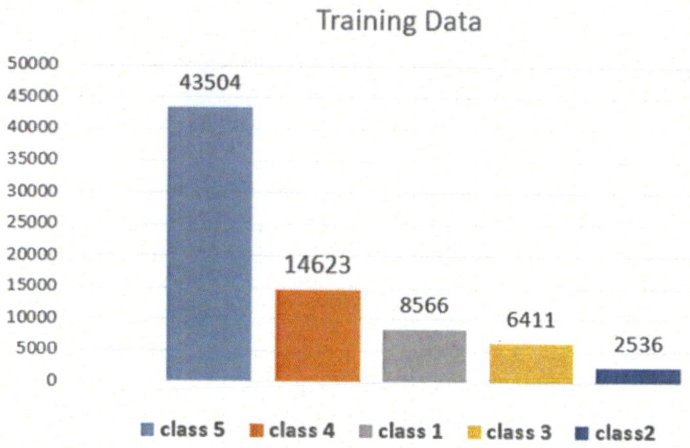

Fig. 4 Distribution of classes in the training data

4.1 Model Optimization

To further improve the performance of the classifier, feature importance will be investigated in this section along with a hyperparameter search.

4.2 Feature Selection

It is preferable to reduce the training features into a collection of variables that are as information rich as possible while training a machine learning model. This is because of three factors. (1) By limiting the number of features, the dimension, sparsity, and statistical significance of the training result are all reduced. Too many features mean more training samples are needed to achieve adequate representation. (2) Reducing sparsity helps to avoid over-fitting (the more features there are, the more complex the model becomes, and thus the greater the likelihood of over-fitting). (3) The effective use of compute power while reaching acceptable performance [19]. Feature selection is done using the SelectKBest() method, which is imported from the sklearn feature selection and used for feature selection and dimensionality reduction on sample sets according to the k highest scores [20]. The performance of the classifiers after feature selection is shown in Fig. 6.

4.3 Resampling

The target class-label is imbalanced There are a number of resampling techniques available for dealing with unbalanced classification issues [21–23]. Undersampling and oversampling are the most popular resampling techniques [24].

The two techniques were used to treat the imbalanced, then the models were trained again. Before feeding the data to the model for training, resampling is performed to ensure that all classes are represented fairly. Undersampling is a method for achieving data balance in which instances from the majority classes are deleted at random. With contrast, in oversampling, examples from the underrepresented classes are repeatedly sampled at random until there is equality across all classes. The distribution of classes before and after resampling is shown in Figs. 4 and 5. Resampling is done by a Python package namely imbalanced-learn which is offering several resampling techniques commonly used in datasets showing strong between-class imbalance. It is compatible with scikit-learn [25]. Figure 6 summarizes the performance of the models before and after applying feature selection and resampling techniques.

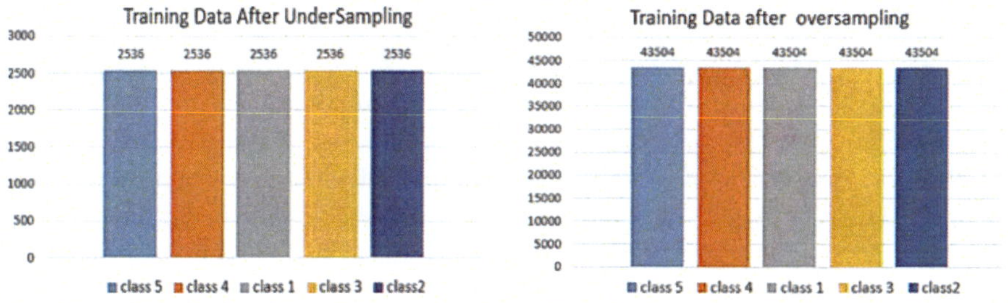

Fig. 5 Distribution of classes before and after resampling

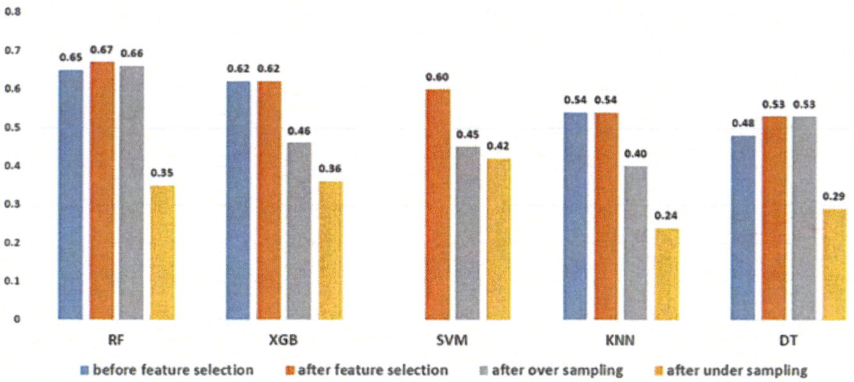

Fig. 6 Performance of the models before and after feature selection, oversampling, and undersampling techniques

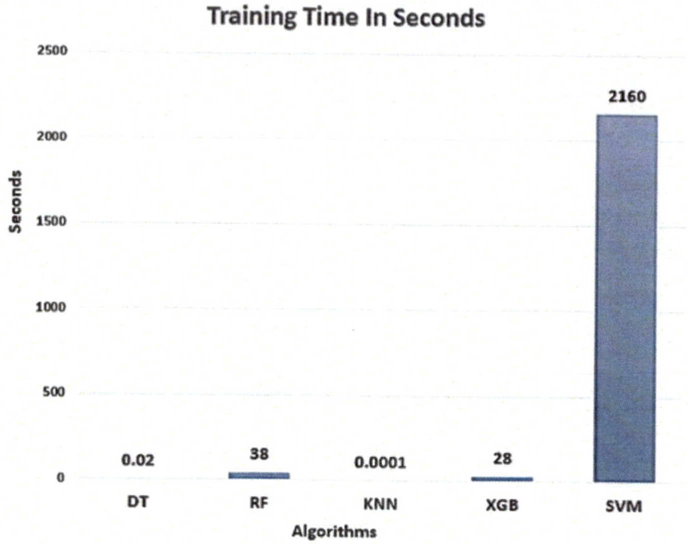

Fig. 7 Training time for the models in seconds after feature selection

As observed in Fig. 6, feature selection enhanced the performance of models (RF) and (DT). SVM model is worked after feature reduction, while (XGB) and (KNN) models not affected by decreasing the number of features. The performance of the models has not improved after applying resampling techniques. According to the result, (RF) model can predict the review score by $F1$-score 0.67 better than (XGB) with $F1$-score 0.62, (SVM) with $F1$-score 0.60, (KNN) with $F1$-score 0.54 and (DT) with $F1$-score 0.53. SVM has the longest training computation time. On the other hand, DT has the fastest training computation as shown in Fig. 7. The performance of the current study overcomes the earlier work [7], which used (XGB) algorithm and obtained an $F1$-score of 0.60. The current study on the other hand obtained an $F1$-score of 0.62 for (XGB) and an $F1$-score of 0.67 for (RF) after feature selection.

As observed in Fig. 7, the time for (DT) and (RF) models is still less than 1 s before and after feature selection, which means (DT)and (RF) models have the fastest training computation time regardless of the number of features, although accuracy improved after feature selection but didn't require more time.

4.4 Key Drivers of Customer Satisfaction

The following features have been discovered, as shown in Fig. 8, to have an impact on customer satisfaction: 'wd_delivery_time_delta', 'order_freight_ratio', 'geolocation_lng',', 'geolocation_lat', 'average_product_value', 'freight_value', 'total_

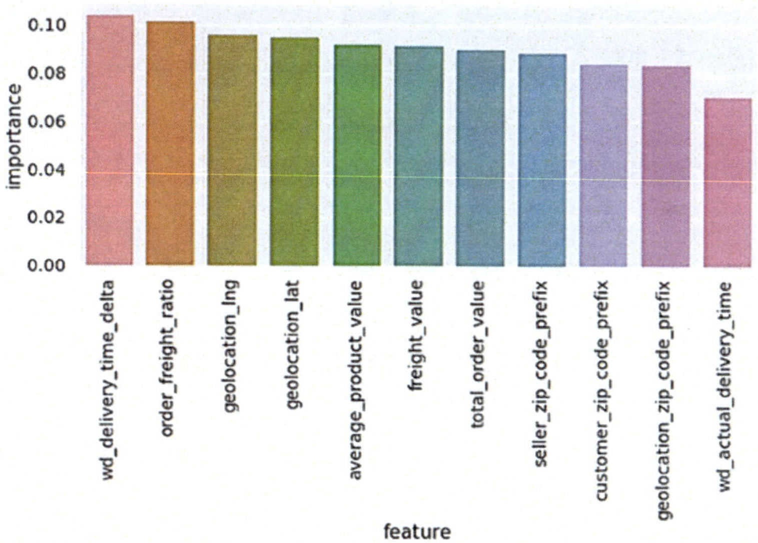

Fig. 8 Important features form Random Forest model

order_value', 'seller_zip_code_prefix',customer_zip_code_prefix', , 'geolocation_zip_code_prefix', 'geolocation_lng', 'wd_actual_delivery_time'.

During the training phase, the (RF), and (XGB) models found out which variables were most important features (Figs. 8, 9), and their relative importance was the same for all algorithms. The 'wd delivery time delta' variable is the most crucial, followed by the 'order freight ratio' variable When there is a big difference between the expected delivery date and the actual delivery date, or when freight costs make up a big part of the total cost, Brazilian customers are not satisfied, Therefore, importance should be paid to these variables.

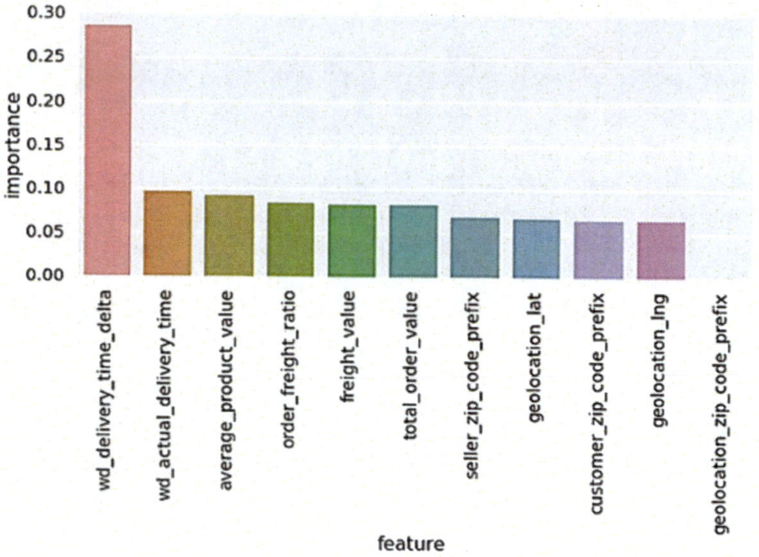

Fig. 9 Important features form XGB model

5 Conclusion

There is now more competition among online retailers thanks to the rapid expansion of online shopping made possible by COVID-19. Businesses in today's market need to focus on customer happiness in order to stay competitive. This study aims to improve the services offered by e-commerce by analyzing the opinions of previous buyers to determine their degree of satisfaction. Predicting the rating prior to the user submitting a review score that represents customer satisfaction was accomplished using five different classification algorithms: Decision Tree (DT), Random Forest (RF), XGBoost (XGB), support vector machine (SVM), and K-Nearest Neighbors (KNN). Methods were compared and contrasted to find the best one. The $F1$-score of the review score may be predicted 0.67 more accurately by the (RF) model compared to the other models. Location of the customer and seller, delivery time for the product, the mean price of the product, total cost of the order, and freight ratio are singled out as crucial factors in an effort to better understand their impact on customers' overall happiness.

6 Future Work

The performance enhancement of the models is recommended as a future study of the current work. One concept for future research is to approach this topic as a binary classification problem using deep learning algorithms and try more techniques to address imbalanced problems.

References

1. Novitasari D, Napitupulu BBJ, Abadiyah S, Silitonga N, Asbari M (2022) Linking between brand leadership, customer satisfaction, and repurchase intention in the e-commerce industry. Int J Soc Manage Studies 3(1):280–289
2. Gunawan I (2022) Customer loyalty: the effect customer satisfaction, experiential marketing and product quality. KINERJA: Jurnal Manajemen Organisasi dan Industri 1(1):35–50
3. Liu X, Kao Z (2022) Research on influencing factors of customer satisfaction of e-commerce of characteristic agricultural products. Proc Comput Sci 199:1505–1512
4. Metsai AI, Tabakis IM, Karamitsios K, Kotrotsios K, Chatzimisios P, Stalidis G, Goulianas K (2022) Customer journey: applications of AI and machine learning in E-commerce. In: Interactive mobile communication, technologies and learning. Springer, Cham, pp 123–132
5. Khalid A, Lee O, Choi M, Ahn J (2018) The effects of customer satisfaction with e-commerce system. J Theor Appl Inf Technol 96(2):481–491
6. Bouzakraoui MS, Sadiq A, Alaoui AY (2020) Customer satisfaction recognition based on facial expression and machine learning techniques. Adv Sci, Technol Eng Syst J 5:594–594
7. Kotsokechagia M (2021) Predictive model for customer satisfaction in e-commerce
8. Moon NN, Talha IM, Salehin I (2021) An advanced intelligence system in customer online shopping behavior and satisfaction analysis. Curr Res Behav Sci 2:100051
9. Hussain Y (2019) Predicting customer satisfaction with product reviews: a comparitive study of some machine learning approaches (Master's thesis)
10. Zhao Y, Xu X, Wang M (2019) Predicting overall customer satisfaction: Big data evidence from hotel online textual reviews. Int J Hosp Manag 76:111–121
11. Brazilian E-Commerce Public Dataset by Olist. Brazilian E-Commerce Public Dataset by Olist | Kaggle, /datasets/olistbr/brazilian-ecommerce. Accessed 19 Dec 2022
12. Milo T, Somech A (2020, June) Automating exploratory data analysis via machine learning: an overview. In: Proceedings of the 2020 ACM SIGMOD international conference on management of data, pp 2617–2622
13. Mahesh B (2020) Machine learning algorithms-a review. Int J Sci Res (IJSR).[Internet], 9, 381–386.
14. Parmar A, Katariya R, Patel V (2018) A review on random forest: an ensemble classifier. In: International conference on intelligent data communication technologies and internet of things. Springer, Cham. pp 758–763
15. Taunk K, De S, Verma S, Swetapadma A (2019) A brief review of nearest neighbor algorithm for learning and classification. In: 2019 International conference on intelligent computing and control systems (ICCS). IEEE, pp 1255–1260
16. Abdulqader DM, Abdulazeez AM, Zeebaree DQ (2020) Machine learning supervised algorithms of gene selection: a review. Mach Learn 62(03):233–244
17. XingFen W, Xiangbin Y, Yangchun M (2018) Research on user consumption behavior prediction based on improved XGBoost algorithm. In: 2018 IEEE international conference on big data (big data). IEEE, pp 4169–4175
18. Grandini M, Bagli E, Visani G (2020) Metrics for multi-class classification: an overview. arXiv preprint arXiv:2008.05756

19. Cai J, Luo J, Wang S, Yang S (2018) Feature selection in machine learning: a new perspective. Neurocomputing 300:70–79
20. Feature selection. (n.d.). Scikit-learn. Retrieved October 6, 2022, from
21. Supervised learning. (n.d.). Scikit-learn. Retrieved March 25, 2022, from
22. Haixiang G, Yijing L, Shang J, Mingyun G, Yuanyue H, Bing G (2017) Learning from class-imbalanced data: Review of methods and applications. Expert Syst Appl 73:220–239
23. Rout N, Mishra D, Mallick MK (2018) Handling imbalanced data: a survey. In: International proceedings on advances in soft computing, intelligent systems and applications. Springer, Singapore, pp 431–443
24. Mohammed R, Rawashdeh J, Abdullah M (2020) Machine learning with oversampling and undersampling techniques: overview study and experimental results. In: 2020 11th international conference on information and communication systems (ICICS). IEEE, pp 243–248
25. Imbalanced-learn documentation — Version 0.10.1, https://imbalanced-learn.org/stable/index.html. Last accessed 25 Oct 2022

Digital Transformation: Network Infrastructure

An HBM3 Processing-In-Memory Architecture for Security and Data Integrity: Case Study

Dina Fakhry, Mohamed Abdelsalam, M. Watheq El-Kharashi, and Mona Safar

Abstract The von Neumann bottleneck is imposed due to the explosion of data transfers and emerging data-intensive applications in heterogeneous system architectures. The conventional computation approach of transferring data to CPU is no longer suitable especially with the cost it imposes. Hence, high-performance data processing mechanisms are needed, which may be demonstrated in bringing computation closer to data. Generating insight where data is stored helps deal with energy efficiency, low latency, storage bus bandwidth and security. In this paper, we explore near data processing focusing on Processing-In-Memory (PIM) to address data integrity through the inclusion of CRC compute units within the DRAM controller as well as address data security through encompassing an Advanced Encryption Standard (AES) acceleration unit. Our experimental results are based against an HBM3 platform. We achieved a computing memory with minimal overhead of commands, area and same power consumption compared to a non-computing one.

Keywords DRAM · HBM3 · Near data processing · Processing-In-Memory

1 Introduction

Throughout the years, the High-Performance Computing (HPC) applications' computational cost has notably declined. The cost of data transfer however did not scale down proportionally with computation cost. The system bandwidth availability represented in I/O, network, and CPU memory hierarchy in addition to low data locality hinder the performance of data-intensive applications. That is mainly caused due to sizeable energy being consumed when moving around such massive data. Another wide range of application domains have also emerged with the evolution of different computing platforms all processing very large datasets. Various applications

D. Fakhry (✉) · M. W. El-Kharashi · M. Safar
Department of Computer and Systems Engineering, Ain Shams University, Cairo, Egypt
e-mail: Dina.walid@outlook.com

M. Abdelsalam
Siemens EDA, Cairo, Egypt

D. Magdi et al. (eds.), *Green Sustainability: Towards Innovative Digital Transformation*, Lecture Notes in Networks and Systems 753,
https://doi.org/10.1007/978-981-99-4764-5_18

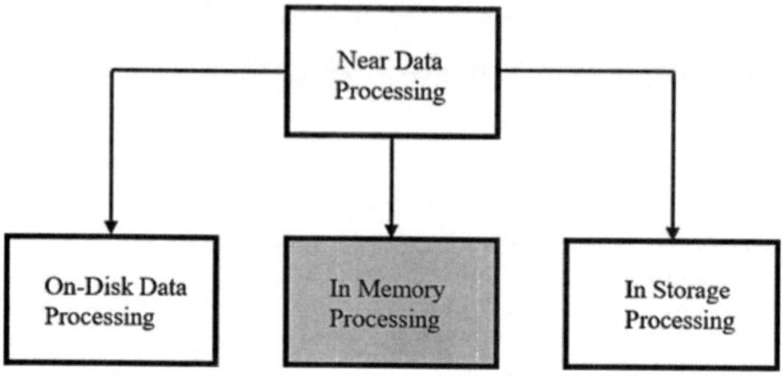

Fig. 1 Near data processing classification

including database acceleration, machine learning and artificial intelligence can perform better and become more scalable if the "move process to data" paradigm is applied. Other applications that can also benefit from such paradigm include compression, encryption and data integrity applications in combination with digital signatures. Several techniques are encapsulated within Near Data Processing (NDP) namely On-Disk Data Processing (ODDP) [1], In-Memory Computing (IMC) or Processing-In-Memory (PIM) [2] and In-Storage Computing (ISC) or In-Storage Processing (ISP).

Figure 1 depicts the different domains of NDP that have been researched in literature.

ODDP associates with the arrangement where secondary storage devices usually Hard Disk Drives (HDDs) are further equipped with memory capacity and computing to process stored data [1] .

On the other hand, PIM mainly associates with equipping RAM with data processing capability [3, 4]. Different innovations and PIM-enablers have been thoroughly discussed in literature like emergence of three-dimensional (3D)-stacked memory dies and resistive memory technologies as well as usage of byte-addressable Non-Volatile Memories (NVM) and usage of the memory cells themselves that reside in a memory chip.

ISP refers to processing data where it resides such that computation tasks are offloaded to the storage devices. With memory hierarchy consisting of storage, main memory and multiple cache levels, the processing location of data is what determines the used technique. The memory system hierarchy is demonstrated in Fig. 2.

In this work, we present an architecture for High Bandwidth Memory 3 (HBM3) memory controller to integrate a CRC computation unit and assess in-memory computing with regard to internal design trade-offs such as area and power consumption which are key factors of interest. Our main aim is not to compare the CRC application itself with other implementations of CRC or discuss how a stacked memory system can be efficiently protected from large-granularity failures by inclusion of error-detecting codes. However, it is mainly discussing CRC as an application example for in-memory computing. Furthermore, we incorporate an AES computation unit

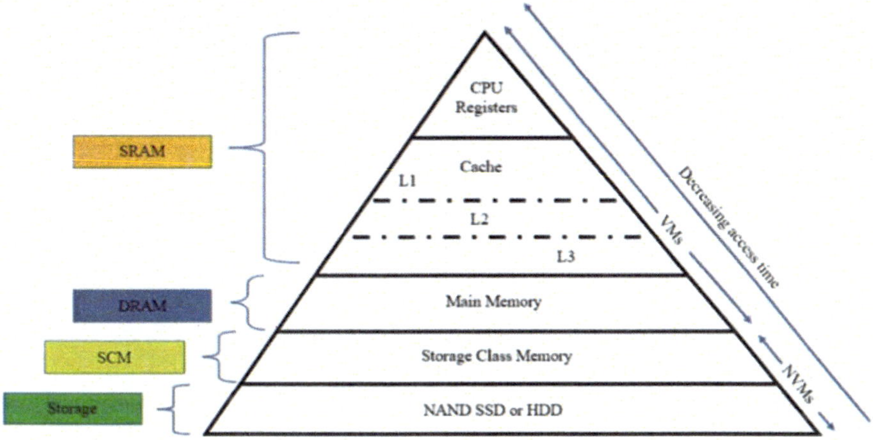

Fig. 2 Memory system hierarchy

within the HBM3 controller with different implementations, one targeting optimized area and another targeting optimized performance and compare the cycle difference and various design trade-offs. We extend the DRAM state diagram to include PIM commands entwining them with standard DRAM commands that allows servicing and operating the standard memory requests during computation.

The paper is structured as follows. Section 2 gives an outline on previous work and PIM architectures. In Sect. 3, we give an overview on HBMs and their applications. In Sect. 4, we provide some background regarding AES and CRC fundamentals since they are part of the PIM architecture we target. In Sect. 5, we present the experimental design. We discuss the extensions made to the HBM memory controller to allow for PIM and evaluate the performance. We conclude in Sect. 6 and propose some future work directions.

2 Background and Related Work

In this section, we explore some of the significant work and architectures concerning PIM as well as some of the challenges faced. Even though computational memories are beneficial to overcome the memory wall, the bandwidth gap between host and storage and to reduce the data movement, they should only be used when needed. Computational memories should be employed whenever the bandwidth is a bottleneck with large data transfers associated or when data delivery can bypass the host. A host CPU however can be thought of as a better processing option if:

- An application is compute-intensive with small data transfer incurred
- Computation itself involves small data
- Algorithm is highly sequential.

Identifying the opportunities from which an application can benefit from PIM has been thoroughly discussed [3].

Various PIM studies have focused on utilizing 3D-stacked memories, mainly Hybrid Memory Cube (HMC) and HBM, to perform the computation tasks where increased capacity and bandwidth can be provided. Ahn et al. have exploited data locality to implement compute units and extend the host processor's ISA with PIM-enabled instructions (PEIs) to trigger near memory computation [5]. They have explored different target applications including large-scale graph processing [6–8] , in-memory data analytics, machine learning and data mining to assess the efficiency of their design.

Another important operation extensively researched with PIM is the multiply-accumulate (MAC) operation. MAC is essential especially for Digital Signal Processing (DSP) and graphics' applications as well as matrix computations for neural networks. Shin et al. evaluated different design trade-offs in terms of energy, area and performance by augmenting the DRAM itself specifically LPDDR4 with MAC units [9]. Others have opted for using HMC [10] and HBM [11] memories to develop PIM frameworks related to MAC. Lee et al. mapped each bank in an HBM2 to a computing unit allowing it to act as a multi-core system [11].

Other PIM design proposals include applications targeting MapReduce [12, 13], machine learning [14, 15], database [16, 17] and cryptography [18].

Several barriers are still hindering the adoption of NDP and restraining their widespread usage. Adoption of new programming paradigms, providing cache coherency and virtual memory support are considered key challenges [19]. Several works addressed cache coherence for PIM with proposals ranging from conventional methods to new approaches [5, 20–23]. Virtual memory challenges, especially because of hosts and PIM modules sharing the same memory, have also been discussed in previous work [5, 19, 24, 25]. Authors have adopted known multi-threading techniques for PIM like OpenMP, MPI and CUDA, OpenCL that may be suitable for platforms with multiple heterogeneous processor cores [8, 25–27].

3 HBM3 and Applications

HBM3 is a DRAM that aims to achieve high-speed, low power operation through vertically stacking memory chips [28]. It provides a wide-interface architecture that is divided into independent channels where a DRAM stack is anticipated to support up to 16 channels. Channels in an HBM3 are not essentially synchronous to one another with each channel subdivided into two individual semi-independent channels known as Pseudo-Channels (PCs). Commands and addresses are being captured on separate row and column buses with the semi-independency of PCs evident in sharing such buses however decoding and executing commands individually. For outputting data, an HBM3 supports a burst length of 8 with each burst comprised of 32 bits. The high bandwidth, low power consumption and great scalability of HBM have made it a great option for network systems, HPC, graphics and data centers.

JEDEC's Task Group TG42_7 has made proposals concerning development of processing memories with Samsung proposing an HBM-PIM architecture.

4 AES and CRC Fundamentals

AES is considered an alternative form of the Rijndael block cipher and is highly used for data protection and security. A fixed block size of 128 bits is used as the plaintext with a variable key size of 128, 192 or 256 bits dependent on whether AES-128, AES-192 or AES-256 is employed, respectively. An AES encryption process goes through some steps for N rounds with a round key being derived from the main key for each of these rounds by going through a key schedule algorithm [29]. For our use case, only AES-128 was implemented and hence the value of N is 10. The 128-bit plaintext is organized as a 4×4 array of bytes (column-major order array of 16 bytes) known as the state array where the output state of each step/round is used as the input for the next step/round or returned as the final result at the end of N rounds. A number of steps make up a single round and they are listed as follows:

- SubBytes
  - Each byte of the state array is replaced with another byte according to a look-up table known as S-Box.
- ShiftRows
  - The bytes of the input state array are circularly shifted to the left by a fixed offset amount (zero, one, two or three positions) according to the row index on which the bytes lie.
- MixColumns
  - A linear transformation where a matrix multiplication over the Galois Field ($GF(2^8)$) between a fixed matrix and the input state matrix is performed to generate the output state.
- AddRoundKey
  - A bitwise XOR operation is used to combine each byte of the state with a round key byte with the round key derived based on the key schedule algorithm.

All rounds are comprised of these four steps with an exception to the final round which does not include the mix columns step.

Figure 3 demonstrates the different steps performed through an encryption process.

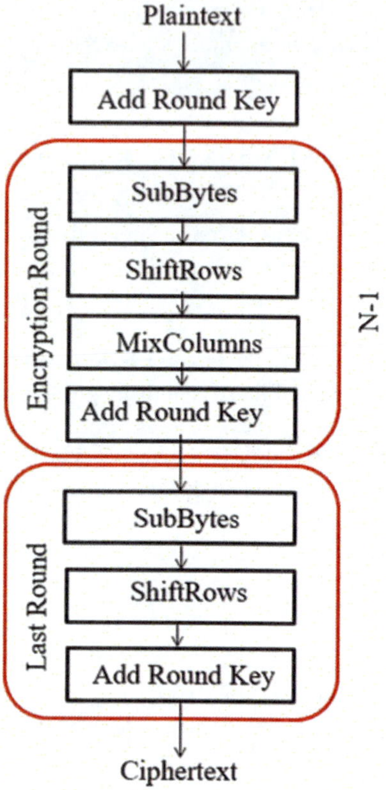

Fig. 3 AES encryption

On the other hand, CRCs are redundancy checks that can be used to ensure data integrity through error detection (but not correction) in digital data. CRC32 is considered one of the most used polynomials due to being widespread in various applications and protocols that include Gzip, SATA, ethernet and others. Several CRC32 polynomials are available including CRC-32C (Castagnoli), CRC-32K (Koopman {1, 3,28}), CRC-32K2 (Koopman {1,1,30}), CRC-32Q however, our main focus is on the normal CRC32 polynomial given by the following equation:

$$x^{32} + x^{26} + x^{23} + x^{22} + x^{16} + x^{12} + x^{11} + x^{10} + x^{8} + x^{7} + x^{5} + x^{4} + x^{2} + x + 1$$

5 Experimental Design

In this section, we discuss the extensions made to the conventional HBM3 state diagram to add new PIM commands for triggering in-memory computation and introduce the platform used to test our design.

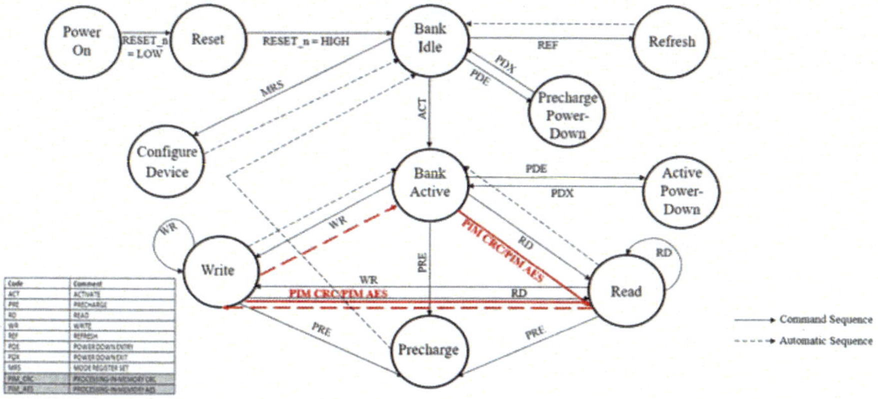

Fig. 4 Modified DRAM state diagram

5.1 HBM3 State Diagram Extension

An HBM3 DRAM features DDR commands entered on both rising and falling clock edges. Commands are either row or column commands covering different functionalities such as activate, precharge, refresh, read and write. DRAM commands issued by the processor are converted using a memory controller into standard read and write commands. A state machine is used to provide a simplified demonstration of the accepted state transitions and the related commands to control them. We augmented the standard state machine to include PIM commands and illustrate their behavior as depicted in the modified state diagram in Fig. 4.

The only addition to the main state machine are the PIM CRC and the PIM AES commands. The datapath for the PIM command whether CRC or AES can be looked at as one similar to a read and a write command combined, hence there was no need for additional states. The starting point for our new PIM commands is the same as that of a read command which can be illustrated from the state diagram by the sharing of the "*read*" state. Interspersing PIM commands with standard memory commands has allowed for interleaved servicing of commands without blocking any. However, it is the host controller's responsibility to ensure the minimum command to command delay so that no two memory-accessing commands overlap. During compute cycles, the memory is available for accessing by other standard non-PIM commands. Figure 5 shows interleaved Read-PIM AES commands. A new PIM command is ignored during compute cycles to ensure that compute units are available when its read latency elapses.

Unlike the work presented in [11] where they needed to set a command for PIM writing and another for reading causing some overhead, we were able to automate it in a single PIM CRC or PIM AES command that was added to the command decoder to carry out a CRC or encryption operation, respectively. For the PIM AES

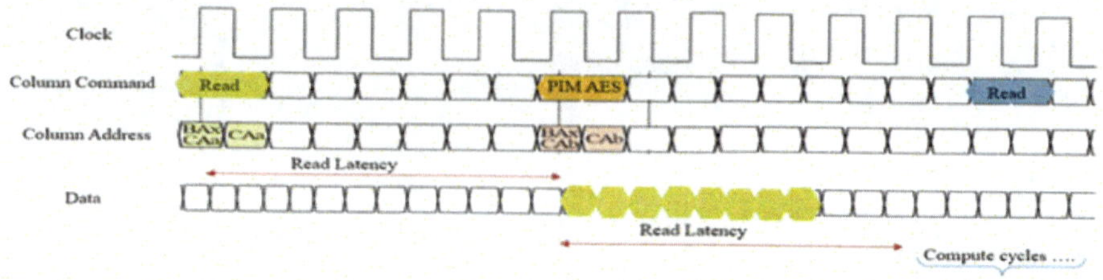

Fig. 5 Interleaved standard - PIM commands

command, it even solely triggers the performance of one step after the other within each round unlike the case presented in [18].

5.2 *HBM3-PIM Architecture*

Before including the PIM computation units in the HBM3's memory controller, the controller was modeled in Verilog HDL and verified using Siemens EDA's Veloce Prototyping System [30].

Figure 6 depicts a simplified block diagram for the modified HBM3 controller architecture which includes a CRC-32 engine and an AES accelerator that act as the PIM computation unit.

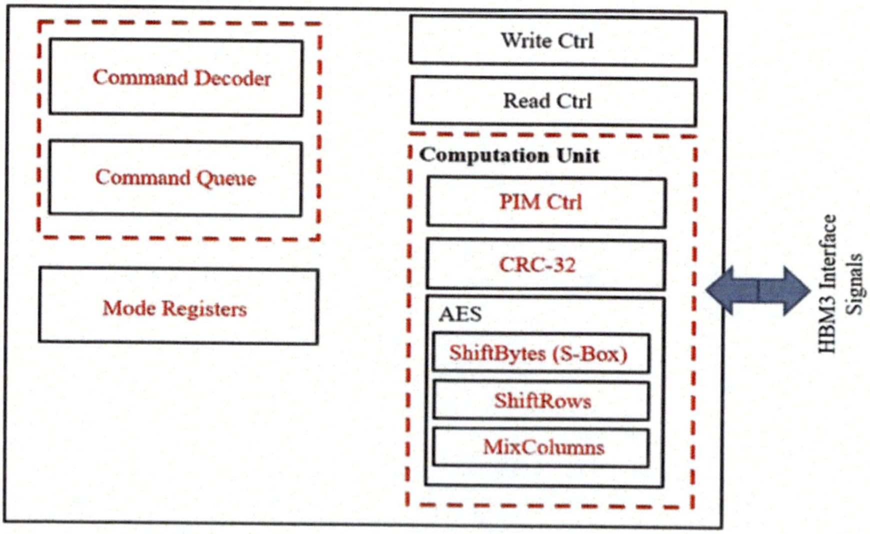

Fig. 6 HBM3 PIM block architecture

CRCs are necessary to maintain data integrity. An optional feature could be calculating the CRC or hash value based on source data then comparing it with a pre-calculated value supplied by the initiator. For our HBM3 controller, we were able to support a memory granularity of 128 bits such that a whole burst of 8 can be fetched in 2 clock cycles. That provided great advantage for supporting AES-128 where we made use of using the first 128 bits as the plaintext and the second 128 bits as the initial round key.

The command decoder was augmented to decode the new PIM CRC and PIM AES commands. Comparing either PIM command with a normal read command, the main difference is the interpretation of what is captured on the interface signals. For a normal read command, interface signals are converted to Stack ID (SID), Bank Address (BA) and Column Address (CA) with the CA being the source address from which data is fetched and returned to host. However, for the PIM command where data needs to be fetched and stored in memory, interface signals are interpreted as source and destination addresses (SRC_CA and DST_CA). We made use of HBM3's **Reserved for Future Use** (RFU) mode registers to:

- Configure PIM-mode entry/exit so as to accept/reject PIM commands in addition to the standard ones
- Load the SID and BA to which the source and destination addresses of the PIM command belong

The use of such mode registers to make use of extra address space is a key enabler to condense reading and writing in a single command. The command queue was also augmented to schedule both standard and PIM commands and generate the needed memory request in time.

Table 1 shows how the decoding of PIM commands differ from standard ones with the read command taken as a comparison where the main difference is how the interface signals are translated. Another difference is the addition of a new opcode that is demonstrated on the rising clock edge C0–C3 in case of the read command which is shown in gray and C0–C2 in case of the PIM CRC or PIM AES command shown in red.

Table 1 PIM command decoding

Command/symbol	Clock edge[a]	Clock cycle[b]							
		C0	C1	C2	C3	C4	C5	C6	C7
Read/RD	R	H	L	H	L	PC	SID0/V	SID1/V	BA0
	F	BA1	BA2	BA3	CA0	CA1	CA2	CA3	CA4
PIM CRC	R	H	L	L	V	PC	V	DST_CA0	DST_CA1
	F	DST_CA2	DST_CA3	DST_CA4	SRC_CA0	SRC_CA1	SRC_CA2	SRC_CA3	SRC_CA4
PIM AES	R	L	H	H	V	PC	V	DST_CA0	DST_CA1
	F	DST_CA2	DST_CA3	DST_CA4	SRC_CA0	SRC_CA1	SRC_CA2	SRC_CA3	SRC_CA4

[a] *R* for Raising and *F* for Falling
[b] *H* for High and *L* for Low

5.3 Performance Evaluation

In order to account for the performance gains obtained by PIM, we compiled and ran the design using Xilinx's XCVU440-FLGA2892-2-E in Siemens' prototyping flow to measure the area and power difference incurred by adding PIM units.

Regarding AES, different implementation approaches were used for modeling the S-Box module resulting in different area and performance results. For realizing a better performance implementation, a look-up table was used to build the S-Box module resulting in a lower number of cycles needed for calculating the substitute bytes however negatively impacting area. On the other hand, for better area realization, we made use of multiplying by 03 in Galois Field GF(256) and computing the affine transformation to calculate the substitute bytes resulting in a lower number of LUTs beings used. We noted that a total of 284 clock cycles were saved in case of the optimized performance implementation in simulation.

Table 2 provides the comparison between HBM and HBM with PIM units added in terms of power and area.

It can be noted that only an 8% area difference is recognized when adding PIM computation units with optimized AES performance and about 4% in case of area optimization whereas nearly no power difference is seen. This shows that low power can still be achieved while simultaneously reducing data movement that would have otherwise been needed if CRC or encryption was to be computed at the host side, leading to a better performance. The speed and latency of the on-board DRAM and its associated controller are also factors that affect the overall performance of the model.

Table 2 Area and power measurements

Design trade-offs		HBM	HBM PIM	
			With AES optimized for performance	With AES optimized for area
Area	LogicLUTs	7380	8369	8187
	LUTRAMs	280	280	280
	FFs	10,900	11,893	11,391
	RAMB18	10	14	14
Total on-chip power (W)		8.800	8.848	8.889

6 Conclusion

The exploding data volumes that require processing and enormous amounts of transportation from the memory system to the processor have led to challenges that triggered investigation of IMC opportunities. We presented some of the different architectures proposed in literature. we addressed data integrity and security by adding CRC and AES compute units within a standard HBM3 controller achieving minimal instruction, area and power overhead. As part of future work, we propose giving the flexibility of adding other CRC functions rather than being limited to CRC-32. We also propose adding the circuitry needed for decrypting the ciphertext and giving the flexibility for computing AES-192 or AES-256.

In addition, some of the future directions are suggested as follows:

- Introducing programming paradigms and prevalent software framework which decreases the integration barrier of NDP elements to optimize data placement and movement within the system is needed.
- Considering applying in-memory and in-storage computing simultaneously as a hybrid integrated architecture to exploit resources and achieve best possible processing speed and energy efficiency [31].

References

1. Mishra M, Somani A (2017) On-disk data processing: issues and future directions. Comput Res Repos
2. Huang D, Qin Z, Liu Q, Podhorszki N, Klasky S (2016) Identifying challenges and opportunities of in-memory computing on large HPC systems. J Parall Distrib Comput 164:106–122
3. Ghose S, Boroumand A, Kim JS, Gómez-Luna J, Mutlu O (2019) Processing-in-memory: a workload-driven perspective. IBM J Res Dev 63(6):3:1–3:19
4. Mutlu O, Ghose S, Gómez-Luna J, Ausavarungnirun R (2020) A modern primer on processing in memory
5. Ahn J, Yoo S, Mutlu O, Choi K (2015) PIM-enabled instructions: a low-overhead, locality-aware processing-in-memory architecture. In: 2015 ACM/IEEE 42nd annual international symposium on computer architecture (ISCA), Portland, OR, June 2015, pp 336–348
6. Ahn J, Hong S, Yoo S, Mutlu O, Choi K (2015) A scalable processing-in-memory accelerator for parallel graph processing. In: 2015 ACM/IEEE 42nd annual international symposium on computer architecture (ISCA), Portland, Oregon, June 2015, pp 105–117
7. Nai L, Hadidi R, Sim J, Kim H, Kumar P, Kim H (2017) GraphPIM: enabling instruction-level PIM offloading in graph computing frameworks. In: 2017 IEEE international symposium on high performance computer architecture (HPCA), Austin, TX, Feb 2017, pp 457–468
8. Zhang D, Jayasena N, Lyashevsky A, Greathouse JL, Xu L, Ignatowski M (2014) TOP-PIM: throughput-oriented programmable processing in memory. In: Proceedings of the 23rd international symposium on high performance parallel and distributed computing, ACM, Vancouver, BC, June 2014, pp 85–98
9. Shin H, Kim D, Park E, Park S, Park Y, Yoo S (2018) McDRAM: low latency and energy-efficient matrix computations in DRAM. IEEE Trans Comput Aided Design Integr Circuits Syst 37(11): 2613–2622

10. Jeon D-I, Park K-B, Chung K (2017) HMC-MAC: processing-in memory architecture for multiply-accumulate operations with hybrid memory cube. IEEE Comput Archit Lett 1
11. Lee WJ, Kim CH, Paik Y, Park J, Park I, Kim SW(2019) Design of processing-inside-memory optimized for DRAM behaviors. IEEE Access 7:82633–82648
12. Pugsley SH, Jestes J, Zhang H, Balasubramonian R, Srinivasan V, Buyuktosunoglu A, Davis A, Li F (2014) NDC: analyzing the impact of 3D-stacked memory+logic devices on MapReduceWorkloads. In: 2014 IEEE international symposium on performance analysis of systems and software (ISPASS), Monterey, CA, Mar 2014, pp 190–200
13. Farmahini-Farahani A, Ahn JH, Morrow K, Kim NS (2015) NDA: near-DRAM acceleration architecture leveraging commodity DRAM devices and standard memory modules. In: 2015 IEEE 21st international symposium on high performance computer architecture (HPCA), Burlingame, CA, Feb 2015, pp 283–295
14. Lee VT, Mazumdar A, del Mundo CC, Alaghi A, Ceze L, Oskin M (2018) Application codesign of near-data processing for similarity search. In: 2018 IEEE international parallel and distributed processing symposium (IPDPS), Vancouver, BC, May 2018, pp 896–907
15. Liu J, Zhao H, Ogleari MA, Li D, Zhao J (2018) Processing in-memory for energy-efficient neural network training: a heterogeneous approach. In: 2018 51st annual IEEE/ACM international symposium on microarchitecture (MICRO), Fukuoka, Japan, Oct 2018, pp 655–668
16. Santos PC, Oliveira GF, Tomé DG, Alves MAZ, Almeida EC, Carro L (2017) Operand size reconfiguration for big data processing in memory. In: Proceedings of the conference on design, automation and test in Europe, European Design and Automation Association, Lausanne, Switzerland, Mar 2017, pp 710–715
17. Xi SL, Augusta A, Athanassoulis M, Idreos S (2015) Beyond the wall: near-data processing for databases. In: Proceedings of the 11th international workshop on data management on new hardware (DaMoN'15), Association for Computing Machinery, Melbourne, VIC, May–June 2015, Article 2, pp 1–10
18. Reis D, Geng H, Niemier M, Hu XS (2022) IMCRYPTO: an in-memory computing fabric for AES encryption and decryption. IEEE Trans Very Large Scale Integr (VLSI) Syst 30(5):553–565
19. Santos PC, Forlin BE, Carro L (2021) Providing plug N' play for processing-in-memory accelerators. In: 2021 26th Asia and South Pacific design automation conference (ASP-DAC), virtual conference, Jan 2021, pp 651–656
20. Santos PC, de Lima JPC, de Moura RF, Alves MAZ, Beck ACS, Carro L (2021) Enabling near-data accelerators adoption by through investigation of datapath solutions. Int J Parall Program 49:237–252
21. Hsieh K, Ebrahim E, Kim G, Chatterjee N, O'Connor M, Vijaykumar N, Mutlu O, Keckler SW (2016) Transparent offloading and mapping (TOM): enabling programmer-transparent near-data processing in GPU systems. In: Proceedings of the 43rd international symposium on computer architecture, Seoul, Republic of Korea, June 2016, pp 204–216
22. Boroumand A, Ghose S, Patel M, Hassan H, Lucia B, Hsieh K, Malladi KT, Zheng H, Mutlu O (2017) LazyPIM: an efficient cache coherence mechanism for processing-in-memory. IEEE Comput Archit Lett 16(1):46–50
23. Boroumand A, Ghose S, Patel M, Hassan H, Lucia B, Ausavarungnirun R, Hsieh K, Hajinazar N, Malladi KT, Zheng H, Mutlu O (2019) CoNDA: efficient cache coherence support for neardata accelerators. In: Proceedings of the 46th international symposium on computer architecture, Phoenix, Arizona, June 2019, pp 629–642
24. Drumond M, Daglis A, Mirzadeh N, Ustiugov D, Picorel J, Falsafi B, Grot B, Pnevmatikatos D (2017) The Mondrian data engine. In: Proceedings of the 44th annual international symposium on computer architecture, Toronto, ON, June 2017, pp 639–651
25. Liu J, Zhao, Hengyu O, Matheus A, Li D, Zhao J (2018) Processing in-memory for energy-efficient neural network training: a heterogeneous approach. In: Proceedings of the 51st annual IEEE/ACM international symposium on microarchitecture, Fukuoka, Oct 2018, pp 655–668
26. Nair R, Antao SF, Bertolli C, Bose P, Brunheroto JR, Chen T, Cher C-Y, Costa CHA, Doi J, Evangelinos C, Fleischer BM, Fox TW, Gallo DS, Grinberg L, Gunnels JA, Jacob AC, Jacob P,

Jacobson HM, Karkhanis T, Kim C, Moreno JH, O'Brien JK, Ohmacht M, Park Y, Prener DA, Rosenburg BS, Ryu KD, Sallenave O, Serrano MJ, Siegl PDM, Sugavanam K, Sura Z (2015) Active memory cube: a processing-in-memory architecture for exascale systems. IBM J Res Dev 59:17:1–7:14
27. Jain S, Ranjan A, Roy K, Raghunathan (2018) Computing in memory with spin-transfer torque magnetic RAM. IEEE Trans Very Large Scale Integr (VLSI) Syst 26(3):470–483
28. JEDEC, High Bandwidth Memory (HBM) DRAM, Standard No. JESD238
29. Daemen J, Rijmen V (2001) AES proposal: Rijndael. Technical report FIPS PUB 197. National Institute of Standards and Technology, U.S. Dept. Commerce, Washington, DC
30. Veloce Primo Enterprise Prototyping Solution. Accessed 20 Oct 2023
31. Kim M, Kim S-H, Lee H-J, Rhee C-E (2021) Case study on integrated architecture for in-memory and in-storage computing. Electronics 10(15)

Improving Digital Agriculture to Achieve Sustainable Development: Analysis and Policy Proposals

Vitalina Babenko, Adolfo Maza, Maryna Nehrey, and Olga Pushko

Abstract Digitalized agriculture can make a significant contribution to achieving the goals of sustainable development. The aim of this paper is to analyse possible areas of digitalization in the agricultural sector. Digitalization of the agricultural sector will contribute to the development of not only the sector, but also the economy, and to the achievement of sustainable development goals. We are assessing the potential for digitalization in the agricultural sector. To increase the efficiency, productivity, and sustainability of digital agriculture, it is necessary to bridge the digital divide, promote cooperation, ensure data privacy and security, support small farmers, and encourage innovation to develop digital agriculture.

Keywords Agricultural sector · Digitalization · Sustainability · Agriculture policy · Information technology · Efficiency

V. Babenko (✉)
V. N. Karazin Kharkiv National University, Kharkiv 61022, Ukraine
e-mail: vitalinababenko@karazin.ua

Kharkiv National Automobile and Highway University, Kharkiv 61002, Ukraine

A. Maza
Universidad de Cantabria, 39005 Santander, Spain
e-mail: adolfo.maza@unican.es

M. Nehrey
National University of Life and Environment Science of Ukraine, Kyiv 03041, Ukraine

O. Pushko
Sumy State University, Sumy 40000, Ukraine

D. Magdi et al. (eds.), *Green Sustainability: Towards Innovative Digital Transformation*, Lecture Notes in Networks and Systems 753,
https://doi.org/10.1007/978-981-99-4764-5_19

1 Introduction

With the world's population expected to reach 9.7 billion by 2050, the agricultural sector plays an important role in providing food. Sustainable agricultural practices are needed to meet the growing demand for food. Digitalization of the agricultural sector is one way to address this issue. The use of digital technologies can help farmers to manage resources more efficiently, increase productivity and profitability, reduce waste and losses, and improve environmental sustainability.

Digitalization of agriculture involves integrating various digital technologies such as the Internet of Things (IoT), big data analytics, artificial intelligence (AI), and precision agriculture into agricultural practices.

The digitalization of agriculture will help increase food production, improve food quality, reduce waste and losses, and enhance food security. By doing this, several of the sustainable development goals proposed by the United Nations could be met. Digitalizing the agricultural sector will contribute to achieving zero hunger (SDG 2). Digital technologies can help manage water resources efficiently, reduce water losses, and improve water quality. This will contribute to achieving the goal of Clean Water and Sanitation (SDG 6). Digital technologies can help promote renewable energy sources in agriculture, reduce energy consumption, and increase energy efficiency. This will all contribute to achieving the goal of Affordable and Clean Energy (SDG 7). Furthermore, digitalization of agriculture will bring closer the achievement of the Climate Action goal (SDG 13), as it can help reduce greenhouse gas emissions.

Against this backdrop, the aim of this work is to assess the efficiency of the agricultural sector in different European countries and then to try to evaluate the potential of the use of digitalization in order to increase it.

2 Literature Review

Digital agriculture has moved beyond its early stages and continues to develop. As a result, academic research is covering an increasingly wide range of challenges and opportunities for the digitalization of the agricultural sector.

An analysis of future visions of digital agriculture from FAO, OECD, and the World Bank was conducted by Klerkx et al. [8]. The authors found that these visions focus mainly on the use of technology to increase food production, without considering other factors that could contribute to food security or address existing inequalities. The authors argue that there is a need to critically assess the use of digital technologies in agriculture and consider their potential impact on ecosystem services and the social context.

Digital agriculture makes the global food system more resilient as it manages the trade-offs between natural and human systems for the economy, environment, and social well-being [2]. However, recent research suggests that the development of digital technologies in agriculture can create win–win synergies and transform agricultural landscapes for sustainable development.

The use of digital agricultural technologies can help redesign agricultural landscapes for sustainability. Yield sustainability maps can be integrated into farm and landscape-level data systems to redesign fields or areas within fields that are unprofitable or environmentally unsustainable and to sustainably intensify high-yielding areas of a field [9].

A review of research on digital agriculture [6] found that there are five patterns of inequality associated with the use of digital technologies in agriculture. These include corporate control over technology development, unequal distribution of benefits, limited access to data and infrastructure, unequal digital literacy, and unequal power and influence over problem-solving. However, these patterns are not caused by digital technologies themselves, but rather reflect existing social inequalities and unequal distribution of power.

Ehlers et al. [5] highlight that digital technologies for agricultural policy are not being fully utilized. Scientists suggest that digitalization can improve monitoring and control, but success also depends on stakeholders' digital competency and suitable institutions. The government can strategically pursue digitalized design options for greater policy legitimacy.

Sustainable farm management can be achieved through current research and state-of-the-art agricultural practices [14]. The authors show that big data technologies are available to manage crops, livestock, and markets, but their adoption is low due to challenges including data privacy and financial investment.

Related to that, creating the infrastructure to provide Internet access, developing/providing the necessary programmes and digital services, and involving rural staff in the use of new technologies are needed to overcome barriers to the digitalization of agriculture [1].

Regan's view is that capacity building for responsible research and innovation (RRI) readiness is necessary to ensure successful innovation in digital agriculture. For scientists to engage in such activities, it is crucial to introduce RRI exercises and dimensions in a careful and effective manner [13].

Studies on the potential of digitalization as well as on the main obstacles and challenges in agricultural policy have been carried out in different countries at platforms. But it also raises fears about data property, privacy, security, and unjust trade behaviours [16]. Digital agriculture can improve farm productivity, process control, and connectivity and improve Australia's food system through data. It can improve global sustainability by connecting farmers, consumers, and intermediaries [4]. In Turkey, it was found that the practice of digital agriculture is limited and not accepted by older farmers and that young farmers need training in digital agriculture to increase productivity. The support of the government is crucial for the creation of an action plan and policy support for digital agriculture [12]. The Ukrainian agricultural sector has significant potential for increased efficiency and development and needs start-ups that use information technology to address the balance of economic, environmental, and social efficiency to increase agricultural productivity [10]. There are several challenges, including the changing ecology of agricultural production and data privacy and security issues, to the digital transformation of agriculture in China. Xie et al. propose to involve small farmers in digitalization by establishing cooperatives and outsourcing agricultural production to service providers [18]. The agricultural sector in Morocco faces a wide range of challenges, and future approaches should take advantage of the opportunities for precision farming [7]. Vietnam has a growing digital agriculture ecosystem with high connectivity and subscription rates, providing opportunities for digital solutions to address food production challenges. Success will require pilots, public awareness campaigns, and supportive policies and investments [3].

In order to achieve these outcomes while avoiding unintended consequences, science has a crucial role to play in the design and use of digital technologies. To achieve this, transdisciplinary research is needed that also addresses socio-ethical issues, data governance, and business models [15].

3 Materials and Methods

To assess the agricultural sector efficiency, we use a method of comparative analysis, in particular Data Envelopment Analysis (DEA). DEA uses linear programming to determine the relative efficiency of an entity by evaluating its outputs against its inputs. Each set of inputs is characterized by a maximum production potential, and the units that produce the maximum output per unit of market inputs are considered the "standard" and create a performance threshold. The analysis is to compare these units in terms of the efficiency of their resource base and their proximity to the productivity threshold.

DEA removes the influence of subjectivity from the assessment by using linear programming and the Pareto optimum as the performance criterion. This method makes it possible to calculate an aggregate indicator for each object under study and to identify priority areas for productivity growth. The first DEA model was developed in 1978, and further developments include a two-vector approach based on the value of production and constant or variable returns to scale.

DEA is suitable for hypothesis testing and multi-country evaluation, for example, when analysing the efficiency of a country's agricultural sector. The method has a wide range of possibilities for assessing efficiency, which makes it possible to analyse the productivity of the agricultural sector and the impact of various factors on it. The proposed algorithm for evaluating the effectiveness of digitalization in the agricultural sector involves solving an optimization problem.

The input-oriented DEA model can be represented by the following Eq. (1),

$$\max \theta = \frac{\sum_{j=1}^{s} u_j y_j}{\sum_{i=1}^{r} v_i x_i}, \tag{1}$$

where θ is the efficiency of the object, x and y are the inputs and outputs, u and v are the weights to be determined to obtain the maximum estimate of the fraction, and s and r are the number of outputs and inputs, respectively.

To estimate the effectiveness of an object, a fractional linear programming problem must be solved by determining the weight coefficients for the inputs (v) and outputs (u), treating u and v as variables. If several objects are considered, the problem must be treated as an optimization problem with constraints. These constraints limit the maximum score fraction to be less than or equal to one; i.e. the ratio of results to costs cannot exceed 100%. To find the weight coefficients for a given object that will lead to the maximum estimate while ensuring that the ratio for the remaining objects is less than or equal to 100%, the model takes the form of Eqs. (2)–(5).

$$(\theta_0) \max \theta = \frac{\sum_{j=1}^{s} u_j y_{j0}}{\sum_{i=1}^{r} v_i x_{i0}} \tag{2}$$

$$\frac{\sum_{i=1}^{s} u_i y_{ij}}{\sum_{i=1}^{r} v_i x_{ij}}, \quad j = \overline{1, n} \tag{3}$$

$$v_1, v_2, \ldots, v_r \geq 0 \tag{4}$$

$$u_1, u_2, \ldots, u_s \geq 0. \tag{5}$$

The weighting factors used in the model cannot be negative. To calculate the efficiency of each object, the optimization problem must be solved n times, where n is the number of objects. When selecting the data (inputs and outputs) for each object, it is important to ensure that numerical values are available that they are positive and that they reflect the analytical or managerial sense of the components included in the relative efficiency assessment. In addition, smaller amounts of input data and larger amounts of output data are preferable, and the units for inputs and outputs may vary.

4 Results

The potential of digital technologies in the agricultural sector

The agricultural sector digitalization can play an important role in its development. The potential of digital technologies in the agricultural sector is considerable. Despite the high costs of digitalization, the impact of digitalization can be very significant.

The agricultural sector digitalization will help to increase the efficiency, productivity, profitability, and sustainability of agriculture. The agricultural sector can implement digital transformation in the following areas:

1. Precision farming.
2. Data-driven decision-making.
3. Digital platforms in agriculture.
4. Supply chain traceability.
5. Smart agriculture.
6. Access to markets.

The use of precision agriculture will help increase the efficiency of the agricultural sector. Precision farming technologies will provide assistance in reducing overlap and optimizing the application of fertilizers and pesticides, reducing waste, cutting costs, and increasing yields and crop quality. By collecting and analysing data on factors such as soil moisture, temperature, and nutrient levels, precision farming can help farmers optimize growing conditions. Precision farming also helps to reduce environmental damage and conserve water resources. The use of precision agriculture offers economic benefits by reducing fertilizer and pesticide costs and labour costs by optimizing tasks such as planting and harvesting.

Farmers can use data-driven decision-making to increase the profitability of their business. By collecting and analysing data on weather, soil conditions, yields, and market prices, farmers can gain valuable insights to help them make more informed decisions.

Farmers can use data-driven decision-making tools in the following ways

1. Optimizing the cost of growing crops: By collecting data on soil health, weather conditions, and crop growth, farmers can make more efficient decisions about the use of inputs such as fertilizer and pesticides. This will help reduce costs and increase yields.
2. Planning according to market needs: By tracking market prices and trends, farmers can make informed decisions about what crops to plant and when to sell their produce. This helps maximize profits and minimize losses.
3. Improving logistics: By tracking data on crop growth and harvest time, farmers can optimize their logistics and transport strategies to reduce costs and minimize waste.
4. Risk management: Analysing data on weather and market conditions will help farmers identify potential risks and take steps to mitigate them. This will help avoid unexpected losses and increase profitability in the long term.

The digitalization of the agricultural sector will contribute to transparency and accountability throughout the supply chain. For example, the traceability of agricultural supply chains can be improved through automated data collection. The use of sensors, drones, and satellite imagery can collect real-time data on crop growth, harvest, and transport. The data can be stored in a centralized database and made available to all participants in the supply chain, providing a transparent and reliable record of the origin and movement of products. By analysing the data obtained, it is possible to obtain information about the actual situation and potential problems and to take action to resolve them before they become major problems. This can help reduce waste, improve quality, and increase efficiency throughout the supply chain. The use of blockchain technology provides a secure and transparent platform for recording and sharing data throughout the supply chain. By using blockchain to track the movement of agricultural products from farm to fork, stakeholders can gain greater transparency across the supply chain and ensure that products are authentic and meet required standards. Mobile applications that provide real-time information from fields, markets, elevators, and logistics services are also highly effective in the agricultural sector. All these tools help to make the agricultural sector more efficient, profitable, and sustainable.

The use of digital platforms in agriculture brings additional benefits to farmers, governments, consumers, and other stakeholders. Better access to information, increased transparency, improved traceability, simplified administrative processes, and improved access to financial services—all these can be achieved thanks to digital platforms. The use of digital platforms at the government level can streamline administrative processes such as licensing, permitting, and certification; help them obtain government subsidies and grants; and access financial services such as credit and insurance, which will help them manage risks and invest in their businesses.

By using technologies such as the Internet of Things (IoT), artificial intelligence, and big data, it is possible to optimize the production of agricultural crops, improve the efficiency of resource use, and increase yields. Smart agriculture technologies can help farmers monitor crop growth, soil moisture, and weather conditions in real time. Smart agriculture can help conserve natural resources and promote sustainable agriculture. By using smart agriculture, farmers can work towards a more sustainable and efficient farming system. Smart agriculture also has the potential to address food security issues by increasing yields and improving food quality.

Digital technologies are opening new opportunities for farmers to access markets and customers. By working on online platforms, farmers can expand their market reach, increase their visibility, and build new relationships with customers around the world. Farmers can use e-commerce platforms, social media, targeted advertising, and mobile applications. Working with platforms such as Amazon, Alibaba, and eBay give farmers the opportunity to reach a global audience and increase sales. Facebook, Twitter, Instagram, and Telegram offer farmers the opportunity to communicate with customers, promote their products, and increase brand awareness. By using social media to share their stories and interact with customers, farmers can build a loyal audience and create new sales opportunities. By using targeted digital advertising, farmers can reach customers who are interested in their products and services. Digital advertising can be highly effective because it allows farmers to target specific demographics and interests and track the success of their campaigns in real time. The use of mobile applications also helps to expand access to markets and customers.

Agricultural productivity data envelopment analysis

Let us assess agricultural productivity in European countries using the DEA method. The input is GDP per capita (current US$) [17] and the output is the agricultural

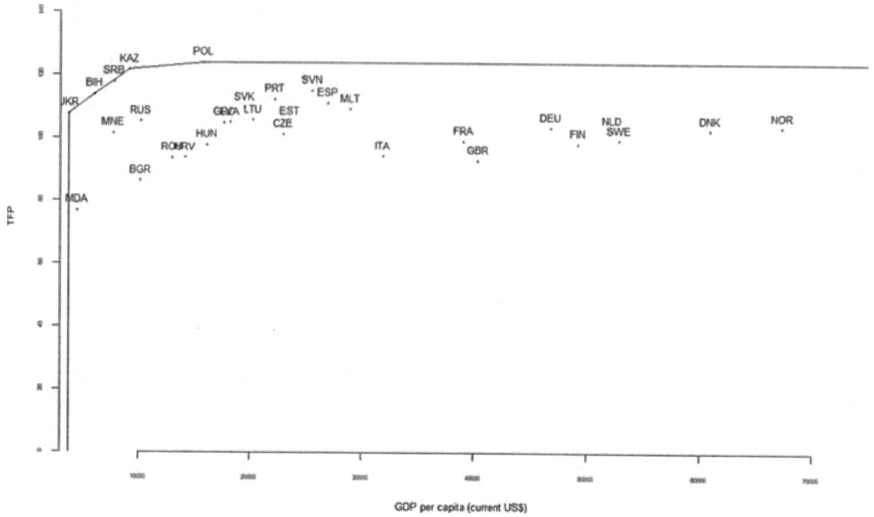

Fig. 1 TFP-GDP data envelopment analysis

total factor productivity (TFP) index [11] (Fig. 1). The results of the analysis show that countries with the largest share of agriculture in the economy have the highest efficiency: Ukraine, Kazakhstan, Serbia, and Bosnia and Herzegovina. The lowest efficiency is mainly found in the Scandinavian countries: Norway, Denmark, Sweden, the Netherlands, Finland, Luxembourg, Switzerland, and Ireland.

The results confirm that developed countries with successful economies focus on other industries rather than agriculture, and we can conclude that the digitalization of the agricultural sector will bring benefits to all countries. For developed countries, the digitalization of the agricultural sector can contribute to increasing the efficiency of agriculture, thus increasing the country's food self-sufficiency and reducing government spending on agricultural support in the country, which is significant in the above-mentioned countries. For countries with a large share of agriculture in the national economy, digitalization of the agricultural sector can be even more effective. Increasing agricultural productivity can significantly increase a country's GDP and contribute to its development.

We also assessed the efficiency of agriculture in terms of land use. We used the DEA method, where the input data is the agricultural area (% of the total area) and the output data is the index of the total factor productivity of agriculture (TFP). The results of the analysis show that Poland and Slovakia have the highest efficiency (equal to 1). Lithuania (0.029), Ukraine (0.034), Bosnia and Herzegovina (0.039), and Spain (0.040) have the lowest efficiency (Fig. 2).

The results of the analysis confirm that most countries have the potential to increase agricultural productivity. And digitalization can be one of the ways to solve this problem.

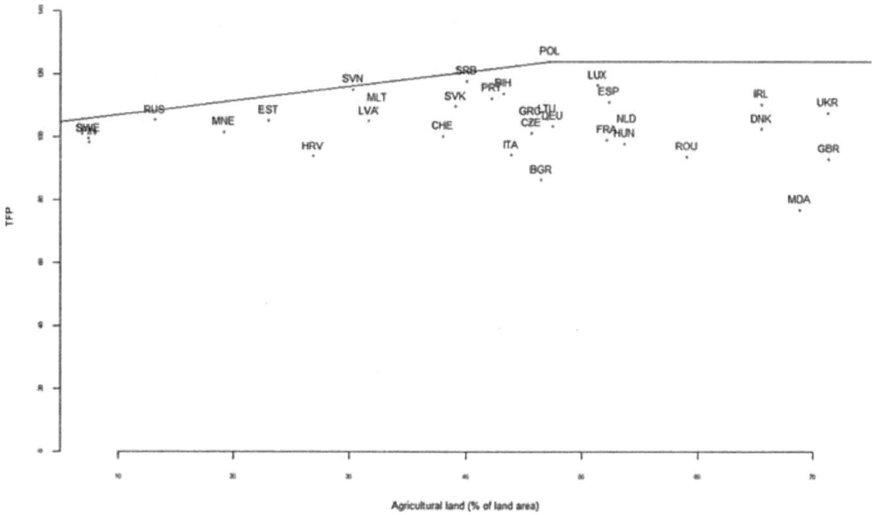

Fig. 2 TFP-Agro land data envelopment analysis

To assess the efficiency of the use of information technology, we performed a DEA with the input data mobile phone subscriptions (per 100 people) and the output data the index of agricultural total factor productivity (TFP) (Fig. 3).

The results of the analysis show that the most efficient countries are: Latvia, Portugal, Serbia, Ireland, and Bosnia and Herzegovina. The least efficient countries are Montenegro, the Russian Federation, Moldova, Italy, and Estonia. As we can see,

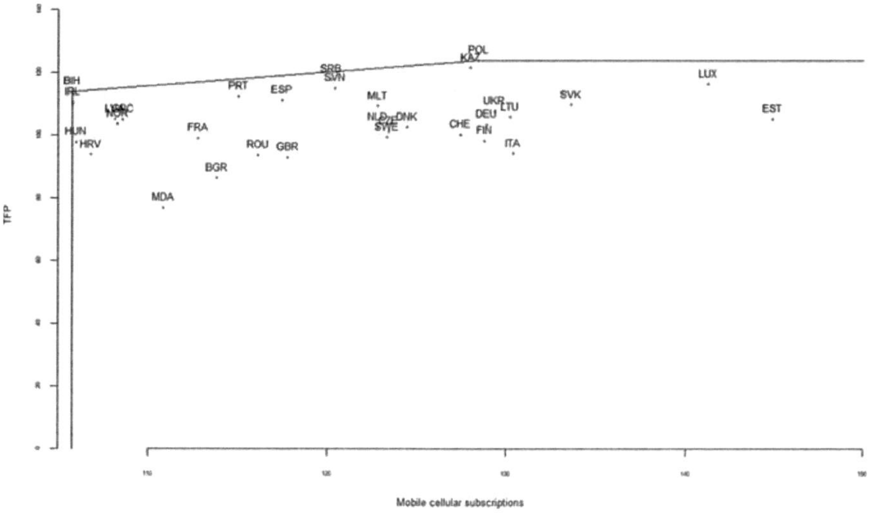

Fig. 3 TFP-mobile data envelopment analysis

this analysis does not allow us to draw any final conclusions about the efficiency of information technology use. Therefore, the next step will be a multi-factor analysis of agricultural efficiency.

To analyse the efficiency of agriculture, we use the DEA method, where the inputs are agriculture, forestry, and fisheries, value added (% of GDP), agricultural land (% of land area), rural population, livestock production index ($2014 - 2016 = 100$), crop production index ($2014 - 2016 = 100$), GDP per capita (current US$), mobile cellular subscriptions (per 100 people), and fixed broadband subscriptions (per 100 people). The output of the model is the agricultural total factor productivity (TFP) index.

The results of DEA models for agricultural total factor productivity show significant differences in the efficiency of resource use across countries. The countries with the lowest efficiency are Italy (0.759), the Netherlands (0.778), Moldova (0.785), and Denmark (0.798). The maximum efficiency is achieved by 50% of the countries surveyed (Table 1).

We conclude that half of the countries have significant potential for the use of digital technologies in agriculture. We mean the potential compared to other countries that have already started to digitalize the agricultural sector. However, the potential for digitalization of the agricultural sector is much greater.

Challenges for agriculture digitalization

Digitalization in agriculture provides significant benefits for agriculture, consumers, the state and society in general. However, some challenges need to be overcome in order to fully realize the potential of digitalization of agriculture. The cost of agriculture digitalization can be high, making adoption of these technologies difficult, especially for small farmers. There may also be additional costs associated with data storage and equipment maintenance.

Moreover, one of the challenges that the agricultural sector may face is the lack of specialists. The digitalization of agriculture requires a range of specialists with experience in different areas. The most important specialists needed for the digitalisation of the agricultural sector are

Table 1 Agricultural total factor productivity data envelopment analysis

Efficiency range	Numbers of countries	%	Countries
$0.7 <= E < 0.8$	4	11	Italy, Netherlands, Moldova, Denmark
$0.8 <= E < 0.9$	9	25	France, United Kingdom, Switzerland, Spain, Germany, Romania, Czech, Greece, Hungary
$0.9 <= E < 1$	5	14	Finland, Lithuania, Bulgaria, Latvia, Portugal
$E = 1$	18	50	Bosnia and Herzegovina, Estonia, Croatia, Ireland, Iceland, Kazakhstan, Luxembourg, North Macedonia, Malta, Montenegro, Norway, Poland, Russian Federation, Serbia, Slovakia, Slovenia, Sweden, Ukraine

1. Agricultural experts: Experts in agronomy, crop production, animal husbandry, and other areas of agriculture are needed to advise on the use of digital technologies to optimize production and increase yields.
2. Data analysts will be needed to manage and analyse large amounts of data generated by digital technologies, such as satellite imagery, meteorological data, and soil analysis.
3. Software developers are needed to create and maintain digital platforms and applications that can help farmers manage their operations, from crop planning to financial management.
4. User experience (UX) and user interface (UI) designers are needed to create intuitive and user-friendly interfaces for digital platforms and applications. This is essential to ensure that farmers can easily access and use these technologies.
5. Internet of Things (IoT) specialists are needed to design and implement sensor networks that can collect data on soil moisture, temperature, humidity, and other environmental factors. These sensors can help farmers make more informed decisions about when to irrigate or fertilize.
6. Business analysts are needed to help farmers identify the most profitable markets for their products, manage their finances, and optimize their supply chains.
7. Project managers will be needed to oversee the implementation of digital technologies and ensure their smooth integration into existing systems.

Data privacy can also be a barrier to the adoption of digital technologies. Farmers need to be confident that their data is accurate and secure. In addition, farmers may lack the technical knowledge to effectively implement and use digital farming solutions. There is also the possibility that individual farmers may resist implementing digital farming solutions due to a lack of understanding or distrust of new technologies.

Policy proposal for agriculture digitalization

The formation of the digitalization policy of agriculture includes several key steps. The process of developing a digitalization policy for agriculture requires, first of all, the definition of objectives. These can be, for example, increasing the productivity of agriculture, reducing waste, increasing the efficiency of resource use, increasing the income of farmers, and achieving one of the goals of sustainable development. A necessary step is to assess the current state of digitization of agriculture in the country and evaluate the available digital tools, infrastructure, and availability of platforms. Further development of the agriculture digitalization policy should identify growth areas. This can be the expansion of access to digital tools and platforms for farmers, the development of new tools and platforms to solve specific tasks, or the creation of digital infrastructure. It is imperative that stakeholder interests are taken into account in the development of agricultural digitalisation policy. Effective policy-making requires engagement with key stakeholders, including farmers, digital service providers, government agencies, and development partners, to ensure that policy is responsive to their needs and priorities.

After the preparatory phase, there is a direct development of policies and regulations. This may include measures to encourage the adoption of digital tools, regulations to ensure privacy and data security, and policies to promote the development of digital infrastructure.

To start implementing the digitalization of agriculture policy, it is advisable to start with a number of pilot projects. This can be the introduction of a policy in a separate region or in a separate sector. It is necessary to evaluate the results of the policy implementation and get feedback from stakeholders. If weaknesses and shortcomings are identified, necessary changes should be made.

Once the policy has been successfully tested and improved, it can be extended to other regions or the whole country. This requires ensuring that the policy is sustainable and can be maintained over time. Throughout the implementation of the policy, it is important to gather feedback from farmers and other stakeholders and improve the policy as needed.

The policy of digitalization of the agricultural sector can be implemented according to various scenarios. However, the basic policy proposals will be as follows:

- Supporting learning and education. The basic element of digitalization is information technology. Insufficient knowledge of information technologies can become the biggest obstacle to digitisation. Governments must therefore support training and education programmes to help farmers and other agricultural stakeholders learn how to use digital tools and technologies effectively.
- Encouraging collaboration and innovation. Governments can foster collaboration and innovation by creating platforms that bring together farmers, researchers, and other relevant actors to share knowledge and ideas. This can include creating digital forums, organizing hackathons or other innovation events, and supporting collaborative research projects.
- Creating an Agritech and Foodtech ecosystem. It is appropriate to support the development of agricultural start-ups that develop innovative digital tools and technologies for the agricultural sector. This can be done through tax breaks, grants, or other forms of financial support.
- Providing subsidies to farmers and other agricultural stakeholders to invest in digital infrastructure such as high-speed Internet, sensor technology, and other digital tools.
- Creating digital farm advisory services that provide farmers with up-to-date information on farming practices, market trends, weather forecasts, and other relevant information. This can be done through mobile applications, SMS, or other digital means.
- Developing regulations for digital agriculture to ensure that digital tools and technologies are used in an ethical and sustainable way. This may include regulations on data protection, cybersecurity, and environmental sustainability.

For the development of an agricultural digitalisation policy, it is advisable to use the following modelling methods:

1. System dynamics. It is advisable to use this approach to model the agricultural sector as a complex system and to study its behaviour over time. The use of system dynamics models allows policy makers to test different scenarios and their impact on the system.
2. Agent-based modelling involves modelling individual agents and their interactions with each other and the environment. It can be used to model different scenarios of digitalization of the agricultural sector and to understand the impact of different policies.
3. Multi-criteria decision analysis can be used to analyse different options and evaluate them according to different criteria. This can help policy makers to make informed decisions that take into account a range of factors.
4. Participatory modelling involves the involvement of interested parties in the modelling process, providing a broader and more inclusive approach to decision-making.

Ultimately, the choice of modelling methodology will depend on the specific needs and objectives of the digitalization strategy for agriculture, as well as the resources and expertise available.

5 Summary and Conclusion

One of the possible solutions to the problems of food security, resource management, and environmental sustainability is the digitalization of agriculture. Using digital technologies can help farmers manage resources efficiently, reduce waste and losses, increase productivity and profitability, and contribute to environmental sustainability. The adoption of digital technologies in agriculture can contribute to the achievement of several SDGs, such as zero hunger, clean water and sanitation, affordable and clean energy, and climate action.

In this paper, we analyse the efficiency of agriculture using the DEA method. We have found that all European countries have significant potential to improve agricultural efficiency by digitalization the agricultural sector.

The development of an agricultural digitalization policy requires a comprehensive understanding of the agricultural sector, the digital landscape, and the needs and priorities of stakeholders. It is important to consult widely and involve all relevant stakeholders in the policy development process to ensure that the policy is effective, feasible, and sustainable.

References

1. Babenko V, Zomchak L, Nehrey M, Salem ABM, Nakisko O (2022) Agritech startup ecosystem in Ukraine: ideas and realization. In: Digital transformation technology: proceedings of ITAF 2020. Springer Singapore, pp. 311–322. https://doi.org/10.1007/978-981-16-2275-5_19
2. Basso B, Antle J (2020) Digital agriculture to design sustainable agricultural systems. Nat Sustain 3(4):254–256
3. Burra DD, Hildebrand J, Giles J, Nguyen T, Hasiner E, Schroeder K, Kropff W (2021) Digital agriculture profile: Viet Nam
4. Cook S, Jackson EL, Fisher MJ, Baker D, Diepeveen D (2022) Embedding digital agriculture into sustainable Australian food systems: pathways and pitfalls to value creation. Int J Agric Sustain 20(3):346–367
5. Ehlers MH, Huber R, Finger R (2021) Agricultural policy in the era of digitalisation. Food Policy 100:102019
6. Hackfort S (2021) Patterns of inequalities in digital agriculture: a systematic literature review. Sustainability 13(22):12345
7. Jabir B, Falih N (2020, April) Digital agriculture in Morocco, opportunities and challenges. In: 2020 IEEE 6th International conference on optimization and applications (ICOA). IEEE, pp 1–5
8. Klerkx L, Jakku E, Labarthe P (2019) A review of social science on digital agriculture, smart farming and agriculture 4.0: new contributions and a future research agenda. NJAS-Wageningen J Life Sci 90:100315
9. Lajoie-O'Malley A, Bronson K, van der Burg S, Klerkx L (2020) The future (s) of digital agriculture and sustainable food systems: an analysis of high-level policy documents. Ecosyst Serv 45:101183
10. Nehrey M, Zomchak L (2022) Digital technology: emerging issue for agriculture. In: The international conference on artificial intelligence and logistics engineering. Springer, Cham, pp 146–156. https://doi.org/10.1007/978-3-031-04809-8_13
11. OECD (2022) Insights into the measurement of agricultural total factor productivity and the environment. https://www.oecd.org/agriculture/topics/network-agricultural-productivity-and-environment/
12. Ozdogan B, Gacar A, Aktas H (2017) Digital agriculture practices in the context of agriculture 4.0. J Econ Financ Account 4(2):186–193
13. Regan Á (2021) Exploring the readiness of publicly funded researchers to practice responsible research and innovation in digital agriculture. J Respons Innov 8(1):28–47
14. Sarker MNI, Islam MS, Ali MA, Islam MS, Salam MA, Mahmud SH (2019) Promoting digital agriculture through big data for sustainable farm management. Int J Innov Appl Stud 25(4):1235–1240
15. Shepherd M, Turner JA, Small B, Wheeler D (2020) Priorities for science to overcome hurdles thwarting the full promise of the 'digital agriculture' revolution. J Sci Food Agric 100(14):5083–5092
16. Verdonk T (2019) Planting the seeds of market power: digital agriculture, farmers' autonomy, and the role of competition policy. Regulating New Technol Uncertain Times 105–132
17. World Bank national accounts data, and OECD National Accounts data files (2022). https://data.worldbank.org/indicator/NY.GDP.PCAP.CD
18. Xie L, Luo B, Zhong W (2021) How are smallholder farmers involved in digital agriculture in developing countries: a case study from China. Land 10(3):245

Cramer-Rao Bound Investigation of the Double-Nested Arc Array Virtual Extension and Butterfly Antenna Configuration

Tarek Abd El-Rahman, Amgad A. Salama, Walid M. Saad, and Mohamed H. El-Shafey

Abstract This paper exploits a sparse array representation, a novel double-nested arc array to generate a multiple virtual nested arc array that forms the virtual non-uniform circular array which has been presented and investigated. The nested arc array has been developed using higher-order statistics, namely using the fourth-order cumulants (FOC), to increase the array aperture. The extended array contains a hybrid of physical and virtual sensors which is distributed in a butterfly shape. The enhanced degrees of freedom (DOFs) in the resultant virtual array make it appropriate for over-and under-determined direction-of-arrival (DOA) estimation. The Cramer-Rao bound (CRB), the number of nested arc arrays, the structure of the double-nested arc related to each other, and the overlap between sensors for such an array are developed and investigated for various arrays. The proposed array performance is superior to that of the uniform one. Further, the proposed array is appropriate for two-dimensional (2D) DOA estimation problems.

Keywords Sparse array · Nested arc arrays · Super nested arc arrays · Fourth-order cumulants · Fisher information matrix · Cramer-Rao bound · Direction-of-arrival

---

T. A. El-Rahman (✉) · M. H. El-Shafey
Faculty of Engineering, Computer and System Engineering Department, Ain Shams University, Cairo 11517, Egypt
e-mail: g17092181@eng.asu.edu.eg

M. H. El-Shafey
e-mail: mohamed.el-shafey@eng.asu.edu.eg

A. A. Salama · W. M. Saad
The Egyptian Technical Research and Development Centre, Cairo 11618, Egypt
e-mail: amgad.salama@acm.org

W. M. Saad
e-mail: walid@sce.carleton.ca

D. Magdi et al. (eds.), *Green Sustainability: Towards Innovative Digital Transformation*, Lecture Notes in Networks and Systems 753,
https://doi.org/10.1007/978-981-99-4764-5_20

1 Introduction

Among the different types of antenna arrays, arc and circular antenna arrays become more reliable used in radar, sonar, mobile, and wireless communication systems [1]. The array configurations in which the elements are placed in an arc and then extended to a circular array are of great interest. They are specified for applications in radio direction finding, air, and space navigation, radar, and projectile body antennas [2], and are of great interest in the design of antennas on that structures. This may be due to reasons of payload space limitations, and aerodynamic reasons when the antenna shape has to adapt to be non-planar to reduce its radar cross-section [3]. For the design of the double-nested arc arrays, one has to adequately choose the number of sensors in the array to have the facility tact of using the indispensable least number of physical sensors only [4]. The design of non-uniform arc arrays with minimum overlap and the maximum number of generated virtual sensors is the main interest of this paper. The nesting concept is a strategy that is provided to enhance the performance of the DOF. The developer tries to increase the DOF using a linear nested array [5], then by extending the nesting concept that depends on the number of the nested array levels, as a statistics order that could be increased to a higher statistic order [6] and then amended using fourth-order cumulants FOC on the linear nested array as [7].

In this paper, two different global optimization methods are used, the nested arc array method [5, 6], and the fourth-order cumulants (FOC) method [4, 7, 8]. The double-nested arc array method is proposed. The double-nested arc array structure and location are used to minimize the overlap, and to fill the gap in the full space with the least number of physical sensors. The FOC and the double-nested arc array are used to increase the total aperture that increases the DOF. The double-nested arc array, the sensors, and the levels need to be carefully localized to get the cardinality best position for antenna elements. The Cramer-Rao bound (CRB) is used to measure the performance [6, 9, 10], and the paper is arranged as follows. In Sect. 2, the co-array system is presented. The virtual sensors generated using FOC and their location are presented in Sect. 3. The configuration specification is in Sect. 4. The bounds on selecting the double-nested arc array location are illustrated in Sect. 5. In Sect. 6, the CRB for the estimated DOA in the FOC domain is obtained. The detected observations are shown in Sect. 7. The CRB by the arced nested arrays with different array structures is explored using simulation in Sects. 8, 9, 10. Finally, the conclusions are illustrated in Sect. 11, appendix in Sect. 12.

1.1 Notation

$\boldsymbol{r}(l), \mathbf{A}(\boldsymbol{\theta}), \boldsymbol{s}(l), \boldsymbol{n}(l) =$ the general matrices,
$\mathbb{C}^{M \times D} =$ the space of $M \times D$ complex-valued matrices,
$\boldsymbol{a}(\theta_d) =$ vector element,

$s(l) \sim N(0, R_{ss})$ = the distribution of the random vector $s(l)$ is normally Gaussian distribution with mean = *0 and* R_{ss} = signal covariance matrix,

cum = the cumulant operation,

$()^*$ = the conjugate operator,

$\text{diag}\left[\sigma_1^2, \ldots.., \sigma_Q^2\right]$ = the diagonal of a matrix,

$E\{\}$ = the expectation operator,

$()^H$ = the complex conjugate transpose of (),

$R_{xx}()$ = the data covariance matrix,

$(\cdot)$ = the dot product,

$\odot$ = the Hadamard product,

$\gamma_{4,q}$ = the cumulant (kurtosis) of the source signal,

$r(l)$ = measured signal,

$v(l)$ = virtual signal,

σ_q^2 = the source power,

Γ = diagonal matrix consists of (FOC),

$\mathcal{M}1_{\text{arc}_1 l_1}$ = sensor number one, arc number one, and nested level number one,

VS_{vz} = virtual sensors generated from physical sensor number v to physical sensor number z,

$\angle$ = angle,

$P_{(,)}$ = permutation operation,

$C_{(,)}$ = combination operation.

2 Co-Array Structure and Signal Concept

By employing the geometry of the array under which the Fisher information matrix (FIM) is guaranteed to be non-singular even in the underdetermined setting (i.e., received signal $\geq$ number of sensors), it is desired to reach the optimal minimum number of used physical sensors to generate the optimal maximum number of virtual sensors. The CRB will provide a fundamental lower bound on the performance compared with any algorithm that aims to localize more sources than sensors [5, 11, 12].

2.1 *Signal Model and Preliminaries*

Consider Q wideband plane waves where $q = 1, 2, \ldots, Q$. impinging on an arc array with

$$M = \left(\mathcal{M}1_{\text{arc}_1 \ell_1} + \ldots + \mathcal{M}m_{\text{arc}_2 \ell_w}\right) + \mathcal{M}_{\text{origin}} \quad (1)$$

where M is the total number of the sparse array, $m = 1, 2, \ldots, M$ is the sensor's number, arc the arc number, and ℓ_w is the sensor's level number w in each arc.

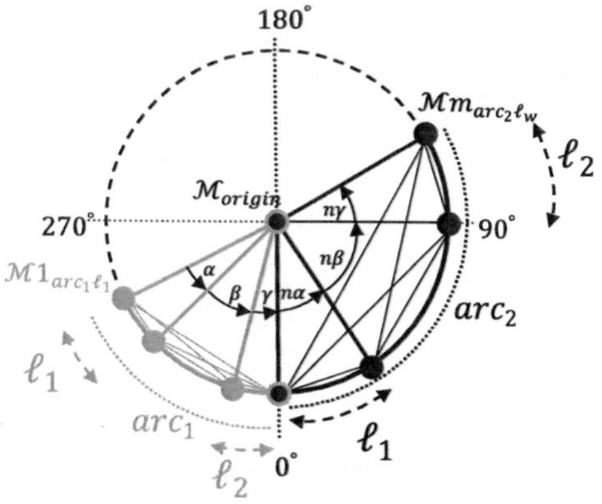

Fig. 1 Double-nested arc array configuration

The number of levels in each arc is identical, the sensors are distributed as one at the center and the other distributed on the merged non-uniform double arc forming the optimum super nested arc. The mth sensor is located at $\boldsymbol{r}_m = [x_m, y_m]^T$, which decomposed to be a sensor at the center of the circuit and the rest at the arc with angle separation (α, β, γ), $(n\alpha, n\beta, n\gamma)$ counterclockwise (CCW) direction. The radius was normalized with respect to the carrier wavelength,λ_o, of the central frequency as shown in Fig. 1. Then the received signal can be expressed as follows [13, 14]

where x, y is a 2D-coordinates, $u(\phi, \psi)$ is the unit vector pointed to the signal source arriving with azimuth angle ϕ and elevation angle ψ, where ϕ_g is the sensor location azimuth angle, r_z is the virtual circle radius, and υ_z the virtual circle number.

The received signal vector $\mathbf{r}(l)$ at the sensor array is expressed as:

$$\mathbf{r}(l) = \mathrm{As}(l) + \mathbf{n}(l), l = 1, \ldots, L \tag{2}$$

where $\mathbf{r}(l) \in \mathbb{C}^{M\times 1}$ is the vector of the observed sensor output, which denotes the lth time snapshot of the received signal at the Mth sensor, $s(l) \in \mathbb{C}^{Q\times 1}$ is the unknown source signal vector, and $\mathbf{n}(l) \in \mathbb{C}^{M\times 1}$ is the vector of additive white Gaussian noise at the Mth sensor. Due to the non-uniform radius concatenations for the virtual circles, the non-uniform distribution of the physical sensors (nested arc array), and the virtual generated sensors (which will be seen later) on each circle, the steering response for each circle which lies on the x-y plane introduces a steering vector, $\boldsymbol{a}(\theta_q) \in \mathbb{C}^{M\times 1}$, which can be expressed as

$$\mathbf{a}(\theta_q) = \left[e^{jk\boldsymbol{u}(\theta_q)^T\cdot r_1}, \ldots, e^{jk\boldsymbol{u}(\theta_q)^T\cdot r_M}\right]^T \tag{3}$$

where $\mathbf{A}|_{\mathcal{C}_\mathcal{Z}} = \left[\mathbf{a}(\theta_1), \mathbf{a}(\theta_2), \cdots, \mathbf{a}(\theta_Q)\right]$, $\mathbf{A} \in \mathbb{C}^{M\times Q}$ is the array manifold matrix for the virtual circle $\mathcal{C}_\mathcal{Z}$, $k = 2\pi/\lambda_o$ is the wavenumber, r_m is the sensor location for the planar array with element on the (x, y) plan at location $r_m = \left[x_m, y_m\right]^T$ where $m \in [1, M]$, and $\boldsymbol{u}(\theta)$ is the unit vector pointed to the signal source. The signal

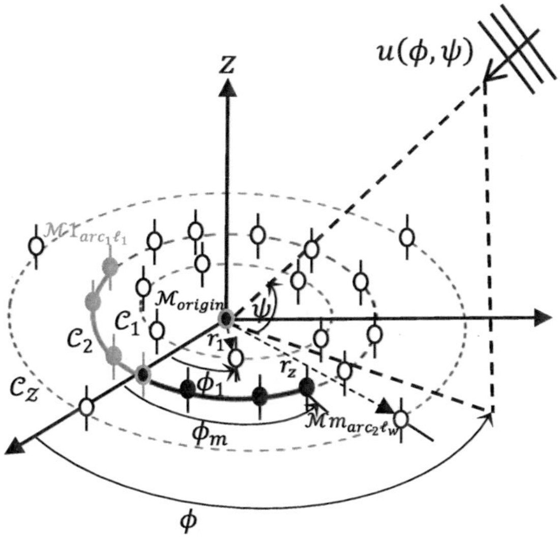

Fig. 2 Double-nested arc array construction (super nested arc array)

arrived with azimuth angle ϕ measured counterclockwise from the x-axis on the x-y plane and elevation angle ψ measured counterclockwise from the x-y plane is as shown in Fig. 2, $\boldsymbol{u}(\theta) = \boldsymbol{u}(\phi, \psi)$,

where

$$\boldsymbol{u}(\phi, \psi) = [\cos(\phi)\cos(\psi), \sin(\phi)\cos(\psi), \sin(\psi)]^T \tag{4}$$

Using the unconditional mode assumption (UMA) [6], the source signal $\mathbf{s}(l)$ is random vector that follows a non-Gaussian distribution. Certain assumptions on the correlation of the source signal, $\mathbf{s}(l)$ are assumed as follows:

(A_1) The source signals: $\{\mathbf{s}(\boldsymbol{l})\}_{l=1}^{L}$ are temporally uncorrelated and are given by $\mathbf{s}(\boldsymbol{l}) \sim N(0, \mathbf{R}_{ss})$, where $\mathbf{R}_{ss} = \mathbf{diag}\left[\boldsymbol{\sigma}_1^2, \cdots, \boldsymbol{\sigma}_q^2, \cdots, \boldsymbol{\sigma}_Q^2\right]$ and $\boldsymbol{\sigma}_q^2$ the power of the $\boldsymbol{Q}$th source signal.

(A_2) The noise, $\mathbf{n}(\boldsymbol{l})$ is spatially and temporally Gaussian process uncorrelated from the source signal, i.e., Gaussian random independent of the source signal, and is given by

$$\mathbf{n}(l) \sim \mathcal{N}\left(0, \sigma_n^2\mathbf{I}\right), E\left[\mathbf{n}(l)\mathbf{n}^H(l)\right] \tag{5}$$

and it is also assumed that the noise power σ_q^2 is unknown. Further, the signal covariance matrix, $\mathbf{R}_{ss}$, where $\mathbf{R}_{ss} = E\left[\mathbf{s}(l)\mathbf{s}^H(l)\right]$ is assumed to be a full rank matrix with rank equals Q and non-singular when the impinging signal $q(\theta)$ wave-fronts are uncorrelated. Furthermore, the eigenvalues are independent (no repetition) so the eigenvectors, are independent too [15]. The received signal vector $\mathbf{r}(l)$ given by:

$$\mathbf{r}(l) \sim \mathcal{N}\left(0, \mathrm{A}(\theta)\mathbf{R}_{\mathrm{ss}}\mathrm{A}^{H}(\theta) + \sigma_n^2\mathbf{I}\right) \tag{6}$$

$$\mathbf{R}_{\mathrm{xx}} = \mathbf{A}\mathbf{R}_{\mathrm{ss}}\mathbf{A}^{H} + \sigma^2\mathbf{I} = \sum_{q=1}^{Q}\sigma_q^2\boldsymbol{a}(\boldsymbol{\theta}_q)\boldsymbol{a}^{H}(\boldsymbol{\theta}_q) + \sigma^2\mathbf{I} \tag{7}$$

where $\mathbf{a}(\theta_q)$ is the steering vector. The direction finder uses the structure in (7) to eliminate the noise component, $\sigma_n^2\mathbf{I}$, whitening the received signal if the noise is Gaussian, then the FOC can be used to suppress its effects.

3 Fourth-Order Statistics

Fourth-order cumulants extended the array aperture by generating the virtual arc array. There are many definitions for the FOC in literature, we utilize the FOC calculation from [4, 16] as follows:

$$\begin{aligned}\boldsymbol{\mu}_{i,j}^{k,l} \triangleq \mathrm{cum}\{\mathbf{r}_i^*(l), \mathbf{r}_j(l), \mathbf{r}_k^*(l), \mathbf{r}_l(l)\} &= E\{\mathbf{r}_i^*(l)\mathbf{r}_j(l)\mathbf{r}_k^*(l)\mathbf{r}_l(l)\} \\ &- E\{\mathbf{r}_i^*(l)\mathbf{r}_j(l)\}E\{\mathbf{r}_k^*(l)\mathbf{r}_l(l)\} - E\{\mathbf{r}_i^*(l)\mathbf{r}_k^*(l)\}E\{\mathbf{r}_j(l)\mathbf{r}_l(l)\} \\ &- E\{\mathbf{r}_i^*(l)\mathbf{r}_l(l)\}E\{\mathbf{r}_j(l)\mathbf{r}_k^*(l)\}\end{aligned} \tag{8}$$

for $(i, j, k, l) \in \{1, \cdots, M\}$, where the received signal vector from array M is $\{\mathbf{r}_k(l)\}_{k=1}^{M}$.

Applying the cumulant-based processing, using the central sensor as a reference and the physical sensors on the arc, as shown in Fig. 3, the cross-correlation can be computed as shown in Fig. 4.

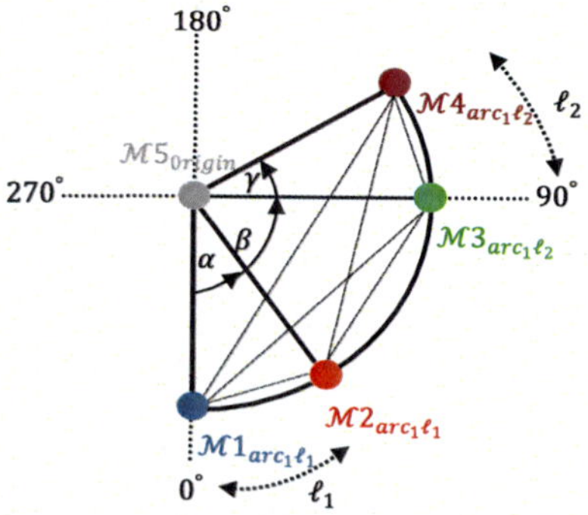

Fig. 3 Five sensors are constructed of a two-level nested arc array with a sensor at the center

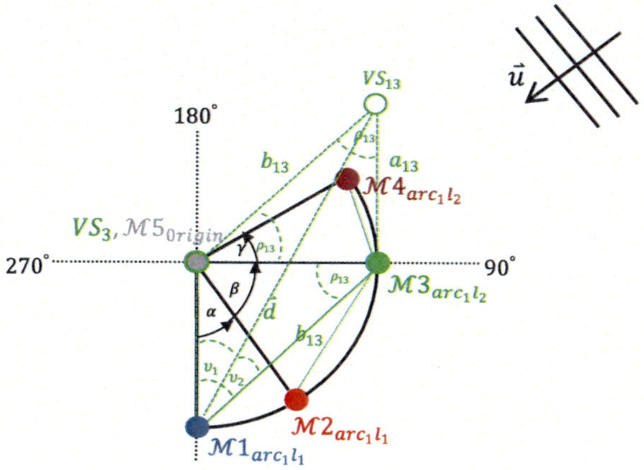

Fig. 4. Arc array's extended virtual sensor localization

3.1 Distribution Concept of the Virtual Array

Simplicity, the proposed merged double-nested arc array (super nested arc array) is applied to one nested arc array structure as shown in Fig. 3. It consists of five sensors, one at the origin and four lying on the arc, forming a two-level nested arc (ℓ_1, ℓ_2). FOC is used to increase the total array aperture.

3.2 Extended Cumulant-Based Virtual Sensor Localization for Physical Sensor $\mathcal{M}3_{arc_1 \ell_2}$

From [4] it was proven that the fourth-order cumulants can be interpreted as the addition of two vectors.

Applying Vector Addition

Since we are going to calculate the virtual sensors generated from the FOC algorithm, which is equivalent to vector addition between the physical sensors, we are going to take one physical sensor and connect it with one of the remaining physical sensors and with the one at the center (origin), which produces two vectors with a known orientation to be added, as shown in Fig. 4.

$$\text{Number of vectors} = C_{((M-1),2)} \tag{9}$$

Inclusionary that the total aperture for the nested arc array can be extended to

$$\begin{aligned}\text{Total aperture extended} =& M + (M-1) + P_{((M-1),2)} \\ &- \text{overlapped condition if occurred}\end{aligned} \tag{10}$$

By applying the derivation to the selected physical sensor $\mathcal{M}3_{\text{arc}_1\ell_2}$ and, similarly for the rest of the physical sensors. Applying the derivation to sensor $\mathcal{M}3$ will be done the same for all the other sensors as shown in the following sections.

($\mathcal{M}1, \mathcal{M}2, \mathcal{M}3, \mathcal{M}4, \mathcal{M}5$) are actual physical sensors measuring the stationary source signal impinging on each sensor according to its location, and ($\text{VS}_{13}, \cdots, \cdots$) a virtual sensor measuring the virtual process of the impinging signal (which measures the process without using a real sensor). So, the actual sensor $\mathcal{M}1_{\text{arc}_1\ell_1}$ measures $\boldsymbol{r}(l)$, which is the signal stationary source, the actual sensor $\mathcal{M}3_{\text{arc}_1\ell_2}$ measures $\boldsymbol{x}(l)$, and the actual sensor $\mathcal{M}5$ measures $\boldsymbol{y}(l)$.

Parallelogram Low Vector Addition

From the array geometry as shown in Fig. 4,

$$\vec{\boldsymbol{d}} = \mathcal{M}\vec{1\mathcal{M}}3 + \mathcal{M}\vec{1\mathcal{M}}5 \tag{11}$$

$$\vec{\boldsymbol{d}} = \vec{\boldsymbol{b}}_{13} + \vec{\boldsymbol{R}} \tag{12}$$

since

$$\boldsymbol{r}(l) = \boldsymbol{s}(l) \tag{13}$$

$$\boldsymbol{x}(l) = \boldsymbol{s}(l)\exp\left(-j\vec{\boldsymbol{q}} \cdot \vec{\boldsymbol{b}}_{13}\right) \tag{14}$$

and

$$\boldsymbol{y}(l) = \boldsymbol{s}(l)\exp\left(-j\vec{\boldsymbol{q}} \cdot \vec{\boldsymbol{R}}\right) \tag{15}$$

Virtual process $\boldsymbol{v}(l)$ measured by the virtual sensor VS_{13} will be:

$$\boldsymbol{v}(l) = \boldsymbol{s}(l)\exp\left(-j\vec{\boldsymbol{q}} \cdot \vec{\boldsymbol{d}}\right) \tag{16}$$

1. The magnitude will be:

$$\boldsymbol{d} = \sqrt{b_{13}^2 + R^2 + 2b_{13}R\cos\rho_{13}} \tag{17}$$

2. The directions:

$$\tan\upsilon_2 = \frac{R\sin\rho_{13}}{b_{13} + R\cos\rho_{13}} \tag{18}$$

$$\tan \upsilon_1 = \frac{b_{13} \sin \rho_{13}}{R + b_{13} \cos \rho_{13}} \tag{19}$$

also

$$b_{13} = \frac{d \sin \upsilon_2}{\sin(\upsilon_1 + \upsilon_2)} = \frac{d \sin \upsilon_2}{\sin \rho_{13}} \tag{20}$$

$$R = \frac{d \sin \upsilon_1}{\sin(\upsilon_1 + \upsilon_2)} = \frac{d \sin \upsilon_1}{\sin \rho_{13}} \tag{21}$$

The only known parameter is *R*, and b_{13} is unknown, by using the cosine rule we could get b_{13} and a_{13}.

The (FOC) can be explicated as the addition of two vectors.

$$\mu_{\mathcal{M}1,R}^{\mathcal{M}1,b_{13}} \equiv \underset{13}{\vec{b}} + \vec{R} = \vec{d} = \mu_{\mathcal{M}1,VS_{13}} \tag{22}$$

The cumulants can be used to form the virtual cumulant covariance by modifying (7) to be as (16) in [4], then the cumulant forms of the virtual covariance matrix are:

$$\mathbf{R}_{xx}|_{\text{FOC}} = \sum_{q=1}^{Q} \gamma_{4,q}\, \mathbf{a}(\theta_q)\big|_{\text{FOC}}\, \boldsymbol{a}^H(\theta_q)\big|_{\text{FOC}} = \mathbf{A}|_{\text{FOC}} \boldsymbol{\Gamma}\, \mathbf{A}^H\big|_{\text{FOC}} \tag{23}$$

That cumulant forms the virtual covariance matrix whose signal subspace can depict the virtual extended apertures. The cumulants can be expressed as vector addition, whereas the cross-covariance can be as a single vector when there are *Q* sources uncorrelated, and incoherent. Then the source covariance matrix $\mathbf{R}_{ss}$ is a diagonal matrix where

$$\boldsymbol{\Gamma} = \text{diag}\left[\gamma_{4,1}, \ldots, \gamma_{4,q}, \ldots, \gamma_{4,Q}\right] \tag{24}$$

which is the spatial covariance matrix of the impinging signals, i.e., cross-spectral density matrix, of the virtually extended sensor array and $\gamma_{4,q}$ (the fourth-order source cumulant) is the FOC of the qth source signal and is given by

$$\gamma_{4,q} = \text{Cum}\left\{s_q(l), s_q(l), s_q^*(l), s_q^*(l)\right\} \tag{25}$$

A is the extended sensor array manifold matrix, $\mathbf{A} \in \mathbb{C}^{F \times Q}$ where *F* is the number of sensors after the generated virtual extension and is given by:

$$\mathbf{a}(\theta_q)\big|_{\text{FOC}} = \left[e^{jk\boldsymbol{u}(\theta_q)^T \cdot r_1}, \ldots, e^{jk\boldsymbol{u}(\theta_q)^T \cdot r_F}\right]^T \tag{26}$$

$$\mathbf{A}|_{\mathcal{C}_{\mathcal{Z}},\text{FOC}} = \left[\mathbf{a}(\theta_1), \mathbf{a}(\theta_2), \cdots, \mathbf{a}(\theta_Q)\right] \tag{27}$$

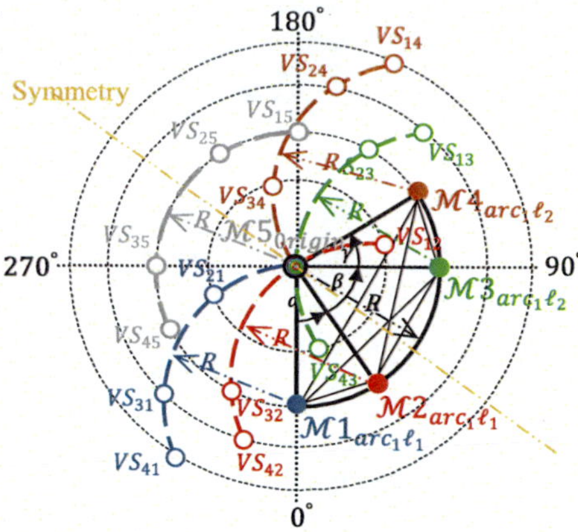

Fig. 5 Virtually generated arcs and circles where the extended cumulant-based sensors (VS_{mm}), generated from the actual physical sensors $\mathcal{M}m$ with the same radius R

$\mathbf{a}(\theta_q)\big|_{\text{FOC}} \in \mathbb{C}^{F\times 1}$ is the steering vector in the cumulant domain with the virtual extends as shown in Fig. 5. Our new approach for the generated virtual sensor direct localization as an element of the virtual arcs has been proved with details in Appendix. It directly takes the shape as shown in Fig. 5.

4 Configuration Specification

After the configuration specification, we are eligible to get the direct complete generated structures as shown in Appendix. It shows that each physical sensor group (five sensors) is responsible for generating a virtual arc and being the central origin sensor for that virtual arc with the same angles sequence, values, and direction as shown in Fig. 5 and with the same radius R from each physical sensor. So according to the number of physical sensors m, the number of the generated virtual arc is shown in Fig. 5, since $m = 1, \ldots, M$.

$$\begin{aligned}&\text{Number of the generated virtual arc} =\\&\text{number of physical sensores} = M\end{aligned} \tag{28}$$

The structures of the non-uniform virtual circles generate from the sensors that have the same radius. The number of virtual circles equals the number of distances between every two physical sensors without repetition for the equidistance, and it is also the number of vectors between every two physical sensors without repetition for the equal vectors, that's measured from the central origin, to be the radius for each virtual circle.

$$\begin{aligned}&\text{The number of virtual circles} =\\&C_{((M-1),2)} - \text{ number of equal vectors}\end{aligned} \tag{29}$$

The generated virtual arcs are shifted with the same sequence number of the physical sensor on the physical arc which is responsible to generate it and to be its central origin individually for each physical sensor as shown in Fig. 5.

5 Constraints on the Physical Arc's Geometry

1. The arc radius $R \leq \lambda/2$, has the same condition for the adjacent sensors.
2. Upgraded the nested arc array structure to double-nested arc array.
3. The separation angle between the levels and between the sensors for the nested arc array are multiples of each other for one arc, also the second arc is multiples of the first one.

6 Cramer-Rao Bound in the Fourth-Order Cumulant Domain

CRB estimates the directions of arrival of $\boldsymbol{Q}$ sources detected by the super nested arc array which has been extended by FOC. The inverse of the total information matrix (CRB) [13] is lower bounds on the covariance matrix. The parameters to be estimated in the cumulant domain DOAs after the noise $\boldsymbol{\sigma}^2$ has been discarded, due to the FOC. Being $\boldsymbol{\theta} = \{\phi_q, \boldsymbol{\psi}_q\}$, signal cumulant, $\boldsymbol{q} = 1, \ldots, \boldsymbol{Q}$. Substituting by $\mathrm{A}(\boldsymbol{\theta})$, and $\mathbf{R}_{ss}$ in Eq. (5) in [9] which was modified to Eq. (6) in [9], the CRB matrix is

$$\boldsymbol{CRB}(\boldsymbol{\theta}) = [\boldsymbol{G}(\boldsymbol{B}, \boldsymbol{\theta})\boldsymbol{P}]^{-1} \tag{30}$$

$$\boldsymbol{G}(\boldsymbol{B}, \boldsymbol{\theta}) = \boldsymbol{J}_\theta(Q)^T \boldsymbol{B} \boldsymbol{J}_\theta(Q) \tag{31}$$

where $\boldsymbol{J}_\theta(Q)$ is the Jacobian matrix of $Q(\theta)$ in (4):

$$\boldsymbol{J}_\theta(Q) = \left[\frac{\partial Q(\boldsymbol{\theta})}{\partial \phi} \; \frac{\partial Q(\boldsymbol{\theta})}{\partial \psi} \right] = \begin{bmatrix} -\sin(\phi)\cos(\psi) & -\cos(\phi)\sin(\psi) \\ \cos(\phi)\cos(\psi) & -\sin(\phi)\sin(\psi) \\ 0 & \cos(\psi) \end{bmatrix} \tag{32}$$

$$\mathbf{B} = \frac{1}{M} \sum_{m=1}^{M} (\boldsymbol{r}_m - \boldsymbol{r}_c)(\boldsymbol{r}_m - \boldsymbol{r}_c)^T \tag{33}$$

From Sect. 2.1 $\boldsymbol{r}_m = [\boldsymbol{x}_m, \boldsymbol{y}_m]$ is the location of the sensor m in the 2D plane, where $\boldsymbol{r}_c$ is the centroid of the arc actual (physical) array and the centroid of the extended virtual aperture array, i.e.,

$$\boldsymbol{r}_c = \frac{1}{M}\sum_{m=1}^{M}\boldsymbol{r}_m \tag{34}$$

$$\boldsymbol{P} = \frac{2LM}{c^2}\sum_{q=1}^{Q}\frac{\omega_q^2}{\sigma_{n_q}}p_q\left(1 - \frac{\sigma_{n_q}}{p_q M + \sigma_{n_q}}\right) \tag{35}$$

L is the snapshot, p_q is the signal power, σ_{n_q} is the frequency noise power, ω_q is the frequency of the impinging signal, and c is the propagation speed. For the 2D DOA estimation as [9], it is parameterized by $\boldsymbol{\theta} = \{\phi_q, \psi_q\}^T$, where $\phi \in (0, 2\pi)$ and $\psi \in (-(\pi/2), \pi/2)$ denotes the azimuth and elevation, respectively. We can see that $\mathbf{G}(\mathbf{B}, \boldsymbol{\theta})$ depends on the source DOA vector $\boldsymbol{\theta}$ and the array geometry, and the source power as a function of frequency.

7 Observation and Detection

The new approach is faster and more direct distribution (localization) for the virtual sensors than that of the conventional way [4], using direct virtual arc generation and then distributing it on a virtual circle. To accomplish increasing the array aperture, and to fill gaps in the space as in Fig. 5, it was proposed to use only two nested arc arrays and not more to avoid the overlapping of sensors on the virtual circles, and for the hardware reduction, which forms a double-nested arc array, and then applies the FOC as shown in Figs. 9, 10, and 11.

8 Simulation Results

The experiments and simulations are composed of

- Apply FOC to extend the array aperture,
- Calculate the location for the extended virtual sensors, and
- Compare the CRB performance of the double-nested arc array for different structures and configurations with the uniform arc array.

9 Results from the Numerical Substitution of Sensors Location

Assuming that there is one source signal impinging the sensor array from an azimuth angle 40° and 80° for the elevation angle with physical arc radius $\boldsymbol{R} = \lambda/2$, and λ is normalized which will apply to all the simulations.

9.1 Applying on a Single Nested Arc

By numerical substitution for the angle (α, β, γ) in Fig. 3 as $\left(30^{\circ}, 60^{\circ}, 30^{\circ}\right)$, the distribution of the sensors is given in Table 1.

- Number of vectors from Eq. (9), $C_{(4,2)} = 6$
- Total extension of the aperture referring to Eq. (10)
  $M + (M - 1) + P_{((M-1),2)} = 21$
- Number of virtual arcs = 5 from Eq. (28)
  The total number of arcs = physical arcs + virtual arcs = 6
- Number of virtual circles = 4, referring to Eq. (29).

10 Numerical Substitution for the Locations of the Double-Nested Arcs Arrays and Their Virtual Extension Forming the Virtual Butterfly Antenna

To make our approach very mild when compared with [4] uniform arc array after being doubled, the FOC must be applied for all cases. Using double-nested arc arrays as in Figs. 6, 9, 10 and 11, the total number of sensors $M = 8$ with two common (shared) sensors, one located at the origin and the other located in common between the two nested arcs. The distance between any two adjacent sensors $\leq \lambda/2$ for all the physical sensors. The signal data sources are as in (2). The incident signal was assumed with heading toward the array with azimuth and elevation angle for all the simulations, with a signal-to-noise ratio (SNR) up to 15dB and assumed 20 snapshots.

Table 1 Sensor's distribution for a nested arc array with angles $\left(30^{\circ}, 60^{\circ}, 30^{\circ}\right)$

Virtual circle	Radius	Virtual sensors	Physical sensors	Angles (deg)
Origin	0		1	Reference
1	0.2588	4		15, 105, 195, 285
2	0.5	6	4	105, 180, 210, 270, 300, 330 0, 30, 90, 120
3	0.7071	4		135, 165, 315, 345
4	0.8660	2		150, 330
Total apertures		16	5	
		21		

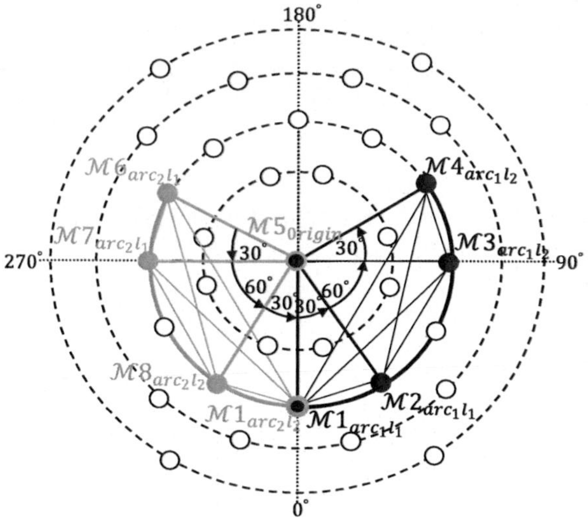

Fig. 6. Double symmetric non-uniform nested arc array with total number of sensors equals eight with two levels nested array each

10.1 Symmetric Non-uniform Double-Nested Arc Arrays

The use of eight's arc array sensors, one at the center and seven distributed on the arc, is shown in Fig. 6 and Table 2. Using Eqs. (9), (10), (28) and (29).

- Vectors numbers: $C_{((M-1),2)_{\text{arc}_1}} + C_{((M-1),2)_{\text{arc}_2}} = C_{(4,2)_{\text{arc}_1}} + C_{(4,2)_{\text{arc}_2}} = 12$
- The total overlapped sensors are:$[(\mathcal{M}1_{\text{arc}_1 l_1}, \mathcal{M}1_{\text{arc}_2 l_2}), (\mathcal{M}2, VS_{78}), (\mathcal{M}3, VS_{75}),$

Table 2 Sensor's distribution for a double-nested arc array arranged as $(30^\circ, 60^\circ, 30^\circ), (30^\circ, 60^\circ, 30^\circ)$

Virtual circle	Radius	Virtual sensors	Physical sensors	Overlap	Angles (deg)
Origin	0		1	1	Reference
1	0.258	8			15, 105, 195, 285, 75, 165, 255, 345
2	0.5	5	7	8	105, 180, 210, 300, 60 0, 30, 90, 120, 240, 270, 330
3	0.707	8			135, 165, 315, 345, 15, 45, 195, 225
4	0.866	4			150, 330, 30, 210
Total apertures		25	8	9	
		33			
Total apertures with overlap		42			

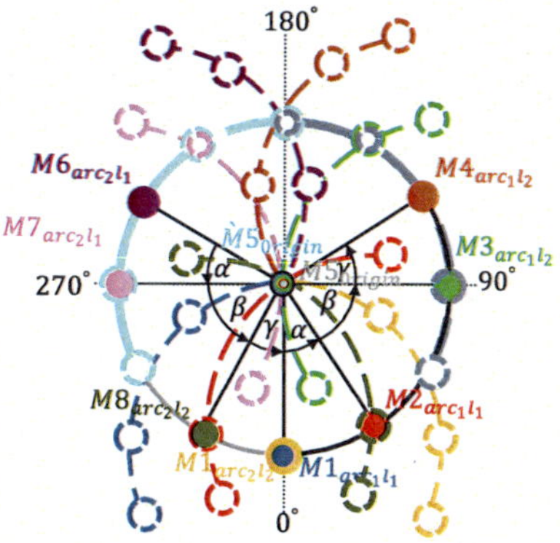

Fig. 7 Virtually generated arcs for a double symmetric non-uniform nested arc array of sensors, named the virtual extended butterfly antenna. overlapped sensors

$(\mathcal{M}5_{\text{origin}}, \acute{\mathcal{M}}5_{\text{origin}})$, $(\mathcal{M}7, VS_{35})$, $(\mathcal{M}8, VS_{32})(VS_{25}, VS_{34})$,$(VS_{15}, \text{V}'\text{S}_{15})$, $(VS_{23}, VS_{85})] = 9$ elements as shown in Fig. 7.

- The total aperture can be extended to: $(M + (M - 1) + P_{((M-1),2)})_{\text{arc}_1} + (M + (M - 1) + P_{((M-1),2)})_{\text{arc}_2}$ − the overlapped sensores $= (5 + 4 + 12)_{\text{arc}_1} + (5 + 4 + 12)_{\text{arc}_2} - 9 = 33$ as given in Table 2.
- The total number of arcs = number of physical arcs + the generated virtual arcs $= 2 + 10 = 12$ as shown in Fig. 7.
- Virtual circles number: $C_{((M-1),2)\text{arc}_1} + C_{((M-1),2)\text{arc}_2}$ − the number of equal distance between two sensors $= 6 + 6 - 8 = 4$ as shown in Fig. 8.

10.2 Non-uniform Double-Nested Arc Arrays

The use of two non-uniform nested arc arrays with one multiple of the other.

- Vectors numbers: $C_{((M-1),2)_{\text{arc}_1}} + C_{((M-1),2)_{\text{arc}_2}} = C_{(4,2)_{\text{arc}_1}} + C_{(4,2)_{\text{arc}_2}} = 12$
- The total overlapped sensors are:

  $$\left[(\mathcal{M}1_{\text{arc}_1 l_1}, \mathcal{M}1_{\text{arc}_2 l_2}), \left(\mathcal{M}5_{\text{origin}}, \acute{\mathcal{M}}5_{\text{origin}}\right),\right.$$
  $$\left.(\mathcal{M}6_{\text{arc}_2 l_1}, VS_{32}), (\text{VS}_{15}, \text{V}'\text{S}_{15}), (\text{VS}_{23}, \text{VS}_{65})\right] = 5 \text{ elements}$$
  as shown in Fig. 9.

- The total aperture can be extended to:
  $$\left(M + (M - 1) + P_{((M-1),2)}\right)_{\text{arc}_1} + \left(M + (M - 1) + P_{((M-1),2)}\right)_{\text{arc}_2}$$
  − the overlapped sensores $= (5 + 4 + 12)_{\text{arc}_1} + (5 + 4 + 12)_{\text{arc}_2} - 5 = 37$ as given in Table 3.
- The total number of arcs = number of physical arcs (double arc) + the generated virtual arcs $= 2 + 10 = 12$.

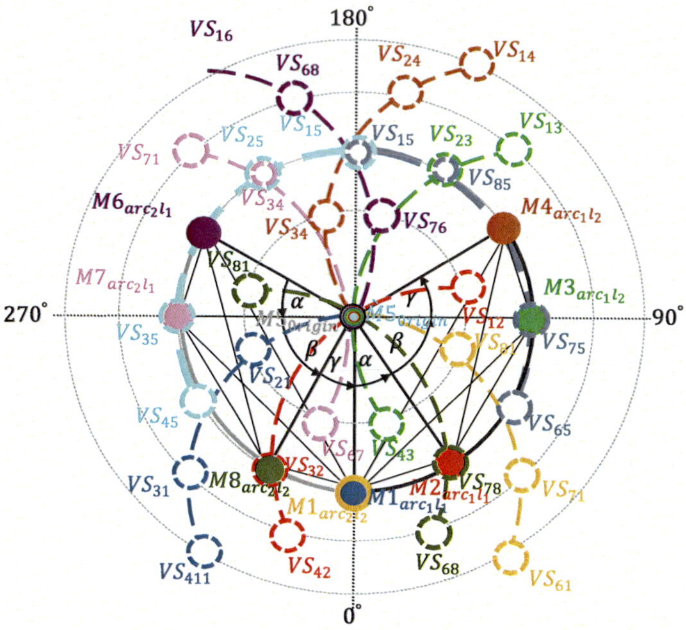

Fig. 8 Virtually generated circles for a double symmetric non-uniform nested arc array of sensors, named the virtual extended butterfly antenna

Fig. 9 Non-uniform double-nested arc arrays with (15°), and (60°) intermediate angles between the nested levels (subarrays), and (7.5°), and (30°) between sensors

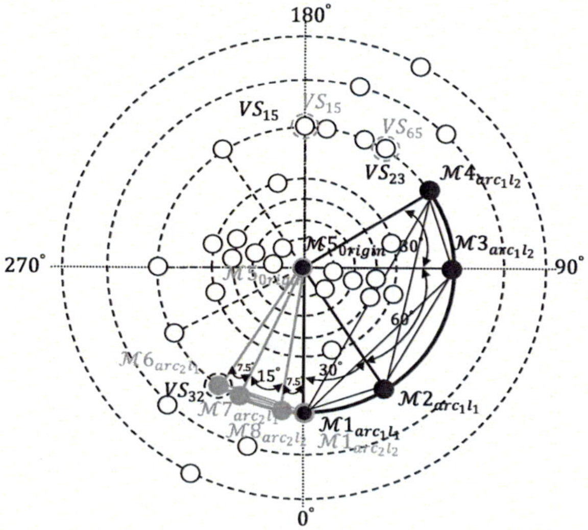

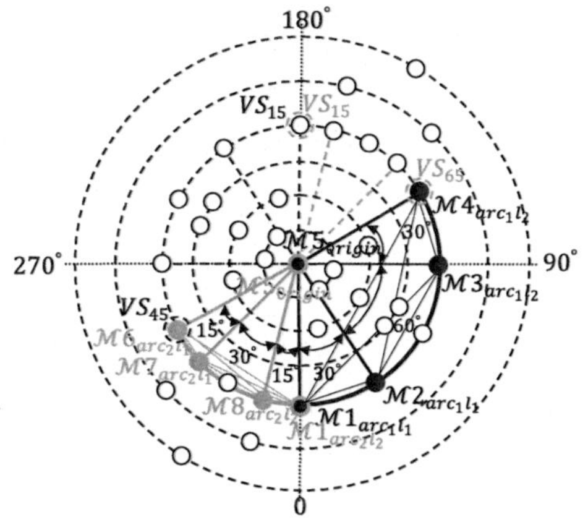

Fig. 10 Double-nested arc arrays where one is multiple (double) of the other with (30°), and (60°) intermediate angles between the nested levels (subarrays), and (15°), and (30°) between sensors

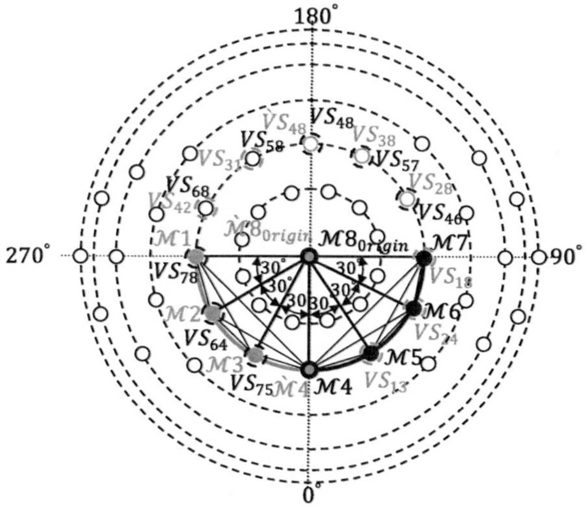

Fig. 11 Double symmetric uniform arc array with uniform distance in between 30°

- Virtual circles number: $C_{(((M-1),2)\text{arc}_1)} + C_{((M-1),2)\text{arc}_2}$ − the number of equal distance between two sensors = 6 + 6 − 5 = 7 as shown in Fig. 9.

Table 3 Sensor's distribution for a non-uniform double-nested arc arrays arranged as $(7.5^\circ, 15^\circ, 7.5^\circ), (30^\circ, 60^\circ, 30^\circ)$

Virtual circle	Radius	Virtual sensors	Physical sensors	Angles (deg)	Overlap
Origin	0		1	reference	
1	0.065	4		63.75, 86.25, 243.75, 266.25	1
2	0.130	2		75, 255	
3	0.195	4		71.25, 78.75, 251.25, 258.75	
4	0.258	6		15, 105, 195, 285, 75, 255	
5	0.5	7	7	105, 180, 210, 270, 300, 157.5, 172.5 0, 30, 90, 120, 330, 337.5, 352.5	4
6	0.707	4		135, 165, 315, 345	
7	0.866	2		150, 330	
Total apertures		29	8		5
		37			
Total apertures with overlap		42			

10.3 Double-Nested Arc Arrays Where One is Multiple (Double) of the Other

The use of two nested arc arrays where the intermediate distance between levels and the separation angles between sensors for the first nested arc is double the measurements of the second nested arc.

- Vectors numbers: $C_{((M-1),2)_{\text{arc}_1}} + C_{((M-1),2)_{\text{arc}_2}} = C_{(4,2)_{\text{arc}_1}} + C_{(4,2)_{\text{arc}_2}} = 12$
- The total overlapped sensors are:

  $[(\mathcal{M}1_{\text{arc}_1 l_1}, \mathcal{M}1_{\text{arc}_2 l_2}), (\mathcal{M}5_{\text{origin}}, \acute{\mathcal{M}}5_{\text{origin}}), (\mathcal{M}4_{arc_1 l_2}, \text{VS}_{65})(\text{VS}_{15}, \text{V}'\,\text{S}_{15})]$ as shown in Fig. 10.

- The total aperture can be extended to: $\left(M + (M-1) + P_{((M-1),2)}\right)_{\text{arc}_1} + \left(M + (M-1) + P_{((M-1),2)}\right)_{\text{arc}_2}$ $-$the overlapped sensores $= (5+4+12)_{\text{arc}_1} + (5+4+12)_{\text{arc}_2} - 5 = 37$ as given in Table 4.
- The total number of arcs = number of physical arcs (double arc) + the generated virtual arcs = 2 + 10 = 12.
- Virtual circles number: $C_{((M-1),2)\text{arc}_1} + C_{((M-1),2)\text{arc}_2}$ $-$ the number of equal distance between two sensors = 6 + 6 = 6 as shown in Fig. 10.

Table 4 Sensor's distribution for double-nested arc arrays arranged as $(15^{\circ}, 30^{\circ}, 15^{\circ}), (30^{\circ}, 60^{\circ}, 30^{\circ})$

Virtual circle	Radius	Virtual sensors	Physical sensors	Angles (deg)	Overlap
Origin	0		1	Reference	1
1	0.130	4		37.5, 82.5, 217.5, 262.5	
2	0.258	6		15, 60, 105, 195, 240, 285	
3	0.382	4		52.5, 67.5, 232.5, 247.5	
4	0.5	10	6	60, 135, 150, 165, 180, 210, 240, 270, 300, 330 0, 30, 90, 120, 315, 345	4
5	0.707	4		135, 165, 315, 345	
6	0.866	2		150, 330	
Total apertures		30	7		5
		37			
Total apertures with overlap		42			

10.4 Double Symmetric Uniform Arc Array

Using eight's arc arrays sensors, one at the center and seven distributed on the arc with identical separation angles between sensors equals 30°. Number of sensors in the first arc $M_1 = 5$, the number of sensors in the second arc $M_2 = 5$, with two common (shared) sensors $M4$, and $M8_{\text{origin}}$ between the two arcs as shown in Fig. 11.

- Vectors numbers: $C_{((M-1),2)_{\text{arc}_1}} + C_{((M-1),2)_{\text{arc}_2}} = C_{(4,2)_{\text{arc}_1}} + C_{(4,2)_{\text{arc}_2}} = 12$
- The total overlapped sensors are:
  $[(\mathcal{M}4, \mathcal{M}'4), (\mathcal{M}8_{\text{origin}}, \mathcal{M}'8_{\text{origin}}), (\mathcal{M}1, VS_{78}), (\mathcal{M}2, VS_{64}), (\mathcal{M}3, VS_{75}), (\mathcal{M}5, VS_{13}),$
  $(\mathcal{M}6, VS_{24}), (\mathcal{M}7, VS_{18}), (VS_{42}, VS_{68}), (VS_{31}, VS_{58}), (VS_{48}), (VS_{38}, VS_{57}), (VS_{28}, VS_{46})] = -$ 13 elements as shown in Fig. 11.
- The total aperture can be extended to:
  $\left(M + (M-1) + P_{((M-1),2)}\right)_{\text{arc}_1} + \left(M + (M-1) + P_{((M-1),2)}\right)_{\text{arc}_2}$
  $-$the overlapped sensores $= (5+4+12)_{\text{arc}_1} + (5+4+12)_{\text{arc}_2} - 13 = 29$
  as given in Table 5.
- The total number of arcs = number of physical arcs + the generated virtual arcs $= 1 + 8 = 9$
- Virtual circles number:

Table 5 Sensor's distribution for a double symmetric uniform arc array arranged as $(30^{\circ}, 30^{\circ}, 30^{\circ})$, $(30^{\circ}, 30^{\circ}, 30^{\circ})$

Virtual circle	Radius	Virtual sensors	Physical sensors	Angles (deg)	Overlap
Origin	0		1	Reference	1
1	0.258	12		15, 45, 75, 105, 135, 165,195, 225, 255, 285, 315, 345	
2	0.5	5	7	120, 150, 180, 210, 240 0, 30, 60, 90, 270, 300, 330	12
3	0.707	4		45, 135, 225, 315	
Total apertures		21	8		13
		29			
Total apertures with overlap		42			

$C_{((M-1),2)}$ – the number of equal distance between two sensores = $(6_{\text{arc}_1} + 6_{\text{arc}_2}) - 9_{\text{both arcs}} = 3$ as shown in Fig. 11.

11 Illustration for the CRB Comparison between Different Double-Nested Arc Array Configurations

By using a fixed number of sensors, normalized wavelength for one incident signal to not affect the configuration, we compared the uniform arc array with different configurations of the double-nested arc array, as shown in Fig. 12, it is found.

1. Double-nested arcs are superior to the uniform double arc arrays.
2. The CRB lower bound occurs when the intermediate angles between the nested levels and between the sensors, are doubled sequentially between the two arcs as shown in Fig. 10 and Table 4.

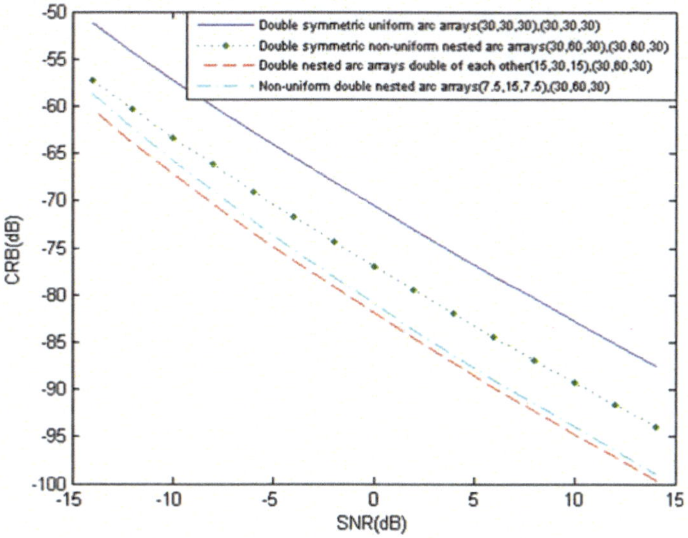

Fig. 12 CRB versus SNR for different double-nested arcs arrays configuration

12 Conclusion

In this paper, a novel sparse array double-nested arc array has been proposed and its lower bound error, CRB, on 2D DOA estimation using FOC has been investigated. The proposed array performance is superior in increasing the degree of freedom and in decreasing the overlapped conditions compared with the uniform arc array. In future work, this configuration will be applied to different DOA methods, trying to reach the optimum number of the nested arc array.

Appendix

Direct and Fast Virtual Sensor Localization

The physical sensor $\mathcal{M}3_{\text{arc}_1\ell_2}$ in Fig. 13 shapes the identified position data of the generated virtual sensor VS_{23} and it provides an element of the generated virtual arc and circle.

Computation Related to the Central Physical Sensors to Configure the Virtual Circle

$$b_{23} = \sqrt{2R^2(1 - \cos(\beta_{23}))} \tag{36}$$

$$\rho_{23} = \cos^{-1}\frac{b_{23}}{2R} \tag{37}$$

$$\mathfrak{z}_{23} = (180 - \rho_{23}) + \alpha \tag{38}$$

Computation Related to the Generated Virtual Arc
As shown in Fig. 13

$$\beta_{23} = \beta \tag{39}$$

virtual arc angles sequence (α, β, γ), which is counterclockwise direction as the physical arc. Applying the use of the sine rule:

$$\frac{a_{23}}{\sin\rho_{23}} = \frac{R}{\sin\rho_{23}} = \frac{b_{23}}{\sin\beta_{23}} \tag{40}$$

$$\frac{a_{23}}{\sin\rho_{23}} = \frac{R}{\sin\rho_{23}} = \frac{b_{23}}{\sin\beta} \tag{41}$$

The generated virtual arc has a radius R exactly like the physical one. So

$$a_{23} = R \tag{42}$$

Similarly, the same derivation for VS_{13}, VS_{43} when the original central physical sensor is shared (overlapped) with a virtual sensor on the generated virtual arc. It takes the same sequence number three to be VS_3 as the physical sensors $\mathcal{M}3_{\text{arc}_1 l_2}$ which is responsible to generate the virtual arc.

These parameters related to the main central physical sensors are used for configuring a virtual circle by the equiradius virtual sensors of the same b_{MM} as shown in Fig. 5. Also, the generated virtual nested arc array VS_{13}, VS_{23}, VS_3, VS_{43} lays with a radius $= R$ for all the virtual senores from the central origin physical sensor $\mathcal{M}3_{\text{arc}_1 l_2}$ and with the same angles sequence values and direction (α, β, γ) as the physical actual arc $\mathcal{M}1_{\text{arc}_1 \ell_1}$, $\mathcal{M}2_{\text{arc}_1 \ell_1}$, $\mathcal{M}3_{\text{arc}_1 \ell_2}$, $\mathcal{M}4_{\text{arc}_1 \ell_2}$ as shown in Fig. 5.

Similarly applying the same calculations on each physical sensor to generate its virtual arc and circle.

Extended Cumulant-Based Virtual Sensors Localization for the Physical Sensor $\mathcal{M}5_{\text{origin}}$

Each of the two physical sensors, one on the arc, and one at the origin can generate a virtual sensor [4], it is configured with the same angles sequence values as the physical actual arc and with a radius $= R$ from the origin sensor $\mathcal{M}5_{\text{origin}}$ as shown in Fig. 5.

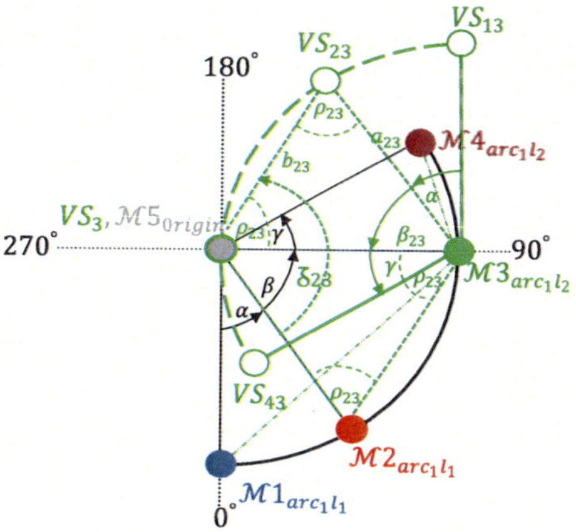

Fig. 13 Arc array's extended virtual sensor VS_{23} localization related to the physical sensor $\mathcal{M}3_{\mathrm{arc}_1\ell_2}$

References

1. Balanis C (1992) Antenna Theory. IEEE 80(1):7–23
2. Albagory Y (2021) An efficient conformal stacked antenna array design and 3d-beamforming. Sensors 21(4):1–29
3. Persson P (2001) Analysis and design of conformal array antennas. 4(4):279–290
4. Doğan M, Mende J (1995) Applications of cumulants to array processing part I: aperture extension and array calibration. IEEE Trans Signal Process 43(5):1200–1216
5. Pal P, Member S, Vaidyanathan P (2010) Nested arrays: a novel approach to array processing with enhanced degrees of freedom. IEEE Trans Signal Process 58(8):4167–4181
6. Koochakzadeh A, Pal P (2016) Cramér-rao bounds for underdetermined source localization. IEEE Signal Process 23(7):919–923
7. Ahmed A, Himed B (2017) Effective nested array design for FOC-based DOA estimation. In: CONFERENCE 2017 IEEE Radar, vol 10.1109, pp 0998–1002
8. Shen Q, Wu S (2016) Extension of nested array with the fourth-order difference co-array enhancement. In: CONFERENCE 2016 Icassp, pp 2991–2995
9. Baysal Ü, Moses R (2003) On the geometry of isotropic arrays. IEEE Trans Signal Process 51(6):1469–1478
10. Benjamin F (1991) On the cramer-rao bound for direction finding of correlated signals. IEEE Trans Signal Process 39(5):1215–1218
11. Choi J, Yoo C (2015) Underdetermined high-resolution DOA estimation: A 2ρ the-order source-signal/noise subspace constrained optimization. IEEE Trans Signal Process 63(7):1858–1873
12. Zhang Y, Qin S, Amin M (2014) DOA estimation exploiting coprime arrays with sparse sensor spacing. In: CONFERENCE 2014 IEEE international on acoustic, pp 2267–2271
13. Doron M, Doron E (1994) Wavefield modeling and array processing, part i—spatial sampling. IEEE Trans Signal Process 42(10):2549–2559
14. Delmas J () Performance bounds and statistical analysis of DOA estimation. In: Array and statistical signal processing, vol 3, pp 719–764
15. Chen Z (2010) Introduction to direction-of-arrival estimation. Artech House
16. Porat B, Frid B (1991) Direction finding algorithms based on high-order statistics. IEEE TSP 39(9):2675–2678

Energy Security of Ukraine in General and in the Field of Electric Power: Theory and Reality

Vitalina Babenko and Boris Pokhodenko

Abstract The article discusses the general principles of forming a methodology for assessing the state of the country's energy security. In view of the differences in regulation mechanisms, the logic of making and applying management decisions, the spheres of energy security regulation have been singled out. The concept of energy regulation in countries with a stable market economy is considered. An analysis of the current state and development strategy of the energy industry of Ukraine in general and the electric power industry in particular were made for the purpose of ensuring the country's energy security. Raised issues require immediate resolution.

Keywords Energy security · Energy sector · Power industry · Energy strategy of Ukraine · Energy balance

1 Introduction

Energy security is one of the priorities of national governments. Each state determines the policy objectives in the field of energy security and the mechanisms for its implementation, taking into account its own peculiarities of the organization of society's vital activities and the model of state management. Management activity in this area is not unambiguous and simple, as it needs to take into account various aspects of the country's life. These are resource, technical, economic, ecological, organizational-administrative, managerial, social, innovative, political, geopolitical, security and other aspects, worldview aspects of energy supply for vital needs and functions of the state, society and economy. In addition, activities in the field of energy security can evolve significantly over time in one and the same country, depending

V. Babenko (✉) · B. Pokhodenko
Kharkiv National Automobile and Highway University, V. N. Karazin Kharkiv National University, Kharkiv 61002, Ukraine
e-mail: vitalinababenko@karazin.ua

B. Pokhodenko
e-mail: boris.pokhodenko@karazin.ua

D. Magdi et al. (eds.), *Green Sustainability: Towards Innovative Digital Transformation*, Lecture Notes in Networks and Systems 753,
https://doi.org/10.1007/978-981-99-4764-5_21

on the dynamics of socioeconomic and scientific and technological development, the transformation of energy market functioning models. Recently, both in Ukraine and in the world, there have been significant changes in the principles of regulation of the energy sector, and the range of methods of using energy as a tool for achieving political goals is expanding. It is this circumstance that determines the variety of theoretical approaches and practices of identifying the object of research (management sphere), its description and methods of assessing the country's energy security state. The need to take into account various aspects of the country's life creates significant obstacles to the development of a unified methodological approach to distinguishing the sphere of energy security.

2 Methodology

A significant amount of scientific research is devoted to the problems of energy management, in particular, the formation of energy policy. H. Mityushkina characterizes the experience of developing energy strategies in the countries of the world with the aim of increasing the level of their energy security; S. Azhnakin studies the problems and prospects of stimulating the attraction of investments in the energy sector of Ukraine [17, 20]. The study of energy management processes in individual regions is covered in the works of O. Amoshi, V. Shevchenko, N. Tkachenko, and Zh. Poplavska. Regional energy development strategies are analyzed in the works of M. Honchar, S. Zhuk, and A. Chubyk; analysis of economic approaches to the use of energy resources is given in the works of N. Belopolsky and D. Turchenko [8, 10, 12, 14]. The motivational function in energetics was studied by Yu. Kostin, K. Uschapovskyi, and A. Minkovich [14, 21]. The problems of strategy formation at energy enterprises are the subject of research by such scientists as O. Sukhodolya, V. Barannik, M. Voropai, N. Savina, V. Haustova, R. Podolets, O. Dyachuk, B. Serebrennikov, and others. O. Dupak, O. Laktionov, and V. Svitalka analyze the implementation of the energy strategy of Ukraine [3, 4, 15, 19, 21]. O. Ivanyuk, N. Podolchak, L. Sereda, and O. Dzioba reviewed energy security management processes. V. Mykytenko considered the problems of increasing the level of energy efficiency national economy, etc. Economists made a significant contribution to the formation of methodological bases for analyzing the energy market of Ukraine, namely: S. Aptekar, O. Balan, T. Birkovich, D. Busarev, V. Geets, G. Darnopyh, A. Dronova, B. Kvasnyuk, O. Kendyukhov, N. Kosar, L. Nazarchuk, I. Nedin, Yu. Prokopenko, V. Tarnavskyi, S. Filippova, I. Chukayeva, A. Shtangret, L. Shostak, and others [5, 6, 8, 13, 18]. The problems of the development of world energy, the use of alternative sources in this field, were analyzed by domestic and foreign scientists, among whom we can single out the scientific works of L. Antonenko, M. Bublyk, V. Verbynskyi, and others [2, 7, 9, 10].

The vast majority of studies on the assessment of the state of energy security are carried out using an integrated approach. The problem of its application is the rather arbitrary selection of groups of parameters, which, according to the researcher, are important for the analysis. The use of a complex approach does not make it possible to develop a universal method for choosing energy security assessment parameters for different countries and specific conditions of functioning of their energy markets [1, 7, 17, 18]. Moreover, the use of such an approach does not allow for strategizing in the field of energy security, since changes during the selection of a set of parameters due to technological transformations and changes in energy market models actually lead to the need to review the methodology for assessing the level of energy security, the selection of modern assessment indicators, and the search for new data for calculations. The dynamics of the processes observed today in the energy markets require the use of systemic approaches.

Currently, there is no single methodological approach to distinguishing the sphere of energy security. Each state uses its own approaches to the definition of the term "energy security", allocation of the sphere of regulation, formation of the management system, and determination of the state (level) of energy security and threats. Usually, the description of the object of research (the field of energy security) is carried out through a set of parameters characterizing the main properties and differences of this object. Sets of such parameters are grouped based on the criterion of similarity on the selected feature. In this way, groups of indicators are most often obtained—economic, political, technological, ecological, social, managerial, etc. The problem of using a complex approach is a fairly arbitrary choice of groups of parameters, which, according to the person conducting the experiment, are quite arbitrary research, are important to analyze in terms of his current knowledge, political preferences, technological development, and security situation. In order to avoid the wrong choice of parameters, the researcher relies on the generalization of the work of other researchers in this field, trying to single out the most important ones.

Having summarized the conceptual approaches to the formalization of the content of the concept of "energy security" in the publications of Ukrainian and foreign scientists, other scientific works, political, strategic documents, and legislative acts of various countries, it is possible to single out certain spheres of regulation, which in a certain context are mentioned by the majority of the subjects of the discussion (scientists, industry experts, politicians, employees, legislators).

Therefore, given the differences in regulation mechanisms, the logic of making and applying management decisions, the following spheres of energy security regulation have been singled out:

- the availability of energy supply is mainly a resource-technical dimension of energy supply of society's needs, which in its simplest form boils down to maintaining the efficiency of systems (equipment) to meet the energy needs of consumers. In foreign English-language literature, this component is mainly associated with the physical resource and technological "availability" of energy (availability/accessibility);
- availability of energy supply refers to the relationship between the resource-technical dimension and the economy, which reflects the profitability of the energy sector (all entities involved in the process of energy supply to society), as well

as the reflection of this activity in the cost of "energy supply" to consumers, society, and state. In English literature, this component is mostly associated with "affordability";

- the acceptability of the energy supply model—the scope of regulation is expanding, combining the accepted models of the organization of the functioning of energy (resource-technical dimension and economic attractiveness) with political aspects of the implementation of state policy (models of energy regulation, influence on adjacent spheres of life). Different countries emphasize certain issues: environmental (pollution, climate change), price (regulation of prices and tariffs, accessibility for consumers), social (protection of energy workers, low-income groups of the population), etc. In English literature, this constituent part is often associated with the "acceptability" (acceptability) of certain decisions depending on the goals of state policy and the principles of organizing the life of society;
- protection of national interests—this area of regulation is often not considered as a separate aspect of energy security policy. Depending on the country, the specifics of its energy sector and foreign policy goals, this aspect is reflected in political decisions at the strategic level (energy security policy) through the identification of priority directions for ensuring national interests in the field of energy.

This sequence of demarcation and expansion of the sphere of energy security is formed on the basis of the complication of management mechanisms and is due to the growth of the number of subjects involved in the relevant spheres, whose interests must be taken into account during the formation and implementation of policy.

The subject of management (researcher), taking into account his practical needs and the goals formed by him, sets the limit of his attention, thanks to which he separates the future object of his management (research).

In countries with an established practice of democratic governance and a mature economic system, the sphere of energy security is generally limited to the sphere of economic regulation of society's vital activities as a management object that combines physical resources, technological equipment, and the economic model of market organization.

In addition, existing global and national political trends bring to the attention of researchers (government officials) current problems related to various aspects of energy use. For example, the issue of environmentally friendly energy use has become especially relevant recently, which was reflected in the effort to add this "current" aspect to the definition of energy security, that is, to the object of research (management).

In general, the focus of energy security on resource, technical, or economic aspects (even taking into account social or environmental aspects) requires a systematic approach to the evaluation of political decisions (taking into account related areas) in relation to the cost of energy supply. An example of this approach is the subordination of political decisions of the governments of countries with a stable market economy to the logic of the economic model and their implementation "only when they are really necessary and in a way that does not distort the internal energy market". That is, all decisions regarding social protection of the consumer, stimulation of energy efficiency, prevention of climate change, etc., are evaluated from the point of view of the possibility of their implementation without distorting the economic logic of energy regulation.

3 Results and Discussion

This approach is logical, but political decisions regarding the functioning of energy markets do not always correspond to economic logic. In Ukrainian in practice, for example, decision-making based on a systemic unifying logic of any nature (the lowest cost for households, complete energy independence, environmental protection, etc.) is not observed. Decision-making on issues of technical regulation, economic efficiency, environmental protection, and social security is inconsistent, logically contradictory, and the achievement of goals (e.g., social protection of consumers) is often carried out at the expense of the progressive deterioration of the technical and financial and economic state of the energy infrastructure, the owner of which is the state.

In particular, the entire energy sector and individual economic entities of Ukraine, as a result of the political attitude to support "state social paternalism", actually worked to their detriment. This mode of functioning was even perceived as completely acceptable within the entire management vertical and was called "planned and unprofitable activity of energy sector entities". Together with the tinization of the economy, the formation of non-transparent relations in the energy markets with the aim of exploiting state assets in private interests, the analysis of the problems of ensuring energy security cannot be adequate to the real situation [16]. At the same time, the theoretical and political concepts of energy security in Western countries do not even contain a conceptual idea regarding the possibility of "planned and loss-making" management. That is why we stand for the need to take into account domestic political aspects when defining our object.

In addition, when formulating the policy of ensuring energy security, it is not considered possible not to take into account the problems related to the use of the energy sector as a "weapon" by individual states to achieve their foreign policy goals. The situation is further complicated by the expansion of the scope of the application of hybrid methods of fighting.

The above requires inclusion in the analysis of issues of ensuring national interests in the energy sector and thanks to the potential of the energy sector. That is, it is proposed to determine the boundary of the object (sphere of energy security) at the level of the sphere of national interests when our object of management will cover such aspects of functioning that are important in the sense of national security (ensuring the realization of national interests).

Summarizing the work of Ukrainian and foreign specialists, it is possible to propose such a definition of energy security, which will set both the boundaries and the main components of the representation of the management object.

Energy security is the ability to satisfy society's needs for energy resources in a technically reliable, economically efficient and environmentally acceptable way, to ensure the sustainable functioning of the national economy in normal and crisis conditions, to protect the sovereignty of the state in the formation and implementation of the policy for the protection of national interests.

The policy of the state of Ukraine in relation to energy security has recently given rise to cautious optimism. Thus, it is recognized that the energy industry of Ukraine is an economic guarantee of state sovereignty, an element of proper governance, a reliable basis for the sustainable development of a competitive economy, and an integral part of the European energy space. In addition, this is confirmed by the "Energy Strategy of Ukraine for the period until 2035" [11].

The following are considered strategic tasks:

- implementation at the state and municipal levels, as well as at enterprises, and continuous improvement of the energy management system, in particular, in accordance with the requirements of standards and international agreements;
- stimulation of energy saving at the level of consumers, formation of energy-efficient consciousness among citizens;
- stimulation of energy efficiency improvement through targeted monetization of end-user subsidies and minimization of subsidy volumes in the future;
- increase in energy efficiency at the stage of electricity and heat generation, reduction of energy losses in the future during its transmission and distribution;
- ensuring the completeness and transparency of the accounting of all forms of energy and fuel and energy resources (electrical and thermal energy carriers, natural gas, etc.);
- formation and regular monitoring of the energy balance of Ukraine and its evaluation according to efficiency criteria;
- reducing the energy intensity of GDP by 2035:
  - compared to GDP (PPP): from the current 0.28 t.e./thousand dollars USA up to 0.13 (per PPP);
  - creation of the necessary energy infrastructure for further sustainable development of the country on a competitive basis.

Energy Independence, Reliability, and Stability of FEK

- ensuring energy independence, including intensive expansion of the resource base and extraction of primary energy resources, domestic capacities for their processing, creation of stocks and reserves, diversification of sources and supply routes, technical and technological rearmament of key enterprises of the industry;
- Ukraine must receive no more than 30% of primary energy resources from one source of supply (for nuclear fuel, the target indicator is determined separately);
- minimization of imports thanks to:
  - intensive increase of hydrocarbon production on the territory of Ukraine;
  - efficient use of resources;
  - intensive development of own resource base;

  - intensive development of RES;
  - optimization of the energy balance;

- restoration of exploration, development, and extraction of hydrocarbon resources on the continental shelf and within the exclusive (marine) economic zone of Ukraine with the creation of favorable conditions for domestic enterprises and foreign investors;
- reduction of dependence on coal supplies of anthracite groups;
- improvement of the critical infrastructure protection system based on the best practices of NATO and EU countries, creation of a crisis management system in the energy complex;
- application of best practices for environmental protection;
- ensuring the reliable functioning of the energy infrastructure, carrying out the necessary modernization measures, reducing the accident rate, and prolonging operation in regular mode;
- ensuring the guaranteed compliance of generating capacities with the volumes and regimes of electricity consumption in the UES of Ukraine, in particular in terms of the availability of regulating capacities;
- integration of the UES of Ukraine into the energy system of the synchronous zone of continental Europe ENTSO-E and regulatory and legal settlement.

Regarding the real state of modern energy security of Ukraine in the electric energy component, the following can be said: Energy security is interpreted as the protection of citizens and the state as a whole from the threats of shortage of energy and energy resources arising due to negative natural, man-made, administrative, socioeconomic internal factors, and foreign political influence factors. When considering this issue, it is necessary to take into account the compliance of electricity production with the needs of the country and the availability of primary energy resources (coal, gas, oil, uranium, as well as sun, wind, water).

How will the economy of Ukraine develop what will its structure be like? If the country will focus on the "heavy" industry (metallurgy, chemistry, defense, and heavy industry), this is one need for electricity. The bet on high-tech products is different. What are the expected rates of economic development and rising living standards of the population?

Electricity is an industry with an extremely long investment cycle. For example, it takes 12–15 years from the idea to build a nuclear power unit to its commissioning. And this is in the presence of consent in society and the presence of political will in the authorities. Thus, the decisions made today will be implemented until about 2035. It is obvious that the strategic planning of the country's electricity industry must be considered for the period up to 2050 with an assessment of prospects until the end of the XXI century (with unconditional clarification of indicators every 5 years). The main types of electricity generation in Ukraine currently include thermal and nuclear power, as well as renewable sources of electricity (RES): hydropower (hydroelectric power stations and hydroelectric storage plants (HAPP); solar power plants (SPP) and wind power plants (WPP). There are other sources of electricity production, but their role is extremely insignificant. Each of these components has problems that affect the country's energy security. The issues of economically and technologically justified combination and joint work of these sources of electricity are important.

In any case, Ukraine will be able to have guarantees of energy security only when the development of its electricity industry is based on the results of a detailed analysis of the compatibility and proportions of various types of generation in the general balance of electricity production needed by the country. The proportions of generation sources in the country's total capacity park must also be justified from the point of view of economics. This question is not just about the tariff and how much the electricity consumer will pay for it. The issue is in the economy, which has a decisive influence on the future of the state.

Many years of thoughtless tariff policy have led to the fact that tariffs for electricity produced by thermal power plants are many times higher than tariffs for electricity produced by atomic power plants. It is impossible to explain this by any technological differences or operating modes. One reason is the greed of the oligarchic-official clan. Owners of coal mines and owners of TPPs are the same persons (oligarchs and the state). It would seem that for such crazy tariffs, we should have the best mines in the world, the safest working conditions for miners, as well as the most modern thermal energy in the world.

For many decades, the electric power industry, power engineering, design and construction, and assembly complexes, as well as scientific support of the electric power industry of Ukraine have been leaders. In the last three decades, many spontaneous, ill-considered, and premature decisions have been made that have led the industry to its current state. The potential of the industry, created in the last century, is being activated, and the renewal of fixed assets practically did not take place. The industry is characterized by staff turnover, a decrease in professionalism, first of all at the management level.

4 Conclusions

In preparation for a radical restructuring, it is necessary to understand in detail the state of the industry and evaluate the technologies that are supposed to be the basis of its development, as well as the risks associated with changes in the structure of generating capacities. It is necessary that the future electric power industry solves the tasks of socioeconomic development of the country, relies on a reasonable balance of different types of generation sources, primarily "green" and nuclear, and solves environmental problems.

Since 1990, three nuclear power units have been put into operation (their construction was started back in the 1970 of the last century), which replaced the three units of the Chornobyl NPP that were decommissioned. Power units are operated in thermal power generation, the youngest of which is almost 40 years old.

We also have many questions about our prospects. It is necessary to build new units of thermal power plants and atomic power plants and to reconstruct electrical networks. Where to get funds? Of course, the Western partners have funds, but they will be loans, and "for the loan—thanks". At the expense of what to repay debts? Increase tariffs for TPP electricity? Where is the next place? If you increase tariffs for APP electricity (recall, their share in electricity production is more than 50%) without reducing TPP tariffs, then the average tariff for electricity in the country becomes unacceptable for business, the population, and the economy of the state as a whole and also the share of WPPs and SPPs with their "exciting" tariff, which will increase to 13% in the future.

But that is not all. We must also build reserve capacities for "support" of wind power plants and thermal power plants. We have to carry out a deep reconstruction of the networks, without reliable operation of which only one name will remain from the electricity market. What economic calculations can we rely on? What are at least indicative financial models for covering the costs necessary to ensure Ukraine's energy security in the near and distant future?

All these questions require immediate answers, including from the scientific community.

References

1. A 2030 framework for climate and energy policies. Green Paper. Brussels, 27 Mar 2013. COM (2013) 169 final. http://ec.europa.eu/energy/green_paper_2030_en.htm. Date of access: 23 Nov 2022
2. Babenko VA (2013) Formation of economic-mathematical model for process dynamics of innovative technologies management at agroindustrial enterprises. Actual Probl Econ 1(1):182–186
3. Babenko V (2019) Formalization of the model of management of the technological innovations. CEUR Worksh Proc 2393:595–602. Online http://ceur-ws.org/Vol-2393/. http://ceur-ws.org/Vol-2393/paper_431.pdf
4. Babenko V, Sidorov V, Koniaieva Y, Kysliuk L (2019) Features in scientific and technical cooperation in the field of non-conventional renewable energy. Glob J Environ Sci Manag 5(SI):105–112. https://doi.org/10.22034/gjesm.2019.05.SI.12
5. Bobrov E (2021) Energy security of the state: monograph. "KROK" University of Economics and Law, Kyiv, 308p
6. Cherp A, Jewell J (2014) The concept of energy security. Beyond the four As. Energy Policy 75:415–421. https://doi.org/10.1016/j.enpol.2014.09.005. Date of access: 25 Nov 2022
7. Energy efficiency as a resource of innovative development: National report on the state and prospects of state policy implementation of energy efficiency/S. F. Ermilov, V. M. Geets, Yu. P. Yashchenko, V. V. Hryhorovskyi, V. E. Lear. NAER, Kyiv (2019), 93p
8. Energy security of Ukraine: strategy and mechanisms of provision/A. I. Shevtsov, M. G. Zemliany, V. O. Barannik [and others]; under the editorship A. I. Shevtsova. Porogy, Dnipropetrovsk (2020), 264p
9. European Energy Security Strategy. Brussels, 05/28/2014. COM (2014) 330 final/ Communication from the Commission to the European Parliament and the Council "European Energy Security Strategy", Brussels, 28 May 2014. https://www.eesc.europa.eu/resources/docs/european-energy-security-strategy.pdf. Date of access: 24 Nov 2020
10. European Energy and Transport (2016) Trends to 2030—update 2005. Eur Commun 146p
11. IEA (International Energy Agency) (2017) World energy Outlook. OECD/IEA, IEA, Paris, p 311

12. Kharazishvili YM (2019) Identification of the level of energy security of Ukraine from the standpoint of sustainable development. Econ Ind 4(88):5–27. https://doi.org/10.15407/econindustry2019.04.005. Accessed 25 Nov 2022
13. Kitskaya LI (2013) Energy efficiency in Ukraine: analysis, problems and ways of promotion. Ukr Sci Ind J Innov Econ 3:32–37
14. Lear VE (2018) National energy security in the context of global goals of sustainable development. East Eur Econ Bus Manag 1(12):77–83
15. Luft G, Korin A (eds) (2019) Energy security challenges for the 21st century: a reference handbook. Praeger Security International
16. Paton BE, Klyuy MI, Korotynskyi OE (2012) Conditions of effective application of solar power systems. Bull Natl Acad Sci 3:48–58
17. Stogniy BS, Kyrylenko OV, Denysiuk SP (2018) Energy security of Ukraine. World and national challenges. Ukrainian Encyclopedic Knowledge, Kyiv, 408p
18. Suhodolya OM (2018) Stability of functioning of the energy system or stability of energy supply to consumers: statement of the problem. Strategic Prior 2:101–117
19. Sukhin EI (2021) Non-traditional energy as a factor of economic security states: autoref. dis… Dr. Econ. Sciences: specialist 04/21/01 "Economic security state". National Security and Defense Council of Ukraine; National Institute of International Security Problems. K., 38p
20. Sukhodolya OM (2019) Problems of defining the sphere of regulation of energy security. Strateg Prior 1:5–17
21. The Energy Independence and Security Act of 2007, USA. https://www.congress.gov/110/plaws/publ140/PLAW-110publ140.pdf. Date accessed: 25 Nov 2022

Digital Transformation Technologies

An Ontology-Based Customer Relationship Management Model for Educational Institutions Using Service-Oriented Architecture

Islam Samih, Mohammed Abdelsalam, and Ibrahim Fathy Moawad

Abstract Increasing satisfaction of customers is the most important goal for any institution. Therefore, customer relationship management (CRM) is founded for this purpose to keep track of the interaction between the organization and their customers to satisfy them and to guarantee their loyalty. In this paper, we introduce a layered architecture model for customer relationship management in Education domain. This model integrates educational domain ontology and service-oriented architecture to improve the business processes and hence to increase the customer satisfaction in the educational institutions. The model design is presented in detail for a huge Egyptian educational institution called the Education City. The paper presents a workflow design for customer relationship management describing the interaction among stakeholders to achieve the different business processes in the Education City. As well as, a detailed ontology design for the Education City is presented to describe and to visualize the different scenarios for the services provided by the Education City. To achieve flexibility and reusability, the model was built based on service-oriented architecture for the possibility of reusing the model in more than one field.

Keywords Customer relationship management (CRM) · Service-oriented architecture (SOA) · Business process management (BPM) · Enterprise service bus (ESB) and ontology

---

I. Samih (✉) · M. Abdelsalam
Business Information Systems Department, Faculty of Commerce and Business Administration, Helwan University, Cairo, Egypt
e-mail: islam.samih21@commerce.helwan.edu.eg

M. Abdelsalam
e-mail: dr.m_abdelsalam@commerce.helwan.edu.eg

I. F. Moawad
Faculty of Computer and Information Sciences, Ain Shams University, Abbassia, Cairo, Egypt

Faculty of Computer Science and Engineering, New Mansoura University, New Mansoura City, Egypt

I. F. Moawad
e-mail: ibrahim_moawad@cis.asu.edu.eg

D. Magdi et al. (eds.), *Green Sustainability: Towards Innovative Digital Transformation*, Lecture Notes in Networks and Systems 753,
https://doi.org/10.1007/978-981-99-4764-5_22

1 Introduction

Customer relationship management (CRM) has become an important tool for increasing the profitability of organizations that enable them to collect data about customers and meet their needs, to achieve customer satisfaction. CRM is a complete business and marketing strategy that combines technology with customer transactions, applied in marketing and service business units, providing flexibility and benefits to organizations [1].

The development of web services using the service architecture model is an acceptable concept in modern technology, which faces most of the complaints of organizations in the absence of an inadequate communication system between the departments of the organization, its employees, and customers which leads to poor communication and poor services provided by the organization.

By focusing on dealing with CRM, organizations can achieve an effective success factor by increasing the satisfaction of users, whether they are customers or employees within the organization. So every organization needs to develop its internal and external communication with its employees and customers to succeed.

The advantages of service-oriented architecture (SOA) are reduced costs, reduced redundancy, improved consistency, and improved efficiency by giving the proposed model implemented by Educational City reusability and flexibility.

In this paper, we present a layered architecture model for customer relationship management in the field of education. This approach combines the ontology for the educational domain with service-oriented architecture to enhance business operations and, as a result, boost customer happiness in educational institutions. Details of the model design for Egypt's enormous Education City educational complex are revealed. The paper provides a workflow design for customer relationship management that illustrates how stakeholders in Education City interact to complete various business operations. A thorough ontology design for the Education City is also supplied in order to describe and illustrate the many situations for the services offered by the Education City. The model was created utilizing a service-oriented architecture for flexibility and reusability so that it could be applied to multiple fields.

The rest of this paper is organized as follows. While Sect. 2 provides an overview of some scientific concepts, Sect. 3 reviews the related work. The proposed model and the business process management are described in detail in Sect. 4 and Sect. 5, respectively. Finally, Sect. 6 presents the CRM-based ontology, and the paper is concluded in Sect. 7.

2 Scientific Background

2.1 *Service-Oriented Architecture (SOA)*

Service-oriented architecture is a software architectural style, where web services are considered as software components that can be available to clients via the application components. The term SOA was used as a reference model to construct and describe the correlations among services in a specific domain [2, 3].

A web service is a part of the software that works with a set of mutually exclusive standards. These standards enable the interoperability of computers regardless of operating systems, hardware platforms, programming language, or network infrastructure. Web services rely on three interconnected XML-based software standards to function: simple object access protocol (SOAP), web service description language (WSDL), and Universal Description, Discovery, and Integration (UDDI), where web services exchange SOAP messages. It can also call services that use multiple interfaces and a description language (WSDL) to implement business processes. The eligibility of a web service to operate anywhere depends on any network without any impact on its performance and the transparency of the network. Web services support the flexibility of websites on the internet. Web services are powerful technologies for practicing service-oriented architecture as the standards and infrastructures that support these technologies are available to make web services based on SOA [4] (Fig. 1).

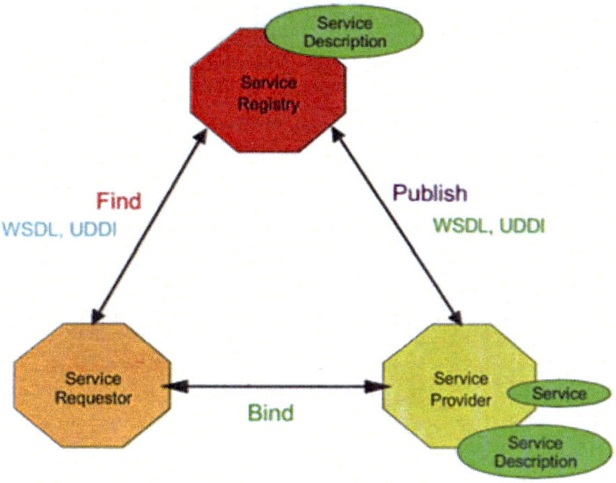

Fig. 1 Basic operations in SOA [4]

2.2 Customer Relationship Management (CRM)

Customer relationship management is a technique used by businesses to better understand their customers' needs and improve customer satisfaction. Customer retention and customer analysis are more crucial than ever before.

Customer knowledge acquisition, particularly potential knowledge acquisition, is significantly more important than customer loyalty acquisition. As a result, CRM goals should be expanded beyond customer loyalty to include gathering and studying customer knowledge as well as transforming customers into value-added collaborators [5].

2.3 Introducing Ontology

Concepts, instances, characteristics, relations, and axioms are used to describe the topic domain. A formal clear specification of a shared conceptualization might be defined as ontology. It is a good way to keep track of things. In remote and heterogeneous information services, ontology also improves semantic search. The purpose of ontology is to allow users and consumers to reuse and analyze domain information. The goal is to create an ontology that represents customer information in a framework that is simple to understand.

There are three stages to the ontology method:

Phase 1: Begin defining the ontology based on client experiences.
Phase 2: Then improving ontology by browsing fresh ideas from customers.
Phase 3: Concludes with ontology consistency checking [6, 7].

Ontology can be built into Customer Support System (CSS) to describe customer information in an appropriate way. This enables CSS clients to get enhanced support for participating in different communities and ensuring that their requests match the services of the organizations where they make the final choices. On the other hand, he highlighted that ontology contributes to the development of searching for information on the web because it creates relationships between approaches and objects, and ontology provides semantic analysis among them [8].

3 Related Work

Ali et al. [9] presented a research paper discussing a problem with Astra Motor Indralaya a company engaged in the sale of products and services. Astra Motor has problems such as the distribution of product information using the brochure causing effort and lack of accuracy in the delivery of information to customers and the problem of booking time which takes a long time for either the motorcycle or parts.

The researchers relied on building a CRM application with the service-oriented architecture of Astra Motor Indralaya. It can market products over the internet so that the information provided will spread more evenly and customers can easily find information about the products being marketed online. In addition, the system built using the black box from the distribution of the questionnaire about the CRM system has been tested and has been designed and built to improve customer loyalty.

The researcher believes that one of the disadvantages of this research paper did not explain the administrative processes within the proposed application and did not use ontology to describe the services provided by the company and the processes within it.

Ali et al. [10] presented a research paper discussing a problem related to customer relationship management where CRM systems aim to monitor and analyze customer behaviors to increase customer satisfaction and loyalty to the company. This made it difficult to interpret large volumes of data from a variety of sources in a short time. Based on this, the research presented a proposed approach to developing a tool that could unify data from different CRM systems and provide explanations about the customer in a common format.

The researchers relied on an ontology-based software tool which has been developed that focuses on the concept of customer in CRM systems. This tool allows companies to react to customer behavior without losing time by analyzing customer attitudes, characteristics, and emotions which are very critical for companies, automatically. Although the tool focuses on the concept of customer in CRM applications and has been developed only to automate processes closely related to this concept, it also provides an infrastructure to automate activities around other CRM concepts with semantic interpretation. The tool was developed using Java and Scala languages. Akka library was used to make the processes in the tool distributed, and Apache Hadoop was used to manage distributed operations on the cloud infrastructure.

The researcher believes that one of the disadvantages of this research paper is that it did not give a model of the data that was interpreted through the used tool, and no tool was used to visualize this ontology.

Shahira et al. [11] presented a discussion paper on the role of cloud computing and big data technology in managing, improving, and developing performance in Egyptian higher education institutions. With how to take advantage of the problem of multiple data stores, the development of multimedia, the spread of the Internet of Things technology, and social media led to the existence of a large volume of structured, semi-structured, and unstructured data.

The researchers relied on the application of cloud-based technology used in the Learning Management System and Student Information System. Where free technology can be provided at low cost with ease of personalized learning, accessibility all contributes to the rapid application of this technology in higher education.

The researcher believes that one disadvantage of this research paper is that the proposed model was not accurately presented, and the layers of the model are not explained with the ontology is not being used to describe the services provided by the proposed model.

Sayed et al. [12] presented a paper that used knowledge management techniques to look at a problem faced by the SOA community in terms of Service Selection, service quality, service configuration, and coordination of different processes between web services. To solve the requirements and difficulties of software engineering, knowledge management has proven to be an effective tool. Knowledge management approaches can help bridge the gap between user needs and web services that can be used to build a software solution.

The researchers in this study used a systematic literature review to identify, estimate, and comprehend all available research on a certain research query, topic domain, or instance of interest. To identify any research gaps in the present study so that proposed portions for further exploration can be enhanced and improved. Propose an analysis for incorporating innovation into research.

The researcher believes that one of the disadvantages of this research paper is that it did not provide an example showing how to apply knowledge management to solve the problem of how to know and choose the appropriate service for the SOA community.

Ahmad et al. [13] presented a research paper discussing a problem about the current situation in modern higher education in Indonesia which is facing a complex situation and which does not interconnect students and lecturers, teachers and universities, and financial institutions.

The researchers concentrated on creating powerful tools with a smoothly connected platform that provides a meaningful service. They claimed that SOA develops a new system for Indonesian higher education that is integrated and modern.

The researcher believes that one of the disadvantages of this research paper did not explain the administrative processes within the proposed system and did not use ontology to describe the services provided by the system and the processes within it.

Francesco et al. [14] introduced a research paper discussing a problem about designing a chatbot model in the educational field to provide support to university students and respond to user questions in a correctly and quickly regarding the courses and curricula offered by universities.

The researchers relied on a system that can recognize inquiries and provide responses to the learner using natural language processing techniques and field ontology. The front-End, Back-Office, Knowledge Base Module, and E-learning BOT Module make up the system architecture.

The researcher believes that one of the disadvantages of this paper is that no tool was used to depict this ontology.

Miguel et al. [15] presented a research paper on the role of how to develop and implement a model to apply to a customer relationship management system in a company that provides consulting services as a case study.

The researchers relied on the reuse and engineering of CRM knowledge resources by proposing a CRM ontology called Curie-O using the OntoUML language.

The methodology used was Design Science Research, for the design and validation of artifacts. The researcher believes that one of the disadvantages of this paper is that no tool was used to depict this ontology.

Hien et al. [16] presented a research paper discussing the problem of building an intelligent conversation model to answer students' inquiries and questions to one a university regarding an advanced course in programming.

The researchers introduced a model based on the ontology technique to know the relationships, called Rela-Scripts to meet the requirements of the students.

The researcher believes that one of the defects of this research paper is that the researchers did not use any of the ontology tools to apply them to the proposed model.

4 The Proposed Model

The proposed model depends on the customer's data, transactions, and desires input, to achieve customer satisfaction and improve service quality.

As shown in Fig. 2, the proposed model consists of five layers which can be summarized as follows.

The Data and Knowledge Layer Is Divided into Two Types

Data warehouse: contains homogeneous data such as customer data, transaction data, and business processes within a model. It also contains heterogeneous data from different systems and applications such as social media applications for the possibility of analyzing this data.

Knowledge base: it contains information and data about the services provided by the institution and knowledge base that developed to answer customer questions through the smart agent.

The business services layer: contains the modules of major functionalities of the model which include customer data registration, event reservation, item reservation, complaint management, and online payment. It is the Communication Logic Layer that acts as the link to the different modules of the platform since it enables the communication between the Data Layer and the Services Layer.

The application layer: for browser-specific web applications. It is a point of interaction between the CRM system and the user. It is used to collect data, show results, deliver outputs, check results, and check the progress of operations.

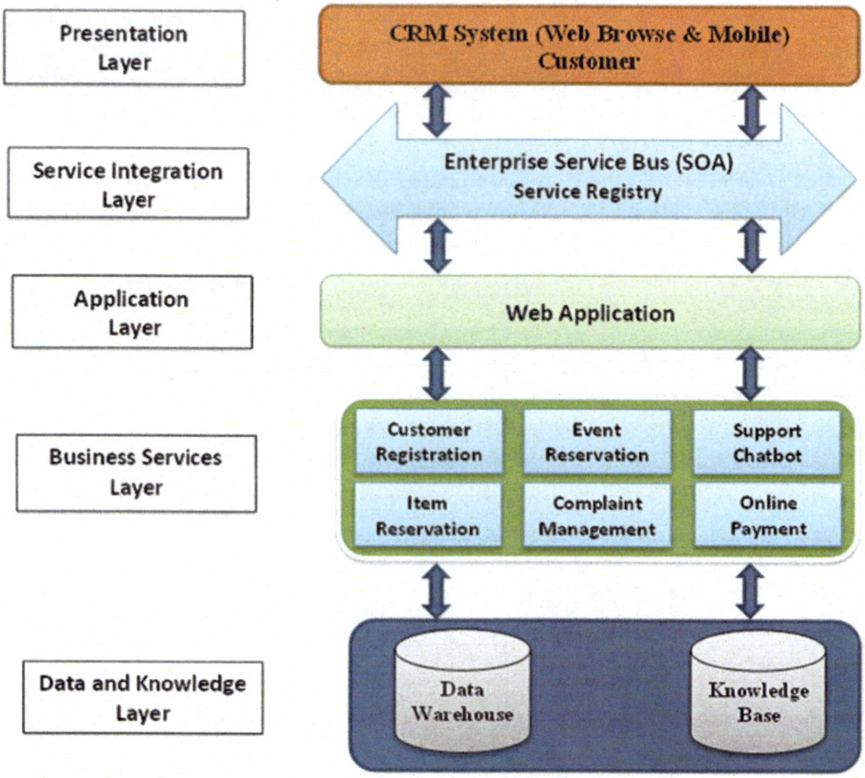

Fig. 2 Proposed model

The service integration Layer: is used to ensure the interaction between a system and other information systems, where the role of an SOA registry is to provide quick and easy access to connect and work between different applications.

The basic concept of enterprise service bus (ESB) architecture is to integrate different applications by placing a communication bus between them and then help each application from talking to the carrier. This separates the systems from each other and allows them to communicate without relying on or knowing other systems [17].

The presentation layer: the presentation layer is the system interface that deals with customers' interactions and transforms data into the form that the application accepts.

5 Business Process Management (BPM)

Organizations contain huge operations and coordination between the work teams to carry out the work properly. This requires the use of business process management for the implementation of operations within organizations successfully [18].

Figure 3 refers to a drawing of an illustration of the business process management model among work teams within the Educational City and takes more than one scene as follows.

- The customer registers the reservation activity and item data.
- The reservation request is sent to customer service Application approval department.

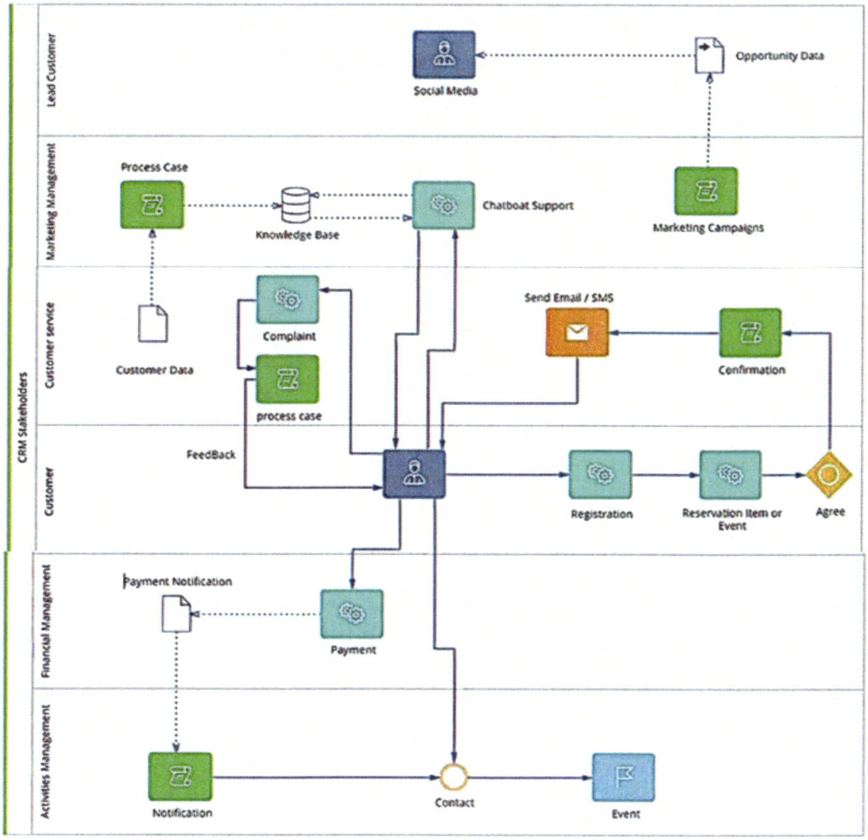

Fig. 3 Workflow model for Educational City

- Customer service is directed to confirm approval of Reservation request.
- The customer pays the value of the reservation, and a payment letter is sent from the Finance Department to the Activity Department.
- The activities department contacts the client to coordinate the establishment of the activity.
- After the activity ends, in the event of a complaint or evaluation of the service provided, the customer sends it to the Customer Service Department for analysis and treatment.
- The Marketing Department analyzes existing customer data from the Customer Service Department to achieve full knowledge of customers and their needs to achieve customer satisfaction.
- The marketing department follows up the questions that are directed to the smart agent to learn about the requirements of existing and new customers and help them with their requests through chatbot support.
- The Marketing Department targets the potential customer and publishes and analyzes marketing campaigns within the social networking site to attract new customers.

6 CRM Based on Ontology

The proposed ontology for educational institutions relies on the belief that all actors involved in CRM could benefit from an ontological investigation of this field, aimed at providing a core set of formally described concepts and relations, useful both for describing CRM processes and for specifying the functionality of CRM applications.

A well-formed CRM ontology would support communication and interoperability both in intra-organization and in CRM processes. The paper discusses in detail the services and customer relationship, as well as the corresponding business knowledge items (services requested and customer records).

This research aims to develop an ontology-based knowledge representation for CRM system that was built using the Protégé editing environment. It has an excellent user interface that eases the process of building or editing ontologies. Figure 4 represents the class hierarchy of ontology.

In Fig. 5, the ontology-based CRM system for Educational City has four main classes:

- Customer: contains types for customers.
- Service: contains item and event classes.
- Teamwork: contains activities team, customer service team, and marketing team, classes.
- Account: contains ID number, contact, and bill.

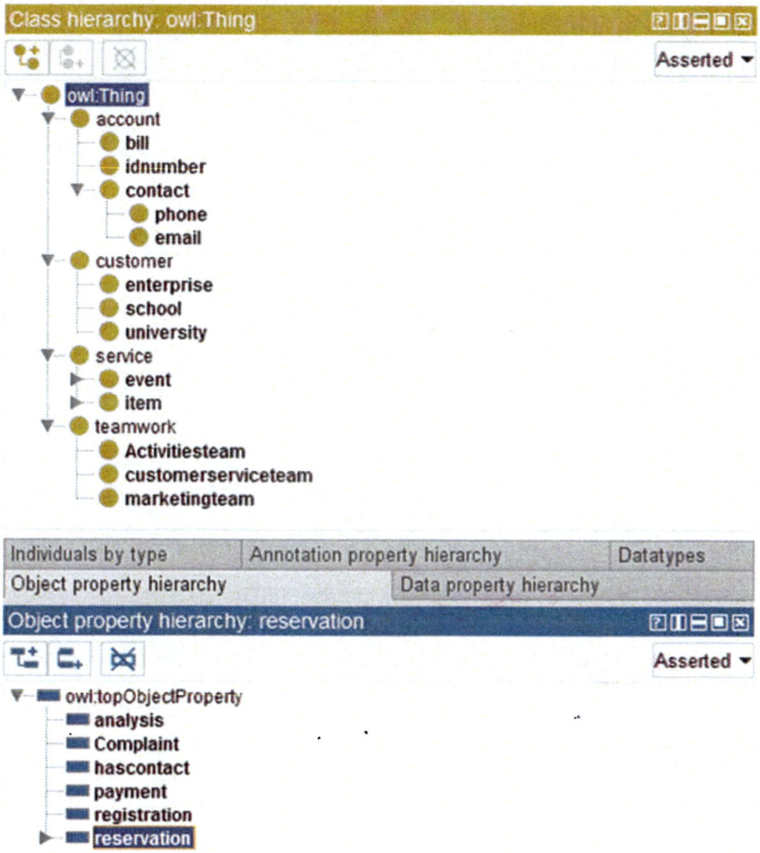

Fig. 4 Class hierarchy of ontology

Each one of these classes may have its subclasses according to the structure of the CRM system for Educational City.

6.1 Ontology Visualization

The purpose of ontology visualization is to display the concepts and structures of ontologies at the level of Ontology Web Language (OWL) constructs rather than displaying the knowledge covered by the ontology. Target users are ontology developers or engineers and other user groups who want to analyze and understand ontology to use it for data annotation and other tasks [19].

Fig. 5 Ontology classes

The ontology of the CRM system for Education City is visualized using the OWLGrEd tool, where Fig. 6 represents all components of the ontology of instances, classes, relations, and properties in general and in full.

Figure 7 illustrates a specific part of ontology to visualize the ontology of CRM system for Educational City using OWLGrEd as follows.

Table 1 shows all the data and their classification according to the previous figure.

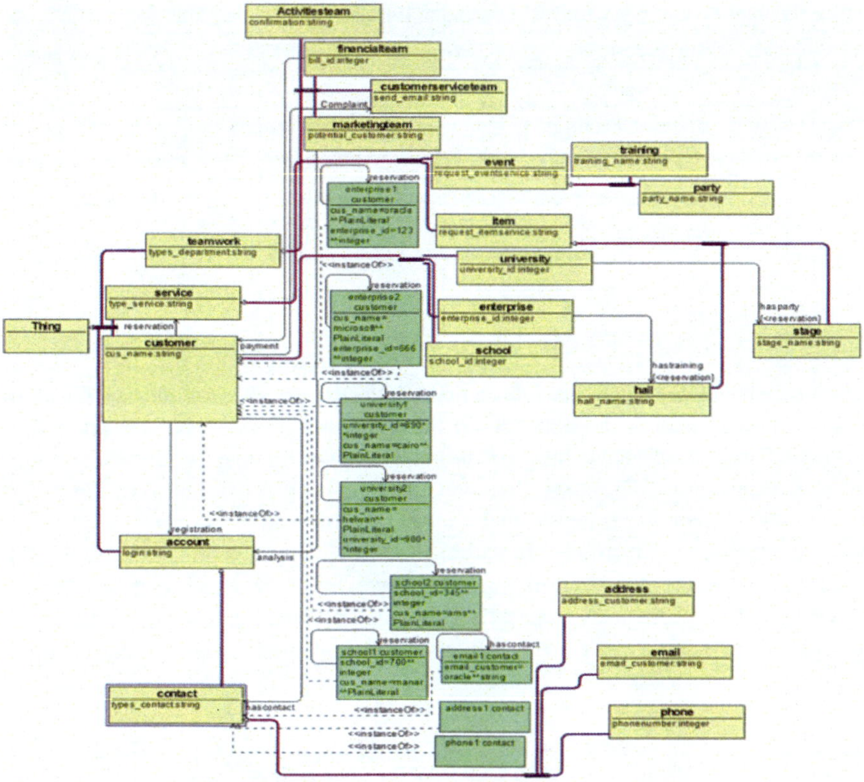

Fig. 6 Complete ontology visualization

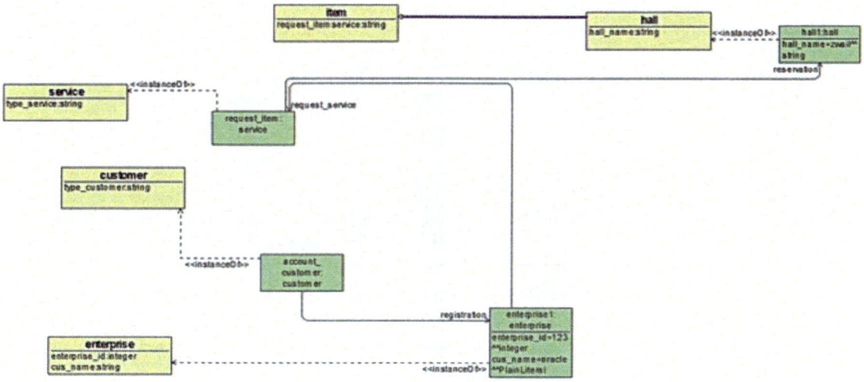

Fig. 7 Ontology visualization for CRM system of Educational City

Table 1 Ontology components for CRM system of Educational City

Class/subclass	Data property	Individuals	Instance of	Object property
Customer	type_customer	Customer	account_customer	Registration
Enterprise	enterprise_id cus_name	123 oracle	Enterprise 1	
Service	request_item	Service	Service	request_service
Item	request_item service	service_item	Hall	Reservation
Hall	hall_name	Hall 1	Zwail	

7 Conclusion

In this paper, we presented an integration model between ontology and CRM to extract the relationships and concepts of the CRM system that has been applied to Education City to describe business processes within the organization and the services provided to customers. The study also showed the possibility of integrating ontology-based knowledge management with business process management to successfully implement and improve processes within the CRM system with the creation of a knowledge base developed and managed by stakeholders to identify organizational knowledge points in business processes within the organization. The proposed model was built using a service-oriented architecture to implement and segment business processes as web services for flexibility and ease of use.

References

1. Talet AN (2012) KM process and CRM to manage customer knowledge relationship management. In: International conference on economics, business and marketing management, vol 29, pp 60–67
2. Yin X (2016) Study on a dynamic E-business application framework based on web service-based SOA. Int J Secur Appl 10(1):55–64
3. Farsi M, Badawy M, Abdelmoneim N, Ali HA, Abdulazeem Y (2019) QoS-aware framework for performance enhancement of SOA in enterprise it environments. J IEEE Access 7:107699–107715
4. Niknejad N, Hussin ARC, Amiri IS (2019) Literature review of service-oriented architecture (SOA) adoption researches and the related significant factors. J Springer Nat 1–24. Print ISBN 978-3-030-12099-3, Online ISBN 978-3-030-12100-6
5. Lee M-C (2014) Customer relationship management and knowledge management enhancing on innovation competence: case study. Int J Manag Inf Technol 10(1):1784–1800. ISSN 2278-5612
6. Alkubaisy SK (2014) Cloud customer relationship management (CRM) ontology. M.S. thesis. Middle East University, Amman, Jordan
7. Alromima W, Moawad IF, Elgohary R, Aref M (2015) Ontology-based model for Arabic lexicons: an application of the place nouns in the Holy Quran. In: 2015 11th international computer engineering conference (ICENCO). IEEE, pp 137–143
8. Šaša A, Krisper M (2011) Ontology-based knowledge management in service-oriented systems. JIOS 35(1):105–118

9. Ibrahim A, Turrahma A, Ruskan EL (2019) The implementation of customer relationship management (CRM) with service oriented architecture (SOA) to improve customers' loyalty". J Phys Conf Ser 1196:1–7. https://doi.org/10.1088/1742-6596/1196/1/012026
10. Tiryaki AM, Atalay M (2020) An ontology based approach for next generation customer relationship management systems. Int J Appl Math Electron Comput 8(04):282–288 https://doi.org/10.18100/ijamec.802695
11. Ali SS, Al Nawaiseh AJ, Helmy Y, Moawad IF, Khalil EE (2020) A proposed model based on cloud computing technology to improve higher education institutions performance. Int J Adv Res Eng Technol (IJARET) 11(9):612–628. ISSN Print: 0976-6480 and ISSN Online: 0976-6499. https://doi.org/10.34218/IJARET.11.9.2020.062
12. Abbas SF, Shahzad RK, Humayun M, Jhanjhi NZ, Alamri M (2019) SOA issues and their solutions through knowledge based techniques—a review. IJCSNS Int J Comput Sci Netw Secur 19(1):8–21
13. Fajar AN, Nurcahyo A, Sriratnasari SR (2018) SOA system architecture for interconected modern higher education in Indonesia. In Selection and peer-review under responsibility of the 3rd international conference on computer science and computational intelligence 2018. Proc Comput Sci 135:354–360. Elsevier Ltd., Available Online at https://www.sciencedirect.com
14. Colace F, De Santo M, Lombardi M, Pascale F, Pietrosanto A (2018) Chatbot for E-learning: a case of study. Int J Mech Eng Robot Res 7(5):528–533
15. Fernández-Cejas M, Pérez-González CJ, Roda-García JL, Colebrook M (2021) CURIE: towards an ontology and enterprise architecture of a CRM conceptual model. Bus Inf Syst Eng 1–29. https://doi.org/10.1007/s12599-022-00744-0
16. Nguyen HD, Tran T-V, Pham X-T, Huynh AT, Do NV (2021) Ontology-based integration of knowledge base for building an intelligent searching chatbot. Sens Mater J 33(9):3101–3123
17. Fayoumi AG (2018) From service-oriented architecture to cloud computing. Glob J Comput Sci Technol B Cloud Distrib 18(1):54–62. 10.17406, Version 1. Online ISSN: 0975-4172, Print ISSN: 0975-4350
18. van der Aalst WMP (2013) Business process management: a comprehensive survey. ISRN Softw Eng 2013:1–37, Article ID 507984
19. Dudáš M, Lohmann S, Svátek V, Pavlov D (2018) Ontology visualization methods and tools: a survey of the state of the art, vol 33, e10, pp 1–39. Published online by Cambridge University Press available at https://doi.org/10.1017/S0269888918000073

Use of the Semantic Wiki Resources for Open Science Support

Rogushina Julia

Abstract One of the main functions of Open Science is open access to scientific publications and other research results. Information resources used by Open Science have to comply with the FAIR principles and be findable. We propose to use semantic Wiki technology to develop scalable resources with complex structure that support information retrieval and navigation on knowledge level. In this work we analyze expressiveness and scalability of the Wiki technology, integration of the Wiki markup with ontological analysis and explore its practical use on example of the portal version of the Great Ukrainian Encyclopedia. We consider specifics of Open Science resources and their typical information objects and propose some solutions for their Wiki-based support.

Keywords Semantic Wiki · Ontology · Open science · FAIR

1 Open Science and Data Representation

Open Science, as defined by the Organization for Economic Cooperation and Development (OECD), is primarily directed at the availability (in digital format, without restrictions or with minimal restrictions) of the results of scientific research [1]. Such approach should provide open access to scientific publications, open experimental data and open peer review of scientific research and open software, open electronic educational resources, etc. Open Science reflects modern fundamental changes in the research process caused by the transition from traditional methods of publishing research results to the dissemination of research results based on modern information technologies [2]. Modern scientific research requires the use of reliable and up-to-date information, and open access in electronic form to the research results such as scientific publications and experimental data makes such work more effective.

R. Julia (✉)
Institute of Software Systems of the National Academy of Sciences, Kyiv, Ukraine
e-mail: ladamandraka2010@gmail.com

D. Magdi et al. (eds.), *Green Sustainability: Towards Innovative Digital Transformation*, Lecture Notes in Networks and Systems 753,
https://doi.org/10.1007/978-981-99-4764-5_23

The paradigm of Open Science is the world scientific community attempt to solve the problem of scientific irreproducibility [3] (that is defined as an inability to repeat experiments of other researchers and reach the same result) by the following basic principles of scientific research as Open access (research results and online cost-free scientific publications should have free access) and Open data (data should be freely available to everyone for use and reanalysis and publication without restrictions from copyright, patents or other control mechanisms.

Currently, Ukraine make a number of arrangements to create a national plan for Open Science that ensure a more open approach to the organization of scientific processes and the dissemination of knowledge, aimed at the development of cooperation between scientists and society. Open science can give Ukrainian scientists and the national scientific system more opportunities to cooperate with colleagues from other countries and integrate their researches.

The EOSC portal was created in 2016 to overcome heterogeneity and unify the scientific infrastructure. Now EOSC is a collection of data repositories and related infrastructure needed to support scientific research [4]. One of the directions of EOSC development is to ensure the search, availability, interoperability and reuse of digital objects that can contain both the results of experiments and the texts of scientific publications. A set of requirements for such data is denoted by the acronym Findable, Accessible, Interoperable, Reusable (FAIR).

Information resources (IRs) that comply with the FAIR principles implement the various functions of data searching, obtaining and representation not by users but by the information system itself, and scientific research organizations and educational institutions, as well as individual researchers, are involved in the management of scientific data. In addition, these principles take into account the specificity of information related to the process and results of scientific research. The FAIR Guidelines do not require any standardization or specific technology to support FAIR principles that act as a general guide for creating data according to their search functionality, accessibility, compatibility and reuse.

The Go FAIR initiative [5] contains the basic principles of improving search capabilities, ensuring access to data, their compatibility and, most importantly, reuse. Core principles of FAIR reflect various aspects of data requirements [6]:

1. Findable: in order to use data and to have access to it, it is necessary to find pertinent information among other information objects (IOs). Therefore, FAIR has a number of requirements for metadata that describe each IO and for tools that support automated analysis of metadata. The expressiveness and structure of metadata are the basic factors in finding of answers that satisfy user needs.
2. Accessible: data and metadata can be retrieved by their identifier using a standardized open communication protocols that assumes authentication and authorization if necessary.
3. Interoperable: used data and metadata need in integration with other information and in interaction with applications or workflows for analysis, storage and processing, use generally accepted representation standards and reference to other data and metadata.

4. Reusable: data and metadata should be detailed with a set of precise and relevant attributes that have a clear and accessible license for use, and conform to community standards relevant to domain.

According to FAIR, the functions of searching, obtaining and representation of data are not implemented by users, but by the information system. At the same time, FAIR considers not only about the data and metadata themselves, but also about algorithms and tools for their management.

The ultimate goal of FAIR is to optimize the reuse of data and their combination in various tasks by detailed description of all important characteristics of data and metadata by unambiguous and relevant attributes in accordance with the domain standards of the scientific community.

All these attributes of data and metadata are interconnected, but at the same time independent and separable. Each of them defines a set of requirements that are transferred to external resources, tools and services in order to ensure their reuse by third parties, including those characteristics that are not directly related to science. At the same time, different groups of users can obtain access to different subsets of data attributes due to gradation of their level of authorities. Combining the metrics of the object description and management of authority level provide a high degree of adaptability of the data and metadata representation in the information systems.

Currently, many European research infrastructures use the FAIR concept to provide access to their scientific data, but most of them focus on general metadata sets that are not designed to support problem-oriented searches with use of domain knowledge relevant to research sphere. The EU Open Science infrastructure integrates a large number of specialized open repositories and data warehouses that provide tools and mechanisms for content management, but use different metadata schemes. EOSC applies the joint development of standards and the implementation of Open Science practices through interaction with research communities, as well as the adaptation for use by scientific communities of such technical components of the FAIR ecosystem, the metadata framework for FAIR digital objects.

One of the Open Science directions is the supply of open access to research results and scientific and technical information through the use of cloud platforms for conducting fundamental and applied research by a wide range of scientists. In addition to a wide range of specialized data analysis services used for research process, one of the basic functions of the Open Science such infrastructure is the support of various forms of information retrieval at the meaningful level. Such semantic search needs in attributes of information objects connected with concepts of research domain and organizational structures (national and international). This component should ensure the reuse and integration of the results of previously performed research in various fields, the results of experiments and calculations, etc.

One of the prerequisites for this is the creation of portal versions of scientific and technical resources with open access that can be realizes as electronic libraries, branch directories and encyclopedias, archives and scientometric databases, etc. Their effective use needs for means of navigation and search into these resources at a meaningful level, using formalized knowledge representations of those subject domains from external knowledge sources to provide the semantic markup of information objects. Semantic markup has to supplement universal standards of metadata presentation such as Dublin Core [7] with domain-specific elements that provide additional search functionality.

Great potential for such retrieval is provided by semantically marked Wiki resources that can use domain ontologies as a source of structural elements of the semantic markup that is relevant to personalized information needs of individual users. Such IRs are not an alternative to electronic libraries with open access to publications and portals of scientific institutions and communities, as well as services that ensure the operation of the infrastructure of Open Science, but they can become a more convenient means of searching for information that is presented in this infrastructure.

2 Problem Formulation

In the broadest sense, the scientific problem related to this study is that the search for information in an open information environment requires the use of formalized knowledge about domain of the user interests. For complex domains of modern scientific research branches, such knowledge has a large volume, a complex structure and largely depends on the specifics of scientific areas and the subjective preferences of individual researchers. Therefore, the availability of open scientific resources and the basic accessibility of the data stored in them is a necessary, but not sufficient, condition for the effective reuse of research results. It causes need in additional, more flexible and personified means of indexing and retrieval of information objects that can be explicitly defined by various interested parties of scientific research.

In order to improve the quality of information support of Open Science, we propose to conduct research and experimental developments in the field of intelligent information systems and resources based on the semantic expansion of the Wiki technologies and the application of ontological analysis methods. The justification of this choice requires an analysis of the functionality compliance of the semantic Wiki resources with FAIR requirements, examination of their practical scaling possibilities and analysis of the Wiki-based semantic search expressiveness.

3 Wiki Ontologies as an Instrument of Knowledge Formalization

Semantic extensions of the Wiki technology (for example, the Semantic MediaWiki plug-in (SMW) [8] for MediaWiki [9]) allow the creation of IRs that meet the FAIR principles. But conformity with these requirements needs in integration of the Wiki technologies with knowledge management tools and ontological models to provide formalized knowledge representation of the relevant domain: built-in Wiki functionality can use this knowledge but cannot fix explicitly all knowledge characteristics. One of the most used approaches for interoperable representation of domain knowledge is ontological analysis [10]. Domain knowledge formalized in various ways provides basis for semantization on all steps of information processing [11]. Formal model of domain ontology O_{domain} most generally is defined as a triplet:

$$O_{\text{domain}} = \langle T, R, F \rangle, \quad (1)$$

where T is a finite set of domain concepts divided into a set of classes T_{cl} and a set of instances of classes T_{ind}; R is a finite set of domain relations between domain concepts from T; F is a finite set of interpretation functions for concepts and relations of ontology O.

Wiki ontology [12] is a special case of domain ontology that formalizes domain knowledge represented by means of the semantic markup of the Wiki resource. The expressiveness of Wiki ontology has some limitations because this ontology contains only the knowledge that can be obtained directly from the Wiki markup. For example, it cannot define characteristics for object properties and data properties, such as equivalence and intersectionability. Elements of Wiki ontology are defined by categories and semantic properties of the Wiki pages. It is important that Wiki ontology can be generated for whole Wiki resource or for its arbitrary subset (for example, for semantic search result).

The formal model of Wiki ontology is based on (1) where T is a set of Wiki pages (including pages of semantic properties and categories), R is a union of set of semantic properties with type "Page" with hierarchy of categories, and F deals with information acquired from the Wiki resource that can be used for logical inference.

In many cases, semantic extensions of Wiki technologies have built-in means for automatic or automated generation of such ontologies. For example, in the SMW technological environment, a Wiki ontology can be created automatically based on any collection of Wiki pages.

The complexity of the Wiki ontology is determined by the set of semantic properties of the "Page" type used in the corresponding Wiki resource and, correspondingly, by the templates used to enter their values. It explains the importance of the set of such properties and the need for its completeness and consistency: on the one hand, it is necessary to create semantic properties for all domain relations that should be displayed by the Wiki resource, and on the other hand, it is advisable to make this set the most compact and understandable for resource developers who have to use these properties to structure Wiki content.

Therefore we propose to consider the types of relations used for domain representation and classify them according to their characteristics that can be used for logical inference (such as symmetry, reflectivity, transitivity), their range space (categories of Wiki pages where they refer) and domain space (categories of Wiki pages where they are used).

In the process of IR developing this reference Wiki ontology can be enriched on base of semantic markup by information about relations between individuals: if one

Wiki page has semantic link to another page but Wiki ontology does not contain such relation between corresponding class individuals then we have to make one of possible decisions: (1) add this relation to Wiki ontology or (2) remove link marked by this semantic property from the Wiki page or change it by other semantic property.

Resources with the more complex structure and big number of information objects need in ontologies that provide the more precise expressiveness possibilities to represent domain knowledge, but this complexity causes more calculative difficulties for their analyzes. Therefore, the problem of a successful choice of the knowledge complexity level is an important factor for the effective development of semantic Wiki resources.

4 Semantic Wiki Resources and FAIR

Wiki ontologies can be used as an instrument for integration of the Wiki resources with other applications, standards and information models. For example, it is very useful for analysis of correspondence of the Wiki technology with FAIR principles.

Semantic Wiki IRs that allow the creation of semantic data and are based on the use of Semantic Web standards, provide a powerful solution for joint publish and editing of data and metadescriptions, creation of various arbitrary sets of properties based on templates of these metadescriptions. They support representation of information also for machine processed form and for human understanding.

Semantic plug-in SMW proposes built-in capabilities for loading files of various formats and adding to them metadata with a different set of attributes that can be modifies, supplemented and analyzed. SMW supports FAIR principles of IR development. Let's consider how FAIR requirements are fulfilled in semantic Wiki resources (see Table 1).

Scalability of the Semantic Wiki Resource

Many important limitations that can trouble IR functioning aren't shown up at small scale but appear when this resource goes into production at large scale. Other problems of large-scale systems can be caused by transition from legacy software and databases to modern technologies and data representations. Growth of data requires effective data storage. For example, if stored data is broken into pieces to store on multiple machines than it also requires some means to coordinate the processing of parts, and this is an important difference of scale-efficient systems from other ones.

Another group of scaling problems deals with teamwork with data and have to take into account the number of content managers and IR users. Scalable IR needs in more complex data management infrastructure and flexible control tools for data access and data movements. They have to allow for various properties of processed information such as the size of the data itself, the complexity of processing algorithms and the number of software modules that realize these algorithms, the number of simultaneously processed objects.

Table 1 FAIR requirements and SMW

FAIR requirements	Wiki resources
Findable	SMW provides intelligent search capabilities based on the semantic properties and categories of individual Wiki pages that can be considered as a flexible set of metadata available for machine processing
F1. Data and metadata have globally unique and permanent identifiers	F1. SMW data and metadata have global unique and permanent identifiers defined as names of the Wiki page. Structure of semantic properties is formalized by the Wiki ontology
F2. Data is described by detailed metadata	F2. MediaWiki engines have the ability to add detailed metadata to data loaded at the standard MediaWiki repository or to hyperlink at external repositories. Semantic properties and categories can be used also for this
F3. Metadata clearly and explicitly identifies the described data	F3. Metadata clearly and explicitly contain the identifier of the described data as a value of the relevant SMW semantic property with type defines into the property page
F4. Data and metadata are registered or indexed by the search resources	F4. Search engines such as Google and Bing can index SMW metadata
Accessible	The IR user needs the means for access to the search results according to his/her authentication and authorization level, and this information in SMW can be added as separate attributes of Wiki pages
A1. Data and metadata can be retrieved by their identifier using a standardized communication protocol	A1. SMW metadata can be retrieved by identifier with use of free communication protocols such as http or secured https
A2. Metadata is available even if the data is not available	A2. SMW metadata is available even if the data is no longer available because metadata is placed separately on the Wiki pages in a reliable repository (in the MediaWiki DB)
Interoperable	SMW data representation can be integrated with other data, other applications or workflows for analysis, storage and processing by publication of additional semantic properties
I1. Data and metadata use common language to represent knowledge	I1. SMW metadata can use the Semantic Web standards and export results of the semantic search into XML and RDF
I2. Data and metadata use dictionaries that comply with FAIR principles	I2. MediaWiki is an open system that can use, integrate, import (by additional plug-ins) various dictionaries, ontologies in the Semantic Web standard languages
I3. Data and metadata includes qualified links to other (meta)data	I3. SMW data and metadata can include links to other metadata and supports to extend the data property sets depending on the task

(continued)

Table 1 (continued)

FAIR requirements	Wiki resources
Reusable	SMW provides data reuse by mechanisms of semantic properties and categories that support flexible changes to the metadata structure
R1. The targets have to be described in detail with a set of precise and appropriate attributes	R1. SMW can add an unlimited number of attributes and categories to each page to fulfill the FAIR requirement about detailed description of metadata by the set of precise and relevant attributes, their sources, etc.

These problems have be solved for all scalable IRs, but researchers define some specific problems in scaling of applications with AI elements and knowledge-oriented IRs [13] such as scalability of IR in general instead of scale for every separate problem, separation of problems at the platform level, unified data access and multifunctional design.

Analysis of the specifics of creating IRs on base of the semantic Wiki technologies shows that the main factors for the successful scaling of such resources deal with the semantic markup structure, namely, the number of relations between Wiki pages, the correct definition their ranges of values and ranges of definition, as well as with clearly formalized values of these relations.

Some scaling problems we can see only in practice when development of IR invokes significant increase of its elements and complexification of structure. Technical possibility to store big amounts of data is necessary for big IR but does not guarantee its scalability. Choice of technological platform for large-scale IRs needs in analysis of its practical use for resources with big amount of data. Therefore we analyze scalability of the MediaWiki and its semantic plug-in on example of development of the portal version of the Great Ukrainian Encyclopedia (e-VUE) that contains heterogeneous information from many fields of knowledge [14] represented by the large number of the Wiki pages with semantic markup. It provides users with intelligent tools for search and navigation (see Fig. 1).

e-VUE uses MediaWiki technological platform [15] and its extension (SMW). Big number of domain experts from various domains and team of professional scientific editors and knowledge engineers work on creation of e-VUE content. Knowledge representation of e-VUE is based on the Semantic Web standards [16] such as OWL [17], RDF and RDF Schema [18] supplemented by original task-specific software solutions. This IR can be considered as a Semantic Web application that satisfies the main requirements formulated by the Semantic Web Challenges [19].

Since 2020, e-VUE has been registered on the web site of the Semantic MediaWiki product community (https://wikiapiary.com/wiki/Great_Ukrainian_Encyclopedia), which shows the rate of growth in the number of pages, user activity, and the number of edits in this IR (Fig. 2). Google Analytics service also improve the increase of site content and number of active users. This growth causes the need in scaling solutions for IR development.

Fig. 1 e-VUE portal

Fig. 2 e-VUE page on the Wikiapi.com website

This IR is focused on modern scientific understanding of the world picture, the history of human civilization, the contribution of the Ukrainian people, etc.

Table 2 shows how semantic Wiki technology supports means for all main aspects of scaling. Therefore this technology can be used for development of semantically enriched IRs for Open Sciences.

Table 2 e-VUE solutions for scaling aspect

Scaling aspect	e-VUE solution based on the Wiki technology
Data size	Control the size of multimedia elements of IR Provide means of mass import of information from external sources
IO number	Control the total number of Wiki pages and delete unnecessary, wrongly created and duplicate pages Develop TIO templates to avoid an increase in the number of similar names of semantic properties and errors in these names and to simplify the perception of information by users Unify metadata of multimedia IOs to avoid duplication Create templates that integrate content of various Wiki pages by semantic queries
Knowledge base structure	Develop templates of typical IOs to unify content Formalize the KB structure and clearly define the semantics of relations between pages (the built-in capabilities of SMW are not enough for this, and therefore it is advisable to use external knowledge organization systems based on ontologies) Determine the semantics of hyperlinks between IR pages and create the corresponding semantic properties, clearly describing their characteristics Develop appropriate templates for built-in semantic queries used on several pages
Processing means	Determine the expediency of connecting plug-ins that extend the functionality of system and do not install those that are not really needed Develop an adequate content indexing policy that takes into account the frequency of information updates and the number of user visits Create queries without redundant conditions by analyzing the taxonomy of the IR categories Minimize external software tools integrated into the IR (such as page visit counters)
Place of data processing	Analyze the number of semantic queries on Wiki pages and their complexity Reduce the number of semantic queries and IOs on the pages that users visit most often (for example, it is impractical to embed complex queries on the main page of the portal, instead of links to pages with these queries) Timely create backup copies of the IR content and structure, ensure the recovery possibilities Generate content at a fixed time interval and add it to the content for pages with a large number of requests and visitors

5 Typical Information Objects of Open Science

Search services should not only provide access to various types of information (natural language documents and multimedia data) and support the acquisition of knowledge from the information found, but also provide the ability to find complex information objects that meet user requirements. Therefore, an important part of research is the creation of a metadata system for describing not only the information resources of Open Science, but also all its objects and subjects.

In addition to commonly used ontological descriptions, it is advisable to generate personalized ontologies that reflect the views of the report of various scientific communities and individual researchers for the creation of metadata. To integrate these personalized representations, it is advisable to develop methods and software tools to compare and align such ontologies with the common ontology of the scientific community.

Examples of the use of this approach are the integration of national and international qualifications frameworks, the formation of multinational territorially distributed research teams, the search for distance education courses, and the integration of the educational services market with the labor market.

Semantic search requires formalized models of user information needs. Such models can be generated on base of domain ontology. If source of information is represented on base on the Wiki technology then we can use its special case—Wiki ontology that describe structure of typical information objects (TIOs).

SMW supports description of TIOs by semantic templates that contain the set of TIO semantic properties and provide correct input of their values.

Development of TIO system for IR is based on the joint work of a knowledge engineers and domain experts. User can construct reference complex IO (CIO) that use elements of domain ontology (1) [20]. CIO as a set of more than one TIO individuals related to each other by ontological relations and meet the requirements regarding the structure and values of CIO properties. Search services should not only provide access to various types of information (natural language documents and multimedia data) and support the acquisition of knowledge from the information found, but also provide the ability to find CIOs that meet user requirements.

Formal model C of CIO based on such ontology has the following structure:

$$C = \langle T_C, N_C, R_C, \{(t_j, t_k, r_m),\ t_j \in T_{\text{ind}},\ t_k \in T_{\text{ind}},\ r_m \in R\}\rangle. \quad (2)$$

where $T_C = \{t_i, i = \overline{1, p}\} \subseteq T$; $N_C = \{n_i, i = \overline{1, p}\}$; $R_C \subseteq R$.

Then semantic search into the Wiki IR is realized by matching of user request with models of Wiki pages represented by Wiki ontology. This is a special case of matching of CIOs with different structure that contain similar IOs [21].

This metadata model is developed by ontology engineering specialists in cooperation with domain experts, and it provides a base both for marking and for information retrieval into IR without special knowledge and without the need to install additional software. It is important that the same IR can be marked from the various points of view based on different ontologies. Such possibility supports personified retrieval of information for different groups and communities of users.

6 Typical Information Objects of Open Science

An important part of research is the creation of a metadata system for describing not only the information resources of Open Science, but also all its objects and subjects. Such metadata model, using knowledge from external domain ontologies, organizational ontologies, ontologies of professions and competences (such as ESCO), as well as various national and international standards, should become the basis for the semantic markup and retrieval of all types of information generated, stored and transmitted by the Open Science infrastructure. These metadata can be represented on base of SMW as semantic TIO templates.

Structure of such templates (their semantic properties, categories, relations with individuals of other TIOs) can be acquired from appropriate ontologies.

We can separate some groups of the most common Open Science TIOs such as scientific publications that has formal properties—traditional (name, authors, date, etc.) and external (multimedia IOs, subtitles, keywords) and meaningful ones (type of research, type of result representation, etc.), and personalities that contain description of research domain and sphere of expertise and research organization. Other TIOs can be more domain-specific and support semantic search into some research branch with the help of domain ontologies.

Proposed TIO-based approach is aimed to generate integrated set of the metadata elements that supports information retrieval for multi-domain areas with use of personified believes with use of standardized Wiki interface.

In addition to commonly used ontological descriptions, it is advisable to generate personalized ontologies that reflect the views of the report of various scientific communities and individual researchers for the creation of metadata. To integrate these personalized representations, it is advisable to develop methods and software tools to compare and align such ontologies with the common ontology of the scientific community. This will allow, on the one hand, to clearly present the differences in the views of various scientific groups and directions, and on the other hand, it will ensure an effective search for the required information, regardless of which terminological system the developers of individual resources choose for metadata.

7 Summary

We analyze semantic extension of the Wiki technology, provide a number of practical investigations and arrive at a conclusion that it can be used as a base for development of the Open Science information resources that provide advanced facilities for search and navigation on the knowledge level: (1) semantic Wiki IRs require all main principles of FAIR; (2) semantic Wiki IRs are scalable and support analysis of large number of information objects by multiple users; (3) semantic markup of these IRs is flexible and can be integrated with knowledge structures from arbitrary external domain ontology; (4) search and navigation into semantic Wiki IRs can use elements of domain knowledge and support unificated user interface; (5) expressiveness of semantic Wiki technology is sufficient for complex IRs of Open Science. Therefore we can recommend this technology enriched by ontological analysis for support of Open Science retrieval.

References

1. OECD science, technology and industry policy papers no 25 making open science a reality. https://cutt.ly/JcuwQtg. Last accessed 03 Dec 2022
2. Open innovation, open science, open to the world—a vision for Europe. https://ec.europa.eu/digital-single-market/en/news/open-innovation-open-science-open-world-vision-europe. Last accessed 04 Dec 2022
3. Challenges in irreproducible research. Nature. https://www.nature.com/collections/prbfkwmwvz/. Last accessed 20 Nov 2022
4. EOSC multi-annual roadmap 2023–2024. https://eosc.eu/sites/default/files/2022-05/20220523_MAR_02_GL.pdf. Last accessed 05 Dec 2022
5. How to Go FAIR. www.go-fair.org/how-to-go-fair/. Last accessed 06 Dec 2022
6. FAIR principles. www.go-fair.org/fair-principles/. Last accessed 06 Dec 2022
7. Dublin Core Metadata Initiative (2014) Dublin core metadata element set, version 1.1. http://dublincore.org/documents/dces/. Last accessed 05 Nov 2022
8. Krötzsch M, Vrandečić D (2011) Semantic MediaWiki. In: Foundations for the web of information and services, pp 311–326. https://files.ifi.uzh.ch/ddis/iswc_archive/iswc/pps/web/2006/lsdis.cs.uga.edu/iswc06/items/swchallenge_krotzsch.pdf. Last accessed 15 Aug 2022
9. MediaWiki. Available from: https://www.mediawiki.org/wiki/MediaWiki. Last accessed 03 Dec 2022
10. Guarino N, Oberle D, Staab S (2009) What is an ontology? In: Handbook on ontologies. Springer, Berlin, pp 1–17
11. Warren P (2006) Knowledge management and the semantic web: from scenario to technology. IEEE Intell Syst 21(1):53–59
12. Rogushina JV, Grishanova IJ (2021) Ontological methods and tools for semantic extension of the MediaWiki technology. In: Proceedings of the 12th international scientific and practical conference of programming (UkrPROG 2020), CEUR workshop proceedings, vol 2866, pp 61–73. http://ceur-ws.org/Vol-2866/ceur_61-73Rogushina6.pdf
13. Dunning T, Friedman E (2021) AI and analytics at scale. Lessons from real-world production systems. O'Reilly Media. https://www.oreilly.com/library/view/ai-and-analytics/9781492094388/

14. Andon PI, Rogushina JV, Grishanova IY, Reznichenko VA, Kyrydon AM, Aristova AV, Tyschenko AO (2021) Experience of semantic technologies use for development of intelligent web encyclopedia. In: Proceedings of the UkrPROG 2020, CEUR workshop proceedings, 2021, vol 2866, pp 246–259. http://ceur-ws.org/Vol-2866/ceur_246-259andon24.pdf
15. Manual: What is MediaWiki? https://www.mediawiki.org/wiki/Manual:What_is_MediaWiki. Last accessed 13 Nov 2022
16. Davies J, Fensel D, van Harmelen F (2002) Towards the semantic web: ontology-driven knowledge management. Wiley, England, p 288
17. OWL Web Ontology Language. Overview, W3C Recommendation: W3C (2009) http://www.w3.org/TR/owl-features/. Last accessed 22 Nov 2022
18. Brickley D, Guha RV (2022) Resource description framework (RDF) schema specification. W3C Proposed Recommendation. https://www.w3.org/TR/PR-rdf-schema. Last accessed 03 Dec 2022
19. Benjamins VR, Contreras J, Corcho O, Gomez-Perez A (2014) Six challenges for the semantic web. https://oa.upm.es/5668/1/Workshop06.KRR2002.pdf. Last accessed 03 Dec 2022
20. Rogushina J, Gladun A, Schreurs J, Salem A-B (2016) Use of ontological analysis as an instrument of knowledge recognition in complex virtual research environments. Int J e-Learn Secur (IJeLS) 6(1):494–504
21. Rogushina YV, Pryima SM (2017) Ontological approach to qualification matching based on competences: model and methods. Nauk Visnyk NGU 6:162–168. http://elar.tsatu.edu.ua/bitstream/123456789/3194/1/4.pdf

Framework Development for Testing Automation of Web Services Based on Python

Oleg Pursky, Vitalina Babenko, Olexandr Nazarenko, Oleksandra Mandych, Tetiana Filimonova, and Volodymyr Gamaliy

Abstract This work presents the technology of framework development for testing automation of web services based on Python programming language. The main approaches to software development and testing were theoretically analyzed. As a result of the study, a framework for testing automation of web services based on the Python programming language using the Selenium browser driver was created. The framework was built taking into account the features of web development on the principles of ease of understanding, expansion, simplicity, and compatibility. The framework consists of two parts: (1) for the first setting before activation by creating a configuration file; (2) for testing the login pages of web applications. The developed framework is suitable for both ordinary sites and information sites, where the main purpose is to get the necessary information as quickly and efficiently as possible. The framework differs from others in its ease of use while maintaining flexibility for more experienced developers.

Keywords Framework · Testing automation · Web technologies · Software development · Python · Selenium

O. Pursky (✉) · T. Filimonova · V. Gamaliy
State University of Trade and Economics, Kyiv 02156, Ukraine
e-mail: o.pursky@knute.edu.ua

V. Babenko
V. N. Karazin Kharkiv National University, Kharkiv 61022, Ukraine

National University of Life and Environment Science of Ukraine, Kyiv 03041, Ukraine

V. Babenko
e-mail: vitalinababenko@karazin.ua

O. Nazarenko
State University of Intelligent Technologies and Communications, Odessa 65023, Ukraine

O. Mandych
State Biotechnological University, Kharkiv 61002, Ukraine

D. Magdi et al. (eds.), *Green Sustainability: Towards Innovative Digital Transformation*, Lecture Notes in Networks and Systems 753,
https://doi.org/10.1007/978-981-99-4764-5_24

1 Introduction

Software is used in almost all areas of our lives, and huge funds are spent on the development of various programs that are in demand in industry and business, in the entertainment industry, in education, and medicine [1]. The task of reducing the cost of software development and improving the quality of manufactured products is one of the most urgent in the information technology industry [2, 3]. When creating modern applications, programs, or web services, developers face the problem of the quality assurance of their product. In order to check the final product for quality that satisfies customers and users, the developers test the product. As technology advances, this task becomes more complex and routine, so automated testing becomes relevant. Testing automation allows you to significantly reduce the costs of development companies, save time and resources spent on testing, and reduce the risks of releasing a low-quality product to the market [4]. Therefore, test automation technologies are gaining more and more popularity among companies involved in the development of software products. Professor Edger Dijkstra in 1972 concluded that testing programs is proof of errors, but never proves their absence. He also considered erroneous approach to programming as a process of achieving results by trial and error («write code—test—find errors—correct—test—...»), because it stimulates programmers not to think about the task [5].

Quality is not an absolute value; it is a subjective concept. Therefore, testing, as a process of timely detection of errors and defects, cannot fully ensure the correctness of the work of the web service. It only compares the condition and behavior of the product with the specification. Usually, the concept of quality is limited to such concepts as correctness, reliability, and practicality, safety, but may contain more technical requirements, which are described in the ISO 9126 standard. The composition and content of the accompanying documentation of the testing process are determined by the IEEE 829-1998 Standard for Software Test Documentation [6]. The most common form of automation is testing applications through a graphical user interface. The popularity of this type of testing is explained by a number of factors, namely the application is tested in the same way that the future users will use it; no access to the source code is required for testing [6].

The development of modern information web-based systems involves the use of multi-stage software development technology [7], one of the stages of which is the testing of web services. The purpose of the study is the development and practical implementation of a framework for automating the testing of web services based on Python programming language that correlated with ninth goal "Industry, Innovation and Infrastructure".

2 Method Review

There are two main methods of testing automation: at the code level—in particular, the method of software testing, which consists in separately tests each module of the program code; GUI testing—simulating user actions use special test frameworks [8]. We will list the most modern and well-known approaches to test automation.

White box testing (or white box testing) is an approach to software testing that involves the tester's knowledge of the internal structure of the program [9]. This allows him to create tests that specifically test the program in what he thinks are weak points. This is exactly such an approach called "at the level of software code".

Black box testing (or black box testing) is an approach to software testing that assumes that the tester approaches software testing as an ordinary user who has no idea about its internal structure [9]. This method is much faster, because the tester does not need to spend time understanding the program from the inside or participate in its development.

Gray box testing is an approach to software testing that combines white box testing and black box testing approaches [9]. The tester only partially knows the internal structure of the program.

Unit testing (or integration testing) is an approach to software testing, the meaning of which is to divide the program into small indivisible components, each of which must be placed in an environment that will simulate the flow of data that the program could give to this component [10]. Unit testing also has been used to test web services with the framework. Unit testing is an approach to software testing, the meaning of which is to split the program into small indivisible components, each of which must be placed in an environment that will simulate the flow of data that the program could give to this component. Then the behavior of this component is studied and compared with the requirements for its operation. It can also be implemented by checking each simple object work results with the expected behavior.

The development of the test automation framework is very relevant, as it can help improve the quality and competitiveness of web services [11]. In the era of Internet development, web services can simplify life, but testing is a complex task that requires a lot of time and money. The testing framework allows to avoid repetitive work that the machine can perfectly perform instead and saves money and time by creating flexible test templates instead of developers reinventing them every time. That is why the framework mostly focuses on unit testing, which is the base of the testing pyramid. It is also possible to set some integration tests, but the framework is focused on unit testing of the webpages.

As a result of the analysis study, it has been concluded that Selenium WebDriver is the most popular and universal test—development tool [12]. Selenium WebDriver is not a testing framework, but a browser driver that allows automating tasks and making the browser execute any command by imitating the user. Selenium WebDriver is a very flexible cross-platform, it is possible to write tests with it, although it will take a lot because Selenium does not have ready-made templates for writing tests.

Cypress [13] stands out from the competition because it is developed independently of Selenium WebDriver, which is the basis of many testing frameworks. Cypress works as a window on top of the browser itself, which is very convenient and easy to use and install, although it may be less useful for experienced developers. JavaScript is used to develop tests with Cypress. Unfortunately, this framework does not support other programming languages. It supports generating tests for popular browsers such as Firefox, Google Chrome, Edge, but not for less popular ones, or Apple's Safari browser, which is a significant drawback.

Robot Framework is also gaining popularity lately [14]. It is a full-fledged web test development framework that also supports the BDD [15] test approach using the Gherkin language. The framework is compatible with other programming languages and browsers. It is a fairly new framework that is constantly being updated, which can lead to some issues. Robot Framework's code is hard to maintain, bugs are not clear enough, which slows down development, and it cannot run tests on different browsers at the same time.

It is also worth noting that a framework is a very balanced way to speed up the testing. Tests can be developed manually, and even though this approach provides the best solution to the testing problem of an individual application, but it takes a lot of time and resources as well as dedicated specialists to pull it off. On the other hand, there are content management systems that allow the used to manually create a website they want using the drag and drop system to create a design. Such systems already have testing in place, and it is delegated to the CMS itself. But such a way of creating websites is very limited, and while it may be good for a lot of traditional use cases, if someone wants to create a unique website the capabilities of such system will not be enough to create custom website logic and databases integrations or other tools integrations. Thus, a framework that already contains all the basic and widespread solutions to common problems and at the same time allows for custom implementations is a very balanced way to test any web application. It may not be useful for small sites, like personal pages or online CV or may be not suited for big professional applications like Google Maps or Zendesk, but anything in-between will benefit from a testing automations framework.

Today, there are a large number of types of testing. Each of them has its advantages and application possibilities, but there is a generally accepted pyramid of testing that helps in this matter. Figure 1 shows the test automation pyramid model [16].

To begin with, you should understand what test automation is and what tests are considered automated. Such tests are those that can be passed without the participation of a person. There is a large number of creations and launch of such tests, and they can be launched by command, or completely automatically, for example, by schedule or when changes are saved to a remote repository. In Fig. 1, the *X*-axis represents the required number of tests, and the *Y*-axis represents the monetary costs of writing and maintaining the tests. Also, the *Y*-axis is inversely proportional to the speed of testing.

Unit testing is at the base of the pyramid. There should be the largest number of them. These are automatic tests that check the functionality of basic elements, such as clicking a button on the site or the operation of a separate function. For

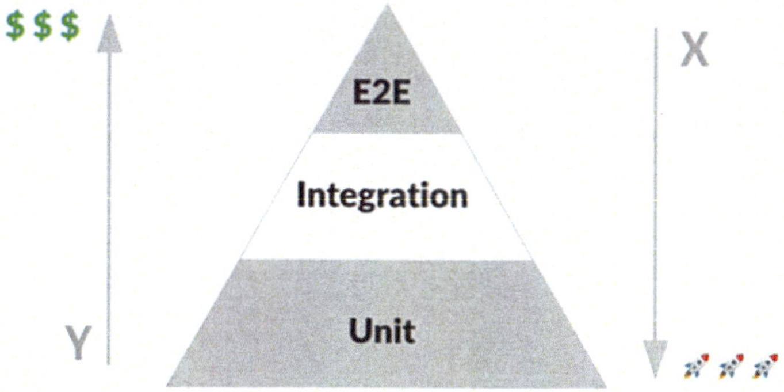

Fig. 1 Testing pyramid [16]

verification, the work of such an element is started with a previously solved data set at the input and correct answers at the output. The operation of the element is checked separately from the rest of the system. The general idea of such testing is to check that modifications to the already written code would not change its behavior and would not lead to the breakdown of already working parts of the site or program. Such tests are very simple and easy to automate if they have already been written. The time spent on one such test is the smallest among all other types of testing.

This type of testing is very useful when many people are working on the project at the same time, and it becomes difficult to follow all the changes and tasks of individual elements. Due to the fact that the product has a large number of unit test elements, there should also be many—each for its own element.

Above unit tests in the pyramid are integration tests (integration testing). In such tests, not individual elements are checked, but software modules, as well as their compatibility and work together, that is, integration. It is necessary to make sure that even in the case of the correct operation of a separate module, it will adequately interact with others and will not issue errors when moving data from one module to another. It is worth noting that in modern realities, such modules are often created by different people or teams, which makes it a very important task to verify the product with integration tests.

When conducting such tests, they can use "plugs"—these are special utilities or modules that can simulate the operation or behavior of other parts of the application.

There are several approaches to integration testing such as: bottom-up integration, where all low-level modules are tested first, only then high-level modules will be tested on them; top-down integration, where it is the other way around: high-level modules are tested first and then low-level modules are connected, if necessary, their functionality can be temporarily replaced with a plug; big bang integration, where all modules are tested simultaneously. This approach can save time, but it can be quite risky, because with a large number of errors, or with incorrectly written tests, because it will be difficult to understand where the errors are and how to eliminate them.

At the highest level of the pyramid are E2E tests. This is short for end-to-end testing, i.e., end-to-end tests and GUI tests. With their help, the finished product as a whole, as well as its user interface, is checked. Such tests take a huge amount of time, are quite complex and expensive, because they test many components from different areas for interaction at once. Often the actions of a normal user are simulated, or manual testing is performed. There are quite a few tools for automating this level of testing. They are used last, after integration and unit tests, and can guarantee good quality of the site or application.

3 Results and Discussion

To solve the problem of creating a framework for testing automation of web services, the method of object-oriented programming has been used. Black box testing is the main way of testing used by the framework. This is the approach that called «GUI testing», because in this approach the tester uses the user interface in a rather creative style in order to detect errors or unforeseen situations. Thus, when using a black box testing, the tester has access to the software only through the same interfaces as the customer or user (front-wnd, browser, desktop form), or through external interfaces. Testing web services are specialized services that allow another computer or process to connect to a system for testing. This is relatively easy to implement via Selenium WebDriver [17]. The framework idea is to automate browser to simulate the end user and also to add some unit tests. The framework is realized in the Python programming language [18]. There are many different programming languages, both specific and universal for many applications. The most well-known universal programming languages are Java, C++, C#, Go, Ruby, JavaScript, and many others. However, the most popular among them according to the TIOBE index, which is an indicator of popularity of a programming language, is Python [18]. Although the TIOBE index does not show the true share of language use, it allows us to see the interest of the programming community in it and the general popularity of the language. Python accounted for 17.08% of all search engine queries in the October 2022 index [19].

The language is interpreted, which means that the code written in Python is not translated immediately into machine code, and this is done by a special language interpreter. This allows the language to be cross-platform, because in order to adapt the program so that it works on a different architecture, it is enough to simply change the interpreter without the need for source code modifications [20]. This is very useful for, for example, web development, because sites must run on any popular platform that supports browsers, and therefore, you need to test the web application on all platforms. Python is also a free language—so its use does not impose restrictions on the programmer, he has the right to use the language for any purpose.

To create a framework for web testing automation, it is very important to know exactly how web applications or web sites are created. We can assume that the site and web application are identical concepts, because the internal structure of almost every web application or website is created using web technologies that implement the client–server system. This works so that the user's request is processed by the web application and is sent over the network to the web server where the site or application is actually located. In response, the web server creates a corresponding web page from the form and sends it back to the web application [21]. The web application in its turn uses the browser as a user interface to display the relevant page or information.

The framework was created using the Visual Studio Community Edition 2019 development environment. Although this environment does not support Python, it can be easily fixed by installing the official Python language support package. This can be done through the official Visual Studio Installer utility. Python version 3.10 was also installed from the official website www.Python.org. The developed framework works by simulating user actions using Selenium WebDriver, which is software that can control any browser actions and is widely used for testing. The tests themselves are implemented using the Python Unittest library.

To expand the functionality, expansion packages were installed using the automatic package installer Package Installer for Python (Pip). One of them is Unittest. Unittest is a package for writing automated unit tests using Python. It is needed to implement the functionality of the framework. This package is one of the most popular testing packages for this language because of its convenience and simplicity. The package includes a TestCase class that can be inherited to create different test groups, a test suite for grouping and running a large number of tests in a queue, and an environment for running tests. However, there are some limitations and standards. For example, the test methods must be formatted in a special way, like adding "test" at the start of the method and using special constructs. This is necessary for the interpreter to successfully find the tests, because anything with a name that does not start with «test» will be ignored. But such a naming practice has quickly become the standard among developers of automated tests, so it is not a big problem in terms of compatibility. Even when using frameworks and libraries that do not require such a rule, tests are still called this way for better understanding. Another important package is ConfigParser. This Python module provides access to the ConfigParser class, which implements the base configuration language for creating configuration files. The language is similar to the .ini format for Microsoft Windows. This is necessary so that the user of the framework can quickly, easily, and clearly provide the necessary data for testing, such as site address, login and password of the test user, and so on. Using this approach is much simpler and more convenient than requesting data before tests, as it would make automation impossible. The configuration file is created by a separate init.py file that must be run before using the framework. The configuration file also includes default data to show the operation of the framework on a test website. The module OS provides extensive functionality for working with the operating system, hence such a name. Although the range of possibilities is incredibly wide, in this case it is needed in order to access the location of the framework files to work properly with the configuration file. The main extension package is Selenium WebDriver. It is responsible for browser management and is required to perform tests. The module is very large, so individual submodules need

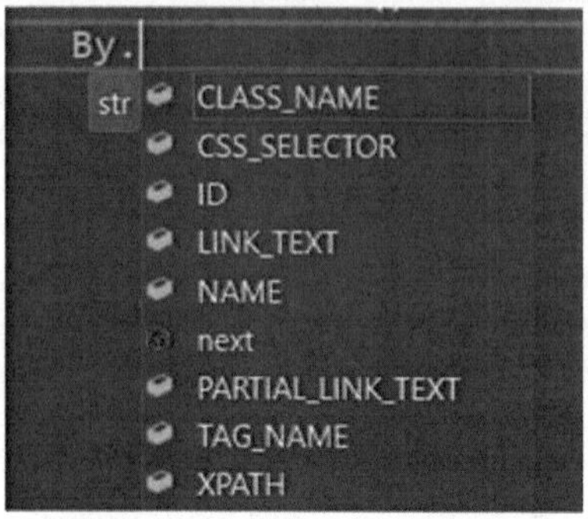

Fig. 2 Example of different locator types in the Selenium «By» submodule

to be imported. The most important ones are «by» to enable the use of Selenium locators, as well as «WebDriverWait» to be able to program the tests that do wait for some responses that take a lot of time.

In order to conduct testing, you need to perform some actions with the elements of the web page. But for this you need to find these elements first. A computer cannot guess where to enter data on a website, unlike a human. Therefore, finding elements is carried out with the help of locators that are implemented with the «By» Selenium submodule. This is a custom element of the Selenium in the form of a Python dictionary [22]. It allows you to find the required HTML DOM element on the page.

The key of the locator dictionary is the imported Selenium «By» element, and the value of the dictionary member is the actual search data of the element, most often represented in a string format (Fig. 2).

The last submodule is «ChromeDriverManager» (Fig. 3), which is needed to run the web driver as a Google Chrome browser. It is possible to use any other popular browsers; you just need to install the necessary packages and import the corresponding submodule. A respective version of a browser instance needs to be downloaded and installed for the Selenium to work with it. Multiple browsers can be installed this way and be controlled with a single Python file for example. It is possible to also automate running the same tests in different browsers.

The framework consists of two parts, Init.py for creating the first configuration file before the first launch and Selenium-testing.py for testing of the login page of web applications (Fig. 4).

As you can see, two sectors and their keys with standard values are created (Fig. 5). At activation, you can see that a new configuration file has been created. The configuration file may then be modified to fit the testing criteria or add more options to the testing process.

As a result of the work, the framework has been developed to automate testing, characterized by high flexibility and ease of understanding. As a web site for demonstration of the sample login page test, https://phptravels.com/ has been chosen, but any site can be tested via the framework (Fig. 6).

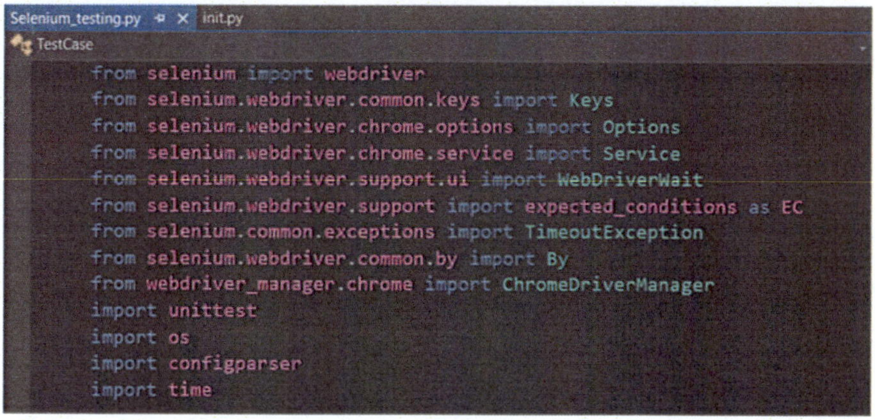

Fig. 3 Importing required submodules

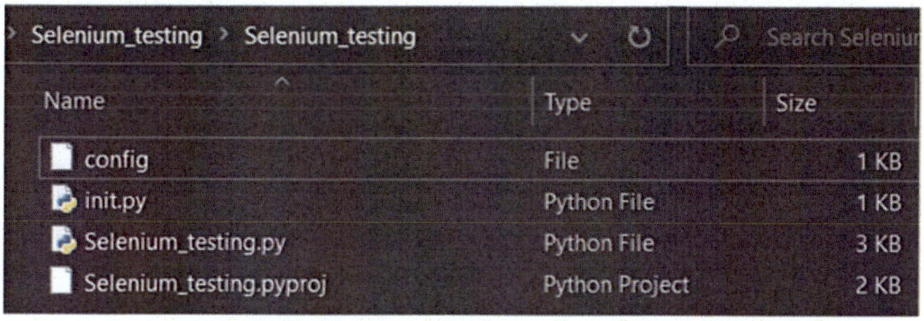

Fig. 4 Structure of the framework

Fig. 5 Contents of the configuration file

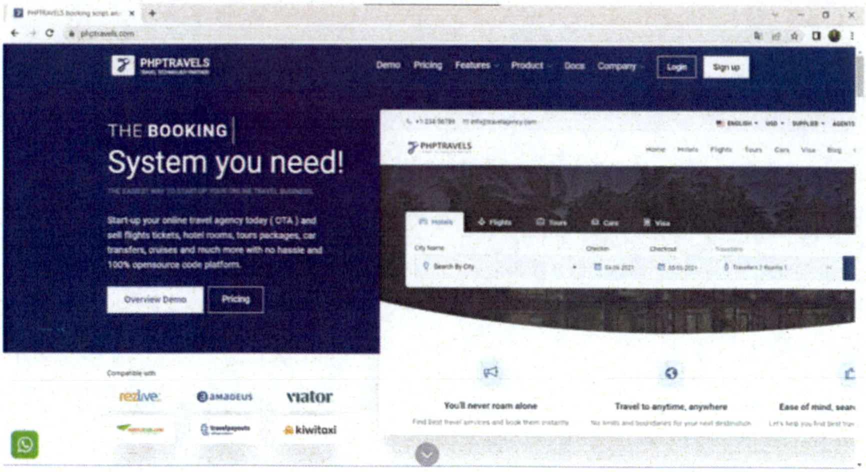

Fig. 6 Website of the layout type https://phptravels.com/

The user authorization form on the site is being tested (Fig. 7). This is an appropriate choice for starting the development of an automation framework, because such a page template is very often used on the World Wide Web and is required for the basic implementation of web application functions.

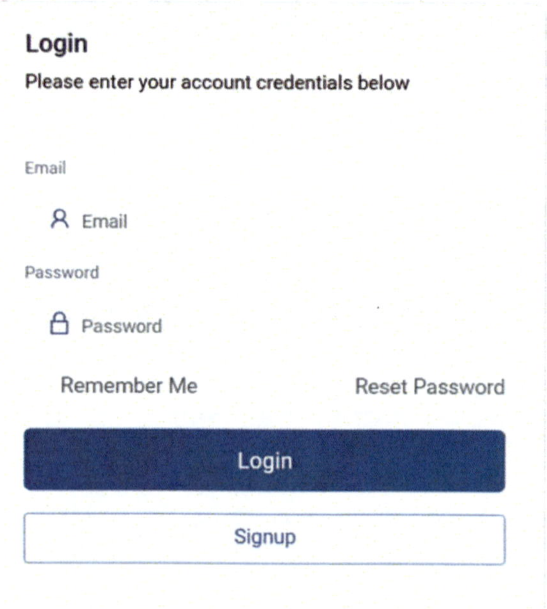

Fig. 7 Site login form displayed in a test instance of Chrome

```python
def test_login(self):
    self.driver.get(ReadConfig.login_page)
    print(self.driver.title)
    email_field = self.driver.find_element(by=By.NAME, value=ReadConfig.element_name_email)
    self.assertTrue(email_field, "No suitable email element found")
    email_field.clear()
    email_field.send_keys(ReadConfig.user_email)
    passw_field = self.driver.find_element(by=By.NAME, value=ReadConfig.element_name_pass)
    self.assertTrue(passw_field, "No suitable password element found")
    passw_field.clear()
    passw_field.send_keys(ReadConfig.user_pass)
    button = self.driver.find_element(by=By.XPATH, value="//button[@type='" + ReadConfig.element_name_button + "']")
    self.driver.execute_script("arguments[0].scrollIntoView();", button)
    self.driver.execute_script("arguments[0].click();", button)

    try:
        wait = WebDriverWait(self.driver, timeout=5)
        submit_completed = wait.until(EC.visibility_of_element_located((By.XPATH, ReadConfig.element_pass_xpath)))
        if submit_completed:
            self.assertEqual(self.driver.current_url, ReadConfig.user_page, "Did not login")

    except TimeoutException:
        self.driver.quit()
    pass
```

Fig. 8 Framework test code example

Firstly, the user/developer should launch the Init.py file to create a sample configuration file. After that by editing it, they can customize the options for testing or add their own via Python code. The user authorization form on the website is being tested. If everything is correct and the authorization form accepts user data and redirects it to the desired page in a timely manner and the key element is successfully loaded on it, then the test will be considered completed.

After that the framework will proceed to the next test. While testing is done, a browser will open in automation mode. This can however be hidden by specifying the parameter in the configuration file (Fig. 8).

The login page test checks for errors in the login form. First, the data from the configuration file is read, and then the transition to the site with the login form goes. Compatibility and working capabilities with the configuration file are provided by the Python «configparser» and «OS» modules. Information about the location of the desired elements is also read from the configuration file. The browser driver then finds the relevant elements using locators and enters test information. Then the form sending is activated, and the page loading timer starts. If the page is not loaded on time, timeout exception will be thrown. In case the page loads successfully, its address will be checked for compliance and then the key verification element will be checked. If successful, the next test will start. At the end, the console will display «OK» if all tests are performed successfully. It is worth noting that even if the page is loaded successfully but takes too long to load, the test will still be marked as failed. Almost every browser action is being checked by Unittest to see whether the information entered is valid. This is done by assert... commands. If an assert fails, then a relevant exception is being thrown. This is the basis of unit testing. Own custom tests can be created by inheriting the unittest.TestCase class and then interacting with the Selenium WebDriver as suitable (Fig. 9).

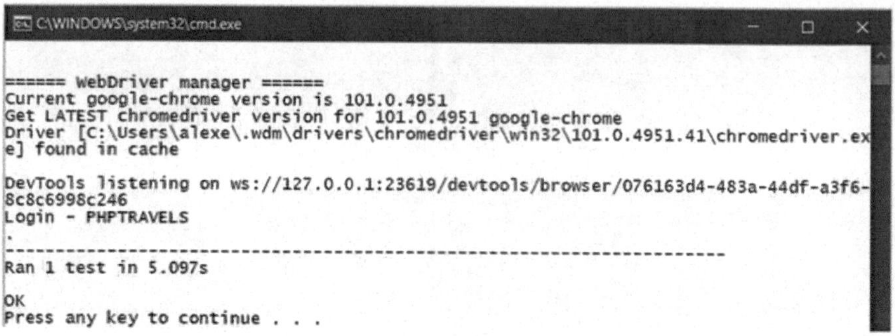

Fig. 9 Test results

The test took only 5 s to complete, including the launch of a web driver, which is a fairly long and resource-consuming task. After completing one test, another test can be launched immediately, thereby it is possible to create test packages that will follow each other in one running instance of the browser, without the need to constantly close and create new instances of browsers. The framework is suitable for both ordinary sites and information sites where the main goal is to obtain the necessary information as quickly and efficiently as possible, such as www.google.com. It is recommended to launch such tests as often as possible. The browser imitates the user input providing real details to the developer on what is happening when interacting with the website. It is also possible to use a different browser for testing, like Firefox or Microsoft Edge for example to ensure the web application works correctly on all platforms. The test was conducted to proof the framework is equally compatible with different browsers. As for its performance, there is little to no difference in time to run the test, and the tests work and show the expected results with all browsers (Table 1).

Thus, any web page can be subject tested, as Selenium WebDriver is used to simulate users' actions. PhP and JavaScript elements of the web page are also subjects to testing. Therefore, all pages that can interact with the user can be tested. The framework has an potential and can be modified to add new features, for example adding automatic test run on different remote systems or virtual containers using own web connection, thus not depending on a third-party infrastructure to run tests.

Table 1 Difference in time to run the test

Browser	Time to run a test (s)	Did the positive test pass as expected?/Did the error test pass as expected?
Google Chrome	5.097	Yes/Yes
Firefox	4.921	Yes/Yes
Edge	5.129	Yes/Yes

4 Conclusions

As a result of the work, the web services testing automation framework has been developed, which allows to speed up the testing of web services. The framework differs from others in its ease of use while maintaining flexibility for more experienced developers. The framework was written in an easy-to-understand Python programming language. Modern and popular development libraries and extension packages have been used, namely Selenium WebDriver, Unittest, and ConfigParser. The extension packages used are lightweight and do not put on the developer rigid constraints, and those constraints meet generally accepted design standards in the developing industry, allowing the framework to be extremely flexible and fit most needs for fast and clear test automation, which can be adapted and supplemented if necessary for individual needs of the project. Unlike others, this framework only implements the important parts of the testing suite, allowing the developers to build their own infrastructure and tests on top of it.

References

1. Kinsbruner E, Bahmutov G (2022) A frontend web developer's guide to testing: explore leading web test automation frameworks and their future driven by low-code and AI. Packt Publishing, Birmingham
2. Patel R (2020) QA analyst/software tester job interview questions & answers-2020: winning the QA/software testing interview. TechCT Solutions Inc., Ontario
3. Pursky O, Selivanova A, Dubovyk T, Herasymchuk T (2019) Software implementation of e-trade business process management information system. In: CEUR workshop proceedings, vol 2546, pp 171–181. http://ceur-ws.org/Vol-2546/paper12.pdf
4. Matsinopoulos P (2020) Practical test automation: learn to use Jasmine, RSpec, and cucumber effectively for your TDD and BDD, 1st edn. Apress, New York
5. Dijkstra EW (1972) The humble programmer. Commun ACM 15(10):859–866. https://doi.org/10.1145/355604.361591
6. Zhan Z (2018) Practical web test automation: automated testing web-applications wisely with selenium web-driver, 3rd edn, Leanpub, Victoria
7. Pursky O, Moroz I, Novikova V, Pavlyshyn S (2021) Stage-by-stage technology for developing of integrated e-trading management system. Int J Bus Inf Syst 38(2):254–280. https://doi.org/10.1504/IJBIS.2020.10023767
8. Khorikov V (2020) Unit testing principles, practices, and patterns: effective testing styles, patterns, and reliable automation for unit testing, mocking, and integration testing with examples in C#, 1st edn. Manning, New York
9. Jorgensen P, DeVries B (2021) Software testing: a Craftsman's approach, 5th edn. Auerbach Publications, Boca Raton
10. McDonough JE (2021) Automated unit testing with ABAP: a practical approach, 1st edn. Apress, New York
11. Angelov A (2020) Design patterns for high-quality automated tests: high-quality test attributes and best practices. Independent Publishers Group, Chicago
12. Gundecha U, Cocchiaro C (2019) Learn Selenium: build data-driven test frameworks for mobile and web applications with Selenium Web Driver 3. Packt Publishing, Birmingham
13. Philip D (2021) Cypress: test automation. Tutorial Points, Madhapur
14. Bisht S (2013) Robot framework test automation. Packt Publishing, Birmingham

15. Lawrence R (2019) Behavior-driven development with cucumber: better collaboration for better software, 1st edn. Addison-Wesley Professional, Boston
16. O'Reilly (2022) The testing pyramid. https://www.oreilly.com/library/view/continuous-delivery-with/9781787125230/f0d4d0ff-8669-43c5-9d78-ed7f361935a7.xhtml
17. Basu S (2020) Selenium with python simplified for beginners—simple, concise & easy guide to automation testing using python and selenium web-driver. Independent Publishers Group, Chicago
18. Ramalho L (2022) Fluent python: clear, concise, and effective programming, 2nd edn. O'Reilly Media, Sebastopol
19. TIOBE Index for October 2022 (2022) https://www.tiobe.com/tiobe-index/
20. Babenko V, Buyak L, Pursky O, Lutsenko I, Perevozova O, Pokhodenko B (2021) The general concept of building integrated information web-systems of E-commerce. In: 2021 11th international conference on advanced computer information technologies, ACIT 2021—proceedings. Deggendorf, Germany, 15–17 Sept 2021, pp 441–444. https://doi.org/10.1109/ACIT52158.2021.9548337
21. Babenko VO, Yatsenko RM, Migunov PD, Salem A-BM (2020) MarkHub cloud online editor as a modern web-based book creation tool. CEUR Worksh Proc 2643:174–184. http://ceur-ws.org/Vol-2643/paper09.pdf
22. Documentation (2022) The Selenium browser automation project. https://www.selenium.dev/documentation/

Author Index

D. Magdi et al. (eds.), *Green Sustainability: Towards Innovative Digital Transformation*, Lecture Notes in Networks and Systems 753,
https://doi.org/10.1007/978-981-99-4764-5